# Maxwell on the Electromagnetic Field

D0021063

Masterworks of Discovery
*Guided Studies of Great Texts in Science*
Harvey M. Flaumenhaft, Series Editor

# Maxwell on the Electromagnetic Field

## A Guided Study

Thomas K. Simpson
Illustrations by Anne Farrell

LCCC LIBRARY

Rutgers University Press

New Brunswick, New Jersey, and London

Second paperback printing, 1998

Library of Congress Cataloging-in-Publication Data
Simpson, Thomas K.
    Maxwell on the electromagnetic field : a guided study /
Thomas K. Simpson.
        p.    cm. — (Masterworks of Discovery)
    Includes bibliographical references and index.
    ISBN 0-8135-2362-1 (alk. paper). — ISBN 0-8135-2363-X (pbk.: alk. paper)
    1. Electromagnetic fields.   2. Mathematical physics.   3. Maxwell, James
Clerk, 1831–1879.   I. Title.   II. Series.
QC665.E4S554   1997
530.1′41—dc20
                                                                            96-8341
                                                                               CIP
British Cataloging-in Publication information available

Copyright © 1997 by Thomas K. Simpson
All rights reserved
No part of this book may be reproduced or utilized in any form or by any means,
electronic or mechanical, or by any information stoirage and retrieval system, without written permission from the publisher. Please contact Rutgers University Press,
Livingston Campus, Bldg. 4161, P.O. Box 5062, New Brunswick, New Jersey
08903. The only exception to this prohibition is "fair use" as defined by U.S. copyright law.
Manufactured in the United States of America

To the memory of Scott Buchanan
who set the goal:
the rediscovery of the liberal arts
in the modern world

BT 12183-A 6-3-99 22.00

# Contents

Series Editor's Foreword     ix

Preface and Acknowledgments     xv

Chapter 1    Beginnings     1

Chapter 2    *On Faraday's Lines of Force*     53
    Maxwell's Text    55
    Interpretive Notes    78
    Discussions    117

Chapter 3    *On Physical Lines of Force*     139
    Maxwell's Text    143
    Interpretive Notes    173
    Discussions    217

Chapter 4    *A Dynamical Theory of the Electromagnetic Field*     252
    Maxwell's Text    254
    Interpretive Notes    287
    Postscript: Maxwell as Experimenter    347
    Epilogue    363
    Discussions    367

Notes     405

Selected Readings     411

Bibliography     417

Illustration Credits     421

Index     423

# Series Editor's Foreword

We often take for granted the terms, the premises, and the methods that prevail in our time and place. We take for granted, as the starting points for our own thinking, the outcomes of a process of thinking by our predecessors.

What happens is something like this: Questions are asked, and answers are given. These answers in turn provoke new questions, with their own answers. The new questions are built from the answers that were given to the old questions, but the old questions are now no longer asked. Foundations get covered over by what is built upon them.

Progress thus can lead to a kind of forgetfulness, making us less thoughtful in some ways than the people whom we go beyond. Hence, this series of guidebooks. The purpose of the series is to foster the reading of classic texts in science, including mathematics, so that readers will become more thoughtful by attending to the thinking that is out of sight but still at work in the achievements it has generated.

To be thoughtful human beings—to be thoughtful about what it is that makes us human—we need to read the record of the thinking that has shaped the world around us, and still shapes our minds as well. Scientific thinking is a fundamental part of this record—but a part that is read even less than the rest. It was not always so. Only recently has the prevalent division between the humanities and science come to be taken for granted. At one time, educated people read Euclid and Ptolemy along with Homer and Plato, whereas nowadays readers of Shakespeare and Rousseau rarely read Copernicus and Newton.

Often it is said that this is because books in science, unlike those in the humanities, simply become outdated: in science the past is held to be passé. But if science is essentially progressive, we can understand it only by seeing its progress *as* progress. This means that we ourselves must move through its progressive stages. We must think through the process of thought that has given us what we otherwise would thoughtlessly accept as given. By refusing to be the passive recipients of ready-made presuppositions and approaches, we can avoid becoming their prisoners. Only by actively taking part in discovery—only by engaging in its rediscovery ourselves—can we avoid both blind reaction against the scientific enterprise and blind submission to it.

When we combine the scientific quest for the roots of things with the humanistic endeavor to make the dead letter come alive in a thoughtful mind, then the past becomes a living source of wisdom that prepares us for the future—a more solid source of wisdom than vague attempts at being "interdisciplinary," which all too often merely provide an excuse for avoiding the study of scientific thought itself. The love of wisdom in its wholeness requires exploration of the sources of the things we take for granted—and this includes the thinking that has sorted out the various disciplines, making demarcations between fields as well as envisioning what is to be done within them.

Masterworks of Discovery has been developed to help nonspecialists gain access to formative writings in ancient and modern science. The volumes in this series are not books *about* thinkers and their thoughts. They are neither histories nor synopses that can take the place of the original works. The volumes are intended to provide guidance that will help nonspecialists to read for themselves the thinkers' own expressions of their thoughts. The volumes are products of a scholarship that is characterized by accessibility rather than originality, so that each guide-book can be read on its own without recourse to surveys of the history of science, or to accounts of the thinkers' lives and times, or to the latest scientific textbooks, or even to other volumes in the Masterworks of Discovery series.

Although addressed to an audience that includes scientists as well as scholars in the humanities, the volumes in this series are meant to be readable by any intelligent person who has been exposed to the rudiments of high school science and mathematics. Individual guidebooks present carefully chosen selections of original scientific texts that are fundamental and thought provoking. Parts of books are presented when they will provide the most direct access to the heart of the matters that they treat. What is tacit in the texts is made explicit for readers who would otherwise be bewildered, or sometimes even be unaware that they should be bewildered. The guidebooks provide a generous supply of overviews, outlines, and diagrams. Besides explaining terminology that has fallen out of use or has changed its meaning, they also explain difficulties in the translation of certain terms and sentences. They alert readers to easily overlooked turning points in complicated arguments. They offer suggestions that help to show what is plausible in premises that may seem completely implausible at first glance. Important alternatives that are not considered in the text, but are not explicitly rejected either, are pointed out when this will help the reader think about what the text does explicitly consider. In order to provoke more thought about what are now the accepted teachings in science, the guidebooks bring forward questions about conclusions in the text that otherwise might merely be taken as confirmation of what is now prevailing doctrine.

Readers of these guidebooks will be unlikely to succumb to notions that reduce science to nothing more than an up-to-date body of concepts and facts and that reduce the humanities to frills left over in the world of learning after scientists have done the solid work. By their study of classic texts in science, readers of these guidebooks will be taking part in continuing education at the highest level. The education of a human being requires learning about the process by which the human race obtains its education—and there is no better way to do this than to read the writings of those master students who have been master teachers of the human race. These are the masterworks of discovery.

James Clerk Maxwell has often been paired with Isaac Newton, the two of them being seen as the twin classic masters of modern physics—Newton, for his principles of mechanics and universal gravitation, giving a mathematical account in which the motions that befall things down here on earth are brought together with those cosmic motions at which earthlings long looked up in wonder; and Maxwell, for his notion of the electromagnetic field, giving a mathematical account of long-unknown invisible processes that pervade nature and give rise to astonishing powers.

After a century and a half of Maxwell's thought at work, we take for granted the pervasiveness and power of the electromagnetic field. The electromagnetic field equations that came from Maxwell's thinking, with the vector mathematics that he helped to develop, are now ordinary tools in the everyday technical life of physicists and engineers. But these tools were the wondrous outcome of an extraordinary inquiry that is worth the attention of every thoughtful person—Maxwell's effort to formulate the notion of a connected whole that is somehow prior to its parts. The question on his mind was something like this: how can we articulate the notion of a structured continuum in which each part of the configuration is sensitive to the motion of every other part, however remote? Maxwell's exploration of that question was a process that took some ten years, in the course of which he produced three papers, as he kept trying new ways to express what he was looking for.

The first of the papers ("On Faraday's Lines of Force," 1855) arose from Maxwell's reading of Faraday's experiments in electricity and magnetism. By imagining the flow of an ideal fluid—an incompressible material with freely moving parts—Maxwell in this paper learns to think mathematically about the forms of Faraday's lines of force. Here Maxwell merely explores the geometry of the continuum that lines of force seem to delineate. In this geometrical mode, which appeals directly to what can be visualized by the reader, mathematics makes its lightest demands; and since the imaginary fluid has no mass, physics makes no demands at all.

The merely geometrical figures of the first paper became physical possibilities in the second ("On Physical Lines of Force," 1861). By a deliberate flight of fancy, Maxwell in this paper shows how space might be filled with a connected physical medium that in obedience to Newton's laws would account mechanically for the known phenomena of electromagnetism. What Maxwell contrives, in order to furnish a physical mechanism capable of doing what is demanded by the mathematical continuum of the lines of force, seems to be a merely playful contraption, rather than the object of a serious theory. However, it does give him a claim to having in principle sketched a conceivable account that makes sense of electromagnetic phenomena. It also gives him a glimpse of the following implication of his elaborate play of thought: by introducing the laws of Newtonian physics, and thus bringing time into the picture, he has raised the possibility that effects are propagated through the electromagnetic medium with the very same velocity that light happens to have. What follows from Maxwell's playful vision is the very serious consequence that light might be a phenomenon of electromagnetic processes.

A mechanical explanation of the sort attempted in the second paper was renounced in the third ("A Dynamical Theory of the Electromagnetic Field," 1865). By using mathematics developed by Lagrange, Maxwell in this paper avoids unverifiable mechanical detail. Here he is seeking an electromagnetic theory that, while coming as close as possible to physical truth, will make no commitment that cannot be justified experimentally. The theory that Lagrangian mathematics enables him to develop is called "dynamical" because it is based on overall *energies* rather than on specific Newtonian *forces*. The field thus emerges as an energy-bearing continuum. This notion of the field makes electromagnetism intelligible; but the existence of the field is physically unprovable, and the ethereal medium in which its energies would reside has properties that are paradoxical. Nonetheless, Maxwell's reasoning culminates in an experiment crucial for the electromagnetic theory of light: the ratio of the unit of electric charge in one system of measurement to that unit in a different system of measurement is revealed to be the very velocity of propagation of waves in the electromagnetic medium—which is, as it turns out, the same as the velocity of light.

As it happened, the accurate establishment of those fundamental units was of the greatest importance to the burgeoning electrical industry at just that time; indeed, Maxwell himself came to the center of attention as an experimenter in the establishment of an immediately related quantity, the British Association, "Ohm," which was another unit of the greatest importance for the new electrical technology. The reading of these three great papers of Maxwell the theorist thus closes with Maxwell the experimenter on view, as fundamental science and expanding technology come together.

Deep currents of thought and massive waves of productive power passed through and emanated from the mind of James Clerk Maxwell, but his own expression of what was on his mind has till now been inaccessible to readers untrained in physics and mathematics. Now, however, such readers will be able to enjoy the company of this great thinker on his way to a thought at the foundation of our physics and technology. In making that enjoyment possible, Thomas K. Simpson has set himself a formidable task: Maxwell did not write on electromagnetism with novices in mind. He did, however, write for human beings who want to think deeply, albeit ones who have the prerequisite learning. For those who lack that prerequisite learning, as well as for those who have it but do not have the time they would need for learning much from Maxwell without help, here is all that is needed for thinking along with Maxwell as he develops a notion that has transformed the world. This guidebook to a body of writings that we need much guidance to read is a most appropriate addition to the Masterworks of Discovery series.

Harvey Flaumenhaft, Series Editor
St. John's College in Annapolis

# Preface and Acknowledgments

This is to be a curious sort of joint venture, in which you, as reader, and I—not as author, but as guide—are about to begin reading together three challenging papers in classical mathematical physics. Our author, in whom we will have for this while a very special common interest, is James Clerk Maxwell. The papers in question are those in which he developed, in three dramatic, sharply distinguished stages, a coherent vision of something he called the "electromagnetic field." We should declare right away that this little book is designed especially for the general reader, even one who may have very little background in mathematics or physics. Though some engineers and physicists might well be interested in meeting the origins of the concepts which are their stock-in-trade, the special purpose of this series is to contribute to a more widespread literacy in the sciences. An abundance of diagrams and other aids will thus be provided to help the lay reader break through the shell of technical skills and vocabularies.

The experience should be facsinating, but it will not always be easy; it might without extravagance be called an "intellectual adventure." Just as the steamship tickets of an earlier era included certain alarming disclaimers concerning wrecks and disasters, so here, at the outset of our voyage, we may wish to come to some agreement about our common purpose, our plan of work, and our expectations—which must not be set too high. Our overall purpose may be stated simply, echoing the vision expressed by Harvey Flaumenhaft in the Series Editor's Foreword: to read these papers of Maxwell's well, and as directly as possible—as though he had written with us in mind. The three papers in question are:

"On Faraday's Lines of Force" (1855)

"On Physical Lines of Force" (1861)

"A Dynamical Theory of the Electromagnetic Field" (1865)

Selections have been made for reprinting here with the intent of laying out a path for a manageable, connected reading through the three. They mark well-defined stages in the shaping of a new paradigm—the *field*—but the reader should not expect a straightforward progression. These are the birth pangs of a new formal order for thought—a dialectical process whose stages may seem at the outset to be more disorderly than orderly. Each rejects the limited and unsatisfactory mode of its predecessor and builds anew on what may seem at first reading to be an arbitrary foundation of its own.

Broadly speaking, the first essay springs immediately from Maxwell's reading of Faraday and is devoted mainly to coming to grips in almost Euclidean terms with the geometry of the strange curvilinear continuum that the lines of force seem to delineate. It thus lays a foundation for thought about the field, but in terms of the flow of an entirely imaginary, massless fluid. Deprived of mass, there is no physics in it, and Maxwell is not yet on a path toward a theory of electromagnetism; yet he has learned to think mathematically about the forms of Faraday's lines of force.

He therefore starts over again in the second paper, "On Physical Lines of Force," in an altogether new direction. This time, he is determined to demonstrate, to himself first of all, that there might *in principle* be a connected, genuinely physical mechanism in space, obedient to Newton's laws, which would account for the known phenomena of electromagnetism. With this, the merely geometrical figures of the first paper would become real physical possibilities. We are still in a realm of imagination—any flight of fancy is admissible, provided only that it contributes to this goal. The result is a brilliant contrivance, marginally successful as physics, but, as he well knows, looking far more like a great comic vision than a firm foundation for serious theory. Yet he has proved his point: he has in hand a sketch of a conceivable account that would make physical sense of electromagnetic phenomena—and there is a strong hint as well that it bears within it the electromagnetic theory of light (the theory that what appears to us as light is actually a phenomenon of electromagnetism).

In the third paper, "A Dynamical Theory," he reverses course once more, renounces any such mechanical explanation, and seeks instead, by means of an abstract mathematical method devised by Lagrange, to avoid all such premature commitment to unverifiable detail. With this Lagrangian theory, which still feels quite new in England in 1865 (called "dynamical" because it is based on overall *energies* rather than on specific Newtonian *forces*), he hopes to build an electromagnetic theory as nearly as possible congruent with physical truth, while making no specific commitment which cannot be justified experimentally. Such a theory would in this way assert all that is known, but nothing that is mere speculation. As he hones these ideas and the field emerges as an energy-bearing continuum, he recognizes that the field concept is in this abstract theory no more or less than a mode of thought—a device of high rhetoric rather than an established fact of science. To the mind's eye, it is immensely appealing, but its existence is physically unprovable. Indeed, he sees the "ether" in which the field energies would reside as a most paradoxical and elusive entity. Yet, as if to certify this intellectual intuition, out of Maxwell's dynamical theory emerges the fully articulated electromagnetic theory of light.

A skeptic might well ask why, in this age in which so many advances press upon us in the sciences, we should go to such trouble to study the labored works of the past. I would not, for example, recommend this path to a student who wished to learn classical electromagnetic theory in an efficient manner. A modern text, employing the streamlined notations Maxwell himself helped to develop and written with the perspective of time, would serve better. We must, then, have some larger or darker purpose in mind.

First, the concept of the *field* is a powerful instrument of thought that is not easy to grasp fully, either in the sciences themselves or in its extrapolation as metaphor. It emerges here as a new paradigm and as it shapes up in these papers of Maxwell's it is loaded with prospects and difficulties, some clearly looming on Maxwell's own horizon. (He was, for example, very much aware of the contradictions lurking in the notion of the field in the way of what was to become relativity theory.) The field is a continuum that is primarily single and whole, and it seems that today we have a special need to find modes of thinking and acting in which *the whole is prior* and the parts significant through their relation to it. The corollary notion of *system*—more urgent upon us as each new unsuspected ecological, technical, or social closure forces itself upon our attention—is very much on the agenda for reconsideration today, taking us back to authors such as Plato, Aristotle, Leibniz, and perhaps even Hegel, who all do look *first to the whole* and who may have been gathering dust on many of our shelves through the long night of atomism.

Second, like other great works, these papers of Maxwell's go beyond their explicit concern to reflect on broader, fundamental questions. In this case, I see Maxwell taking us to issues that would be called epistemological—what are the modes and possibilities for human knowledge?—and to questions of ontology—what is, after all, *out there*? Each reader will take these questions from Maxwell's text in a different way; but they belong to a domain in which we all have a common interest—one which for a long while was branded as "metaphysical" and thought to impede the work of the sciences. If this is a time in which we have tended to lose our bearings, both in the sciences and in society as a whole, perhaps a touch of metaphysics would be a sound prescription. I make no claim that Maxwell had just these issues in mind as I have phrased them: but his thought surely reaches far beyond its formal, expressed limits—and his searching questions bear at least as much on our own time as they ever did on his.

And now: how shall we work together to accomplish the ambitious task of reading these papers? On the one hand, we are inviting the reader who has little working knowledge of either mathematics or physics to share this enterprise. On the other hand, Maxwell at times presumes considerable familiarity with physics, alluding to phenomena and theories of various sorts, and in the later papers uses a good deal of analytic math. It is my interpretive task to bridge the gaps which may arise between the two of you, author and reader, in a generous series of Interpretive Notes to accompany Maxwell's text. I will fill in steps, offer diagrams and explanations, and "teach" just enough in the way of elementary mathematics and physics to keep the way to Maxwell's text open for, hopefully, even the least informed, most disoriented fellow-traveler. In the course of this effort, I will inevitably explain for the benefit of some readers a good many things that others will already know, in some cases quite possibly better than I. Our book is a kind of one-room schoolhouse, with a mixed class. But this I think will present relatively little difficulty. The notes will be unobtrusive, placed in such a way that the reader can readily turn to those that will be helpful and ignore those that will not.

In addition to the specific notes that I have mentioned, some explanations will necessarily be more extensive. These, as opposed to "Notes," will be called "Discussions," and will be gathered separately. These Discussions introduce certain systematic foundations, such as rules of the calculus or the laws of physics, or they spell out derivations in detail.

They will be numbered and reference to them will be made by cues placed within Maxwell's text as well as at appropriate points among the Interpretive Notes. The reader is expected to consult all of this apparatus *ad lib* and, of course, can always come back to a Note or a Discussion later on.

A brief annotated bibliography will be provided to go with each chapter, which will be referred to as we go along. These will take the interested reader to a few of the leading works that Maxwell himself had in view, as well as to a small selection of especially useful texts and to some of the many commentaries on the fascinating history of these matters. Appropriate works for our purposes are classics of the world of texts, a little old-fashioned perhaps and not always still in print. They should be readily at hand in any respectable public library. I have avoided loading our already burdened text with biographies of the scientists whose work Maxwell alludes to as these are available in encylopedias, most especially in the *Dictionary of Scientific Biography,* where, incidentally, there is an excellent overview of Maxwell's life and work. There has been a very active discussion of Maxwell's thought, and methods over the years almost since the moment he wrote—ranging from later enthusiastic appreciation (for example, by Einstein), to rather prompt angry rejection (Pierre Duhem). The books to which I refer will serve to acquaint reader with the principal benchmarks of this discussion over the years.

Overall, however, my intent will be limited to clearing the way to the reading of Maxwell himself. Insofar as I go beyond that I will at times try to give form to questions which Maxwell's text suggests to me, turning them back to you to deal with as you will.

I would especially like to encourage readers, most of all those who have not had laboratory experience in electricity and magnetism, to gather simple items of equipment for themselves, commandeer the kitchen table, and engage in creative investigations of their own devising. The second part of Chapter 1 will be a good source of suggestions, and they abound throughout Maxwell's text as well. Science supply houses advertise in the popular scientific magazines, while the adventuresome can forage in the hardware store or the family junk box. For the faint-hearted, kits put together as introductory challenges for young engineers will serve very well as avenues to Maxwell's science.

For the opportunity to work on this engaging project, I must thank first of all Harvey Flaumenhaft, the series editor, who conceived the vision of a series of guides that would open the way to a new kind of reading of the classics of science. Those who know the work of St. John's College, where both Harvey and I are members of the faculty, will recognize that this project springs from its "Great Books" program, which has shaped the teaching there since 1937. In that sense, the ultimate vision lies with the founders of that program at St. John's, including especially Scott Buchanan, the first dean, to whose memory I have dedicated this volume. It is striking that Buchanan—convinced that the sciences lay at the center of the liberal arts in the modern world—put Maxwell's *Treatise on Electricity and Magnetism* on the first reading lists, though it is not clear whether at that time the college met immediate success in rising to his challenge!

In the same vein, I thank those many members of the faculty and alumni at St. John's, as well as many other friends, who read an early version of this guide and helped greatly with their comments, among them Joseph Cohen, Mark Daley, Dana Densmore, Howard Fisher, Jim Judson, Samuel Kutler, J. Shipley Newlin, Peter D. Paul, Nevitt Reesor, Ralph Swentzell, David Townsend, and John van Doren. I deeply regret that William Gleason, a friend who gave me some of the most heartfelt advice, did not live to see the finished result.

Most particularly, I am indebted to a faculty study group consisting of Harvey Flaumenhaft, Chester Burke, and Marilyn Higuera, who made their way valiantly through the first draft, leaving an Ariadne's thread of invaluable suggestions as they went. Marilyn most nobly gathered these suggestions and made them systematically available to me. I only wish I could be sure that I have done justice to them all. I want to thank as well all those students at St. John's who have undertaken to work with this text, in particular the group who studied it with me in the graduate program at St. John's in Santa Fe in the summer of 1993.

The draft version of this text was prepared under NEH Grant RH-20858–88, and I am most indebted to those at the Foundation who had confidence in the concept of this series. I am deeply grateful to the Harry and Lynde Bradley Foundation for its generous grant to underwrite the publication costs of this volume and thus to ensure that the book can reach the nonspecialist readers to whom it is chiefly addressed. My thanks go to Karen Reeds, my editor at Rutgers University Press, and to others for their patience with this difficult and anomalous project.

Many institutions have provided generous assistance in making illustrations available or in granting permission to use materials in their possession. Dr. Bernard Finn, Director of the Division of Electricity at the Smithsonian, has been especially helpful. Others are acknowledged with gratitutde in the list of credits for the illustrations.

Very special thanks must go to Anne Farrell, the computer graphics artist (and former student), who was willing to lend her skill, through many intense hours of conversation, to create the figures which we hope will help substantially in interpreting Maxwell's vision. Any errors in the figures are entirely my responsibility, but many of the nuances which make them effective are the product of her insight.

Finally, I offer my thanks as ever to my son Eric Simpson, fellow adventurer in this as in many other projects. Several of the photographs were taken during a journey we made together years ago into Maxwell land.

# Chapter 1

## Beginnings

## Meet the Author

At the time he wrote the first of the papers we will be discussing, in 1855, our author, James Clerk Maxwell, was just 24 years old and at the very threshold of his career as a scientist. Instead, then, of beginning with a biography sketching his whole life and works, let us start with a simpler look at this person, young as he is, who is busy preparing "Faraday's Lines of Force." He has just turned for the first time to the serious study of the subject of electricity, and is in the throws of a conviction he has had since reading Michael Faraday's *Experimental Researches in Electricity,* the final volume of which has recently come off the press. Thus we may well in the flow of time place a benchmark in the year 1855. The widely celebrated Crystal Palace Exhibition has taken place just three years earlier, spreading before an admiring world wonders of art and industry—most especially, the mysteries of the newest science, electricity. Thus we may anchor in the year 1855 the slippery perception of what it is to be "modern"—though it is the vision of new possibilities for a philosophy of nature, not a fascination with modernity as such, which interests Maxwell.

In this year, we find young Maxwell a fellow of Trinity College, Cambridge, from which he has recently graduated with high honors. It may well surprise readers today to learn that the centerpiece of liberal education as understood at Trinity College in the mid-nineteenth century is not so much the traditional humanities but rather mathematics. A sound basis in classics and languages is presupposed but the emphasis at the university is on a form of mental discipline that takes mathematics as its focus. This is seen by its proponents as "discipline" in the best and broadest sense—sound preparation of the mind for any of the tasks life

Fig. 1.1. James Clerk Maxwell at Cambridge in 1855, holding a top for experimenting with color mixing.

may present. The "mathematics" involved is itself broadly understood. Founded on the one hand in geometry and the geometrical style adopted by Isaac Newton in the *Principia,* it extends to mastery of powerful methods in differential and integral calculus which are newer and owe their origin rather to the French.[1] Furthermore, "mathematics" in the Cambridge curriculum (understood in something closer to its root sense as the domain of things strictly *knowable,* for which the Greek term is *ta mathemata*) includes without strict distinction both pure mathematics

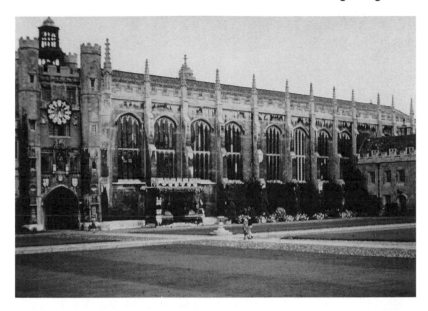

Fig. 1.2. Trinity College, Cambridge.

and a range of topics, such as mechanics, optics, and heat, which would fall today under the heading "mathematical physics." Significantly, in Maxwell's undergraduate years, the science of electricity remains too incoherent to be included, while in the three papers we are to read, Maxwell will give to this new science exactly that intelligibility which will soon justify its admission to the curriculum.

For us in the twentieth century, reflecting on this surprising method of preparing future statesmen and civil servants, it is interesting to consider that the notion of "science" is still poised in 1855 in transition from a branch of philosophy pursued at leisure to the separate, professional study demanded by a restless industrial world. Science is at this point still an appropriate concern of the broadly educated liberal mind, and it is in that spirit that Maxwell is going to work on it. We might well note in passing a striking correlation with our own concern in the present study, which is designed to make accessible to nonspecialists the thinking underlying electromagnetic theory, a domain today normally reserved for experts in engineering and mathematical physics.

Success in acquiring these Cambridge disciplines is judged at the point of graduation in the "Tripos" examinations, a grim week of tests, which

Maxwell confronted in January 1854, in a vast, cold room called the "Senate House." From this ordeal candidates emerge drained of every ounce of energy and knowledge and locked for life in a strict ranking scrutinized with eager anxiety by family, friends, and a critical academic community. Maxwell has done very well in this olympiad of intellectual athletics.

For Maxwell, however, mathematics is no mere academic exercise, but an object of true delight and devotion. Coached in the Cambridge tradition mainly by a private "tutor"—in his case, the celebrated William Hopkins, who has honed many of the best Cambridge minds—Maxwell gained easy competence and, no doubt, concomitant satisfaction in the solution of difficult problems in analytic mathematics. He will use such techniques and build on them in all of his work over the years.

We might well pause to wonder whether this mathematical virtuosity of Maxwell's will not stand ominously between his text and the non-mathematical reader for whom the present study is especially intended. The problem is a real one, but Maxwell, who, as we shall see, comes to Cambridge as a practical Scotsman, retains a deep suspicion of these aristocratic mental "hieroglyphics." He most of all will keep in mind the challenge of expressing the significance of the equations in accessible, well-chosen prose.

Now, following graduation, Maxwell is staying on as a fellow at Trinity, applying himself intensively to several major subjects, one of which is a dedicated study of electricity—a topic that is not yet included in the Tripos regimen so he is able to approach it with an open mind. His reading during this period is voracious and his speculation, revealed in letters and essays, wide-ranging and philosophical. His life has always had a very strong private center. It is generally remarked that he retains rural Scottish manners and patterns of speech which at Cambridge mark him as distinctly odd; but many find his remarks—witty, cryptic, or ironic—worthy of close attention. He has been elected a member of the select "Apostles Club," a traditional society of twelve undergraduates who meet to exchange ideas by reading original essays which are thereupon subjected to searching discussion by the group. It is there that he has come under the influence of Frederick Dennison Maurice, the Christian Socialist, who is by now an important factor in Maxwell's life. Maxwell had watched two years earlier as Maurice was expelled from the Cambridge faculty for writing heretical doctrine and, evidently under

Fig. 1.3. Maxwell's birthplace, 14 India Street, Edinburgh, an elegant Georgian building, today the home of the Maxwell Foundation.

Maurice's initial inspiration, Maxwell taught classes of workingmen during his Cambridge years. Years later, Maxwell will write a remarkable volume, *Matter and Motion,* specifically intended to introduce nonmathematical working people to the concepts of abstract mechanics—an encouragement, surely, to us in our present endeavor!

Maxwell is setting out in effect to forge for himself an intelligible science of electricity. In preparation, he is engaged in mastering the complexities of a variety of formal theories that have recently been arriving in England from the Continent, all of them based on the paradigm of action-at-a-distance, and all incorporating intricate mathematical strategies. He is far more excited, however, by the prospect opened by two other bodies of very current work, pointing in a very different direction. These are, first, the brilliant mathematical insights of William Thomson of Glasgow, a young Cambridge alumnus with whom as a fellow Scotsman Maxwell is already in correspondence;

Fig. 1.4. The fields of Galloway: a view of Maxwell's estate, Glenlair.

and, in sharp contrast, the immense body of work of Michael Faraday, speculation and experiment woven together in the rich fabric of the *Experimental Researches in Electricity,* the third and final volume of which has appeared just in time for Maxwell's reading.

When Maxwell first arrived as an undergraduate at Cambridge in 1850, he brought with him strong marks of Scottish roots of two very different kinds. Though he was born in Edinburgh, attended the Edinburgh Academy, and had been a student at the University of Edinburgh for three years before moving to Cambridge, he had spent the first years of his life in remote Galloway, and the estate at Glenlair was always to be his true home. We must refer the reader to biographies of Maxwell to read of these influences on his life, but it is clear that his scientific work is in fundamental ways shaped by these remarkable polarities in his life—between rural Scotland and sophisticated Edinburgh, on the one hand, and between these Scottish foundations in all their complexity and the distinctly foreign atmosphere of aristocratic Cambridge. As he turns to Faraday's writings, Maxwell must feel some immediate kinship with this brilliant but modest thinker, whose "unmathematical" work is on the whole rather patronized than respected by the aristocracy of science at Cambridge. Max-

Fig. 1.5. Early experiments in hydrostatics: Maxwell navigating in the duck pond at Glenlair.

well will insist that Faraday's intuitions are more genuinely *mathematical* than many of the sophisticated analytic equations.

## William Thomson

To share some sense of Maxwell's starting point in his thinking about electricity, we will do well to look for a moment at the work of Thomson and Faraday. Thomson (known to history by his later title, Lord Kelvin) was born in 1824, and thus is some seven years Maxwell's senior; under the exceptional tutelage of his father, who was a professor of mathematics at the University of Glasgow and the author of texts in differential and integral calculus, Thomson had been able to enter the University of Glasgow as a student at the age of ten. He came under the influence of a powerful French scientific tradition centered on methods of analytic mathematics. He grew up reading the mathematicians Laplace and Lagrange, and has brought this experience to bear on the subject of electricity in striking ways.[2] In particular, he has recognized a formal analogy between equations in electricity and those in heat. Now problems in electrostatics that had remained intractable are easily solved by transferring them to the domain of heat, where it was found that their solutions were already well known. Thomson's revelation of this analogy has been particularly striking to Maxwell, both philosophically in its suggestions

Fig. 1.6. William Thomson, later best known to the world as Lord Kelvin.

for the order of nature and as an opening for thought about electricity. It suggests a way to make Faraday's qualitative insights concerning the magnetic lines of force accessible to mathematical treatment.

It is the universality of Thomson's genius which is most astonishing. In parallel with his leading work in theory, in thermodynamics as well as electricity and magnetism, he excels in the design and perfection of instruments for both the laboratory and the commercial world—we will meet him at the time of Maxwell's third paper as the triumphant engineer of the Atlantic cable. At Glasgow, he has at this point established the first teaching laboratory in Scotland or England, one in which students learn experimental principles and techniques by sharing in his own research.

Maxwell, whose disengagement from the world's terms will be as striking as Thomson's easy mastery of them, is following to a large extent a mathematical pathway which Thomson has laid out. Many of the topics we will encounter in Maxwell's papers have been anticipated in some way by Thomson and much of the mathematics borrowed

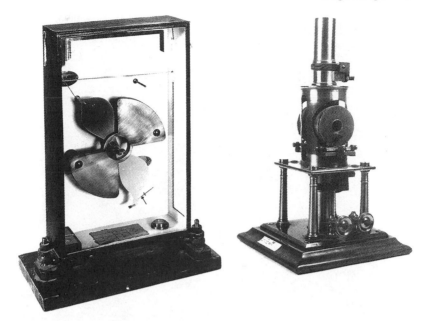

Fig. 1.7. Two of Thomson's instruments: the quadrant electrometer and the cable galvanometer.

frankly from Thomson's papers. Yet as they come into Maxwell's reflective world, these ideas take on altered forms, suffer a sea-change. Maxwell's quiet, reflective way, mixed often with philosophical reservations and penetrating irony, tends to reach some depth of the question in which Thomson is uninterested. The attraction of Faraday for him is far greater than that of Thomson.

## Michael Faraday

It is indeed striking that the very year in which Maxwell picks up the theme of Faraday's work is also the year in which Faraday, forty years Maxwell's senior, has closed his investigations, publishing as we have seen the third and last of the volumes of his immense *Experimental Researches in Electricity*. Maxwell, in turn, in preparation for his entry into the forum of thought on electricity, has just read through all three of these volumes—heroic work, certainly, for Faraday writes in a style which aggregates in a massive, episodic account the detail of his efforts in hundreds of varied investigations at the laboratory bench. He seems to

Fig. 1.8. Michael Faraday.

leave out hardly a turn of his concurrent thinking, which never rests as
the experimental work proceeds. Maxwell has been reading many other
authors, but his reading of Faraday has overbalanced all the rest, not only
in sheer magnitude but in its grip on his imagination. For as we shall see
Maxwell dedicates this first of his efforts, titled "On Faraday's Lines of
Force," to an extensive interpretation of the one most characteristic,
leading theme throughout Faraday's *Researches*. For Maxwell as for
Faraday, the *line of force* is not just a powerful concept capable of
drawing many aspects of electrical science into a single vision; it points

Fig. 1.9. Faraday lecturing before the Prince of Wales at the Royal Institution.

to a view of science and the world which was most characteristic of Michael Faraday and deeply appealing to Maxwell. Maxwell has found in the "line of force" the path to which he is ready to dedicate his own efforts.

We can quickly sketch Faraday's formal biography as its elements are very simple. The beginnings of his career were extremely modest and during the long course of his work he has held a single post at the Royal Institution. At age 64 in our year 1855, Faraday is, as we have seen, finishing his productive scientific work at the moment when Maxwell, aged 24, is beginning his: the quest for the electromagnetic field is passed in the course of one year from Faraday's hands to Maxwell's. In one of history's magic moments, Faraday in 1855 handed the *Experimental Researches* directly to their ideal reader.

The son of a blacksmith who died while he was young, Faraday grew up in a state close to poverty. He had received only the most elementary primary education and had been apprenticed at an early age to a bookbinder. Beyond the rudiments of reading, writing, and calculation he was self-educated, learning from books as they passed through the bindery. He learned his electricity from an article in the *Encyclopedia Britannica* at the shop. He soberly and it seems scrupulously followed the advice contained in a handbook by Isaac Watts, a young man's companion entitled *The Improvement of the Mind.*

His poverty was never, however, abject. His family belonged to a secure dissenting sect, followers of the eighteenth-century preacher Robert Sandeman, which was a small, closely knit, and supportive congregation dedicated to the concept of community in Christian faith. There can be little doubt that this quiet faith has consistently informed Faraday's scientific work.

No summary can convey either the content or the character of Faraday's writings. As perhaps the ultimate example of their genre—the literature of probing scientific inquiry organized episodically in linked series—they are a texture of penetrating speculation and intelligent experiment so interwoven that it is virtually impossible to isolate one significant topic from another. In effect, this is the very point of his work: a relentless search for the wholeness of nature hidden beneath the variety of the phenomena. The unbroken weave of the *Experimental Researches* reflects the coherence of the world Faraday is seeking; it does not fragment into neatly divided and labeled compartments. We may get some sense of the nature of "science" as Faraday conceives and practices it if we look just at his concept of the "lines of force" to which Maxwell's first paper is directed. Faraday was impressed by this notion early when he called the patterns traced out by iron filings, as others did, the "magnetic curves." Toward the end of his work, these patterns have come to bear nearly the whole burden of his thought on these topics and he speculates increasingly about the possible "physical existence" of these lines.

For Maxwell, the lines are destined to become fundamental concepts of a coherent science. But what sort of use does Faraday himself make of them? Not, I think, as elements of a scientific *theory* in any formal sense. As we have seen, Faraday was almost totally uneducated in any sort of formal mathematics. It is difficult to grasp the significance of this fact for science. However brilliantly he has succeeded in self-education in other areas, he had apparently never felt it necessary, or perhaps possible, to acquire the mathematics he had missed. The result is that he does not have before him with any vividness that universal paradigm of a *reasoned deductive system* founded on axioms and postulates.

Is this lack a *defect*? There is at least the possibility that Faraday's innocence of formal mathematics is not merely accidental ignorance but deliberate choice. Note his reaction when confronted with a brilliant

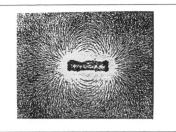

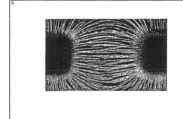

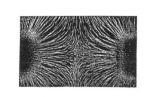

Fig. 1.10. Magnetic curves traced by iron filings.

mathematical argument of André-Marie Ampère's: "I regret that my deficiency in mathematical knowledge makes me dull in comprehending these subjects. *I am naturally skeptical in the matter of theories* and you must not be angry with me for not admitting the one you have advanced immediately."[3] This "skepticism of theory" turns him away from *theory* almost on principle. He tends to regard mathematicians as operating on a height, while his own work, as experimentalist, is close to nature and to fact. It would be a mistake, I believe, to overlook the element of a dissenter's pride that mixes with humility in his characterizations of his more modest work. Upon the discovery of a fundamental phenomenon of magnetic rotations, he wrote to a friend: "It is quite comfortable to me to find that experiment need not quail before mathematics, but is quite competent to rival it in discovery; and I am amazed to find that what *the high mathematicians* have announced as the essential condition of rotation . . . has so little foundation."[4] There is more than a hint of a critical, possibly moral note in this rejection of mathematics and theory, suggesting a fundamental connection with his Sandemanian convictions. Faraday has remained in a phrase he uses without, I think, any hint of apology, an "unmathematical philosopher."

What then *is* Faraday's concept of science? It is evident that he is no *mere* empiricist, cleverly providing data for theoreticians to work up into

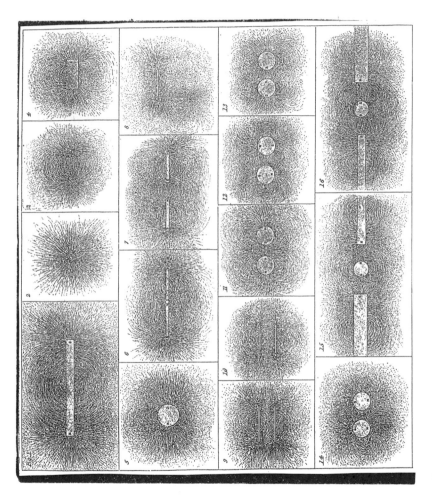

Fig. 1.11. A page of magnetic configurations from the *Experimental Researches*.

theories—although many theorists, patronizing him as a "discoverer," regard his work as no more than tributary to their own. On the contrary, Faraday himself possesses one of the most fertile and insistent of *speculative* minds. In a certain sense, he is constantly producing new hypotheses and he is constantly reasoning from them. Yet the result of this, reported in the thousands of paragraphs of the *Experimental Researches* and pursued privately in the still more numerous paragraphs of his diary, is not "theory"—not something one *knows* by way of intellectual grasp so much as something one has *met* as an element of God's creation.

It is possible to be a thinker without being a theorist. The *Experimental Researches* are dense with questions—Faraday's method is that of unremitting inquiry. The very notion of a "series" of researches is, in a sense, that of a chain of linked questions and answers. But the underlying question is always: what really *exists* in nature? The practical form that this question takes is that of the *test:* what will be *revealed* if I do *this*? Can I produce some visible or tangible evidence, which will be the sure sign of the existence of a suspected power or state?

Writing in December 1854 about the magnetic lines of force, Faraday has carefully drawn a distinction between the limited powers of mathematical physics and his own search for "the one true physical signification" of the phenomena:

> What we really want is not a variety of different methods of representing the forces, but the one true physical signification of that which is rendered apparent to us by the phaenomena . . . if there be such physical lines of magnetic force as correspond (in having a real existence) to the rays of light, it does not seem so very impossible for experiment to touch *them;* and it must be very important to obtain an answer to the inquiry concerning their existence, especially as the answer is likely enough to be in the affirmative.[5]

Throughout the *Researches* Faraday has sought what he calls "contiguity" in nature: the lines of force fill the space between two magnets attracting one another. When Maxwell joins his efforts to those of Faraday, he seems to be drawn not only to a promising mode of explanation but to a concept of science itself, opting, despite his own powers in analytic mathematics, for a science centered not on such formalities but on an intimate understanding of the ways of nature in some simpler sense. He seems to share with Faraday the belief that it is the work of science to lead us to see better, as fact, what lies before us in the creation.

Two years after Maxwell had delivered the paper we are about to read first, "Faraday's Lines," which he had sent to Faraday, he received a letter in which Faraday expresses a plea that I think thereafter becomes a guiding thought for Maxwell:

> There is one thing I would be glad to ask you. When a mathematician engaged in investigating physical actions and results has arrived at his conclusions, may they not be expressed in common language as fully, clearly, and defnitely as in mathematical formulae? If so, would it not be a great boon to such as I to express them so?—translating them out of their hieroglyphics that we also might work upon them by experiment. I think it must be so, because I have always found that you could convey to me a perfectly clear idea of your conclusions.[6]

Maxwell is in fact deeply sympathetic to Faraday's concepts and methods and to his view of science and the world. Later, at a point at which the electromagnetic theory of light is becoming apparent to Maxwell, he writes to Faraday, to whom this success is ultimately due:

> When I began to study electricity mathematically I avoided all the old traditions about forces acting at a distance, and after reading your papers as a first step to right thinking, *I read the others, interpreting as I went on, but never allowing myself to explain anything by these forces.* It is because I put off reading about electricity till I could do it without prejudice that I think I have been able to get hold of some of your ideas.[7]

It appears that Maxwell from the beginning of his work in electricity—that is, from just that point at which we are beginning our own reading—devotes himself altogether to Faraday's point of view, both of the nature of science and of the nature of the world, for the two must ultimately mirror each other. He shaped the world of the *Researches* according to his own abiding convictions. It is perhaps the triumph—at once scientific and literary—of the papers we are about to read that Maxwell will effect the translation of Faraday's thought into a new, formal guise with such gentleness and understanding that Faraday's concept of nature and of the proper work of science will come to realization, intact, in the result.

## "The Present State of Electrical Science"

Maxwell begins "Faraday's Lines" with the remark that "the present state of electrical science seems peculiarly unfavourable to speculation."

Fig. 1.12. Faraday in later years.

He appears to assume that his reader will be more or less aware, as he has just become, of the "state" of the science and the problems which it presents. In order to coordinate our reading with Maxwell's own inquiry, then, let us very briefly review the "state of electrical science" in our benchmark year, 1855, the date in which the paper was read to the Cambridge Philosophical Society.

Readers for whom matters of electricity and magnetism are new may take comfort in the fact that Maxwell, too, is approaching this systematically for the first time—a domain in which he says, perhaps a trifle less than candidly, he has "hardly made a single experiment." For our part, we will by no means attempt to confront all the complexities of this not-yet-integrated science—to do so would be to immerse ourselves in just that morass that Maxwell, who has explored it, finds so little

Fig. 1.13. Magic of the arc lamp. Scene from the ballet *Elektra*. Prior to the invention of the incandescent lamp, the term "electric light" referred to this arc light.

conducive to understanding. It will be enough for us to get an initial glimpse
of a few of the principal elements of the science about which Maxwell will
be writing. Later, we can fill in details as they become important to us and
keep track of a few major developments in the science as they occur.

The year 1855 belongs to a dramatic period in the development of the
science of electricity. A century has elapsed since the first work of such
pioneers as Benjamin Franklin and Joseph Priestly—years of intense
effort on the part of many of the most competent natural philosophers
and mathematicians, in a wide range of areas of research.[8] In each of
these islands of investigation, a body of information, hypothesis, and
technology has proliferated. In some areas theoretical developments are
far advanced: theories have been highly elaborated using mathematical
techniques which will remain powerful through the twentieth century.
Such is the work of Ampère and the fundamental theories of the German
researchers whose work, as we saw earlier, Maxwell has been studying.
For the most part, however, these are oases of isolated initiative, each

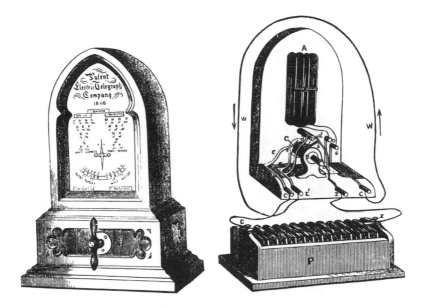

Fig. 1.14. Wheatstone's telegraph, 1845.

with its own methods, terminology, and point of view. Often, even in a single domain alternative theories are in a problematic, unresolved relation to one another.

It is a cautionary thought for us—whose very lives are "wired," for whom electricity is everywhere built into the processes of daily affairs—to consider that in the year 1855, though excitement about the new science is in the air, there are virtually no electrical devices in view. Though Faraday has found a principle for the production of continuous electromagnetic rotation, the first commercial appearance of the electric motor lies twenty years in the future. The concept of the incandescent lamp is even more remote, and the arc light, just finding its way into lighthouse service, is in most instances rather a breathtaking spectacle than a means of practical illumination. Among the few commercial applications of the electric current are electroplating and the astonishing electric telegraph, the latter rapidly proliferating over the globe.

Ten years earlier, the first telegraph lines had been run in England and on the East coast of the United States, where by 1846 they had stretched from Washington to Boston: they are now penetrating the West. Railroads are widely adopting the telegraph and taking the new wires with

Fig. 1.15. Telegraph operating room. In Maxwell's time, telegraphy is rapidly becoming a major industry.

them in their own process of expansion. Driven by the thirst for instant reports of the world's financial markets, news agencies are forming to serve a rapidly growing daily press: the Associated Press in the United States and Reuter's on the Continent (where pigeons are used to fill gaps in the wiring). In 1855, telegraphic cables have just linked Dover and Calais and the British military has run a cable under the Caspian Sea for use in the Crimean War. All this means that practical batteries, a new technology of electric wiring, and instruments, standards, and units of measurement are coming into lively demand. William Thomson is just at the point of a major involvement in the great Atlantic cable project, which will, in Chapter 4, become a central part of our story. As we shall see, the commercial requirements of a ramifying new industry will soon force the adoption of a common terminology, standard units, refined instruments of measurement, and finally, the collection of electrical practice and theory under a single conceptual roof.

If this is the situation with respect to the electrical current and its new-found applications, the case of magnetism is significantly different, for the magnetic compass has long been established as a working tool in both navigation and in surveying—the latter especially in the United States, where major parts of the West are still *terra incognita*. Not only

Fig. 1.16. Construction of a telegraph line over the Missouri river.

is there a continuing demand for improved nautical compasses—a need Thomson will take the lead in filling—but precise mapping of the curiously irregular and changing magnetic field of the earth has for a long while been a matter of high military and commercial priority. Such exact and internationally coordinated measurements have demanded refined instrumentation based on agreed and reproducible standards, built on sound theoretical structuring of systems of units and measurement. This work, at once theoretical and practical, has been carried out exactingly, beginning some twenty years ago, by Wilhelm Weber and Karl Gauss in Göttingen; they organized precise, widespread investigations through a Magnetic Union, whose "Results" were published annually for several years.[9] To a large extent, however, magnetism is at this point still treated as a science and technology complete in itself. Although Weber personally has extended his work to include an action-at-a-distance theory of the relations of magnetism and electricity, the magnetic studies (with their instruments, standards, and procedures) tend to constitute a realm

Fig. 1.17. An early reciprocating electric motor, conceived in the image of the steam engine.

of special interest in their own right without a unifying relation to electricity in a single inclusive science.

We turn now to look more systematically at the state of affairs in our benchmark year in six principal domains. These will be first those of electrostatics, magnetism, and current electricity, each treated separately, and then the chief phenomena which link them: the Oersted effect, electromagnetism, and electrodynamics. Admittedly, as we paint

Fig. 1.18. An electroplating establishment where the generator still uses revolving permanent magnets.

Fig. 1.19. Magnetic observation station of Weber and Gauss. Precise knowledge of the earth's field is becoming essential for navigation.

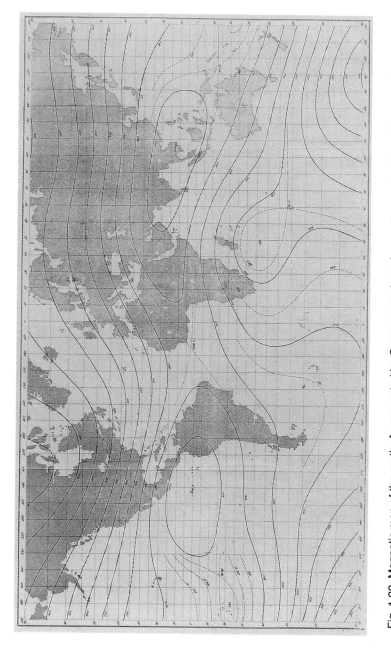

Fig. 1.20. Magnetic map of the earth. As computed by Gauss, a product of very advanced mathematics.

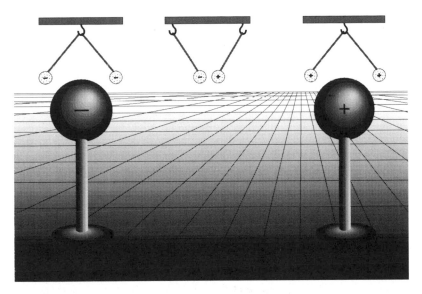

Fig. 1.21. An arrangement of an electrostatic apparatus.

this picture, we will at times compromise our adherence to the style and perceptions of 1855 for the sake of clarity for the modern learner.

## Electrostatics

The phenomena of "static," or frictional, electricity—by which electricity was first known to the ancients and from which it takes its name—has by 1855 been studied by generations of natural philosophers working in the traditions of Newton or Descartes.[10] Basic principles and the law of action of the force between so-called "charged" bodies are well established, and can easily be summarized. Objects are known to become "charged" in two ways, called "positive" and "negative," governed by the rule that "like charges repel; unlike charges attract." This electric charge is most often thought of as a fluid, flowing through conductors or trapped upon insulators. Whether there are in truth two distinct fluids or an excess or defect of a single fluid has not been resolved other than by conventions of parlance. There are some, among them Faraday, who doubt the existence of such a "fluid" at all.

It is easy to generate such charges for personal experimenting or even to conduct inadvertent experiments when walking across a carpet and touching a metal object in the case of dry weather or a heated environment. The paradigmatic experimental apparatus is a ball of pith suspended by a thread, together with a metal sphere on an insulating stand (Fig. 1.21). In the figure, the large sphere on the left has been given a "negative" charge by contact with a rubber rod rubbed with fur; that on the right has been "positively" charged by contact with a glass rod rubbed with silk. Thereafter a set of small, light pith balls have been charged by contact with the spheres and are shown appropriately repelling or attracting one another in dutiful obedience to the rules.

Thus far, we have spoken only in qualitative terms. The quantitative question becomes Newton's: that of the "law of force," i.e., precisely how will the force between two charged bodies decrease as the distance between them is increased? Coulomb showed in extensive, exacting experiments toward the end of the eighteenth century that this law for electric charges is of the same form as Newton's gravitational law governing the force between two masses, such as the sun and a planet: the force *decreases* with the square of the *increasing* distance.[11] In symbols:

$$f \propto \frac{q_1 q_2}{r^2}$$

where $q_1$ and $q_2$ represent the two charges, $r$ the distance between them, and $f$ the force between them.

In this stark "inverse-square law" of attraction or repulsion, we notice that no allowance is made for any process intervening to propagate the effect between the two bodies or of the *time* that that might take; the effect is conceived as mystical and instantaneous. Maxwell's theory will show that there is in fact a process, that time is required, and that this action-at-a-distance equation is thus, strictly, not true. Hence, interestingly, though this law seems in texts even today to constitute a secure foundation for the science of electricity, we might say that the whole effect of Maxwell's work will be to displace it, leaving to it just the role of a limiting equilibrium case in which time can be ignored!

The theory thus being silent about any role for an intervening medium, its point of view is that which has been termed by the tradition "action-at-a-distance." Since the equation accounts exactly for all the observed phenomena and correctly predicts all the resulting motions of the bodies, we must ask whether any more could be demanded by the rational mind? Here, perhaps, is the leading question with which all of Maxwell's work on electricity and magnetism begins.

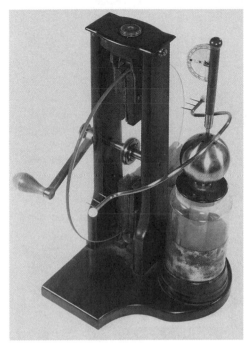

Fig. 1.22. A frictional electrostatic machine, with Leyden jar for collection of the output. At the top, a pith ball with a scale measures the accumulated charge.

A strong statement of a positive demand to confine reason strictly to the evidence alone, a kind of severe parsimony of the rational, is given by Ampère:

> First of all to observe the facts, varying the conditions as much as possible and accompanying this with precise measurement in order to deduce general laws based solely on experiment; then to deduce from these, independently of any hypothesis concerning the nature of the forces which produce the phenomena, the mathematical value of these forces, which is to say the formula which represents them—such is the road which Newton followed. This has been the approach generally adopted by the learned men of France to whom physics owes the immense progress which has been made in recent times, and it is this which has served as guide for me in all my researches into electrodynamic phenomena.[12]

Can we meaningfully go further, to ask *by what means* one body exerts a force on another over a distance? It is Ampère's stark account of science from a mathematician's point of view that stands as the challenge for

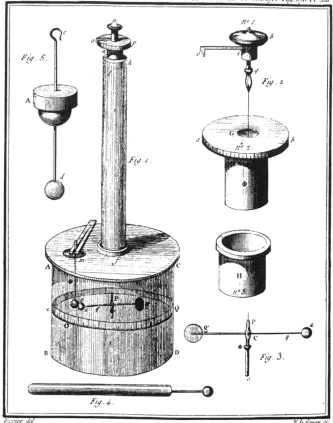

Fig. 1.23. Coulomb's torsional balance, for demonstrating the inverse-square law. Two bodies, one on a rotating arm, are similarly charged by contact. A calibrated suspension wire measures the repulsion beween them as a function of their separation.

both Faraday and Maxwell. We have seen how restless Faraday is with any mathematical surrogate for actual *explanation,* and Maxwell in the papers we are to read follows Faraday in an immense effort to fill the gap with an account of an electromagnetic "field."

One very powerful analytic advance has been the introduction of the concept of *potential:* in the electrical case, the potential of a point at distance $r$ from a charged body is defined as the *work per unit charge* that it would be necessary to do to carry a positive test charge from infinity to that point (Fig. 1.24). This work is then said to be stored in the field as "potential energy."

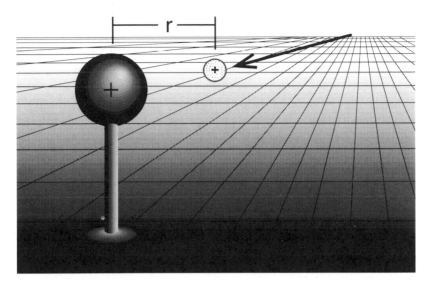

Fig. 1.24. The concept of electric potential. The potential at any point at distance r from the large sphere is the work required to bring unit positive test charge from infinity.

"Energy" is a term we will meet often in Maxwell's papers. In general, *energy* is defined as *the ability to do work.* It may take either *potential* or *kinetic* form: *potential energy* is energy of position, while *kinetic energy* is energy of motion: a ball at the top of a hill has potential energy, which it will exhibit if allowed to roll. A rolling ball on the other hand has kinetic energy, which it will exhibit if something gets in its way. The charged test sphere of Fig. 1.24 has energy by virtue of its position, which it will exhibit if it is released and allowed to fly off as a result of the repulsive force acting on it. In situations such as this, and in the gravitational case, energy is said to be *conserved:* as the ball rolls down hill, the kinetic energy gained is just equal to the potential energy lost, barring friction. When we speak of electric "potential," in the first instance we really mean the potential energy stored in the region of a charged body.

A crucial advantage of the "potential" concept is that since it is a measure of *work* it is not directional. The potential belonging to a point is represented by a simple number, while the *force* acting on a test body at the same point would be represented by a "vector" quantity, one which

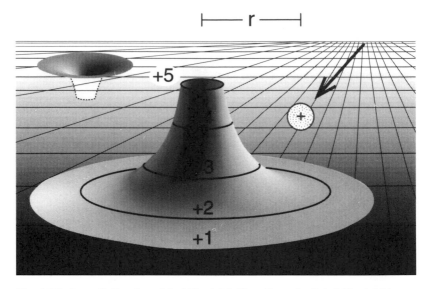

Fig. 1.25. A gravitational model of Fig. 1.24. Here the potential of Fig. 1.24 is represented by the height of the surface in a gravitational model. The "well" in the distance represents the potential of a negative body.

must designate two things: both *magnitude* and *direction*. Use of the "potential" concept thus radically simplifies many computations, as will become apparent as we go along. As we shall see, potentials can be added up as simple numbers. The potential concept thus lends itself to the methods of the integral and differential calculus, which are based on summation, and powerful theorems in this mode had been developed based ultimately on Coulomb's equation for actions of charges upon one another over distances.

The concept of potential and the formal theories for dealing with it were developed in application to problems of gravitational attraction before they were applied to electricity. In fact, a simple gravitational analogy popular in twentieth-century texts may help the reader in thinking about electrical situations. Over small regions on the earth's surface—where gravity may be taken as uniform and parallel (simply, "down")—we do work in carrying an object, or ourselves for that matter, up a hill. The work done is directly proportional to the vertical height we rise. As we have seen it is stored as potential energy, and can be recovered by falling down the hill or simply descending in a more

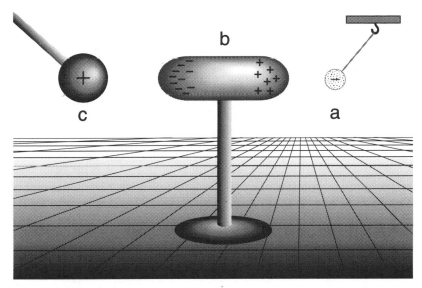

Fig. 1.26. Electrostatic induction.

orderly manner in any direction we choose. An imaginary terrain, with if we wish an accompanying "topographic map," may therefore be constructed to help in visualizing an electrostatic configuration. Thus, we may draw, in Fig. 1.25, a symmetric "mountain" to model the electric potential everywhere in the region shown in Fig. 1.24. Similar, more interesting terrains may be designed to aid in interpreting the potential distributions implicit in other figures we draw—the reader is invited to envision one for the arrangement of Fig. 1.21.

Despite these powerful mathematical techniques, in 1855 many phenomena quite common in laboratory situations remain resistant to formal mathematical treatment. One such situation is depicted in Fig. 1.26. Here b is initially an uncharged metal body on an insulating stand; we may if we wish think of the metallic conductor as containing positive charges which are able to move freely. The positively charged sphere c as it approaches has repelled the positive charges to the far end of the insulated conductor, as proved by the test charge a, leaving the near end negative. We say that b has been charged by "induction": though its total charge remains zero, charges will have redistributed themselves, attracted or repelled by c according to their signs. If b could be taken apart in the middle, the right-hand piece would carry away its positive charge while the remainder would be left as a negatively charged body.

But what exactly will be the resulting distribution of charge on b, shown qualitatively in Fig. 1.26 ? Even with the advanced mathematical techniques available in 1855, the question can be answered quantitatively only for special, convenient choices for the shape of b. Thomson is just in the process of unveiling a powerful method for tackling problems of this kind.

The picture changes greatly if we look at electrostatics for a moment from Faraday's point of view. He is indeed far from the Newtonian tradition. Innocent of mathematics, he does not work with functional relationships such as that expressed in Coulomb's force law; he tends not to think in terms of ratios or proportions. He not only almost never writes an equation, but he never asks the kind of question which has an equation as the natural form of its answer. His attitude, then, toward "Coulomb's law" is one which approaches desperate antagonism: he rebels at the formulation "with a strength VARYING INVERSELY. . . ." The capital letters are his, expressing his outrage at what he considers a blatant violation of the principle of conservation of the force: how can it then "vary"? He understands in some way, we must suppose, the algebraic relation as a description of the effect, but the equation seems to Faraday utterly unjust to the force. "Why, then, talk about the inverse square of the distance?" he asks, commenting on the dismissal of his own theory by the astronomer-royal, Sir George Airy: "I had to warn my audience against the sound of this law and its supposed opposition on my Friday evening. . . ."[13]

To see how Faraday approaches the phenomena that we have been discussing from other points of view, let us look at representative passages from "Series XI" of the *Experimental Researches,* devoted to electrostatics. For Faraday the crucial term is *induction:*

> Amongst the actions of different kinds into which electricity has conventionally been subdivided, there is, I think, none which excels, or even equals in importance that called *Induction.* It is of the most general influence in electrical phenomena, appearing to be concerned in every one of them, and has in reality the character of a first, essential, and fundamental principle.[14]

We have seen this phenomenon illustrated in Fig. 1.26. Though it presents problems for exact calculation, it might seem in principle comfortably in accord with the Coulomb theory of forces acting at a distance according to a mathematical law. But for Faraday it appears as a crucial *alternative* to any theory of forces acting at distances, and such laws leave him supremely ill at ease:

> I was led to suspect that common induction itself was in all cases an *action of contiguous particles* . . . and that electrical action at a distance (i.e., ordinary inductive action) never occurred except through the influence of the intervening matter.

What does he mean in renaming electrical action at a distance "ordinary *inductive* action"? It seems that there is no "action at a distance" for Faraday:

> At present I believe ordinary induction in all cases to be an action of contiguous particles consisting in a species of polarity, instead of being an action of either particles or masses at sensible distances; and if this be true, the distinction and establishment of such a truth must be of the greatest consequence to our further progress in the investigation of the nature of electric forces.

Figure 1.26, then, becomes for Faraday the paradigm for *all* electrical actions. He seems to interpret it in this way: body b intervenes between c and a, by its own polarization transmitting the force that c has to a. It is clear that in simpler direct actions of one body on another, such as those of Fig. 1.21 in which no evidence of induction is apparent, Faraday yet perceives in his mind's eye intervening "contiguous particles" at work. One decisive piece of evidence for this would be, he imagines, induction occurring "in curved lines"—Coulomb's law refers to a force acting *directly along the straight line* between the bodies. Utterly committed to one of science's truly great creative errors, Faraday apparently never came to understand that curved lines of force can arise perfectly naturally in the mathematical theory by what is called the "composition of forces," a concept fundamental to the *Principia,* but unknown to him (Fig. 1.27).

The attention that Faraday's new view draws to the role of a presumed intervening medium as "transmitting" the static electric force has led him to focus on the effects of various interposed substances, for which he has introduced the term "dielectrics," and to define their "specific inductive capacities." He virtually eliminates from his own thinking the concept of the electric "charge" as any sort of substance or fluid—what appears to the world to be a "charge" *on* a conductor is in his view never so: it is in every case a state of affairs, a polarization, of the intervening medium—even if there is no medium to be found in a given case but the vacuum itself.

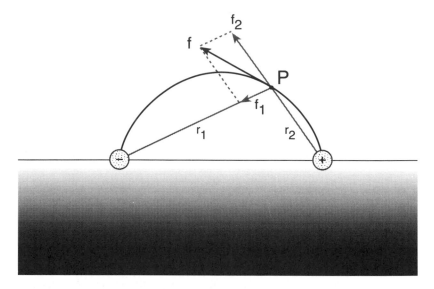

Fig. 1.27. Induction in curved lines.

All the figures we have drawn must, then, be reinterpreted—if we are to try on Faraday's view for a moment—removing the "charges" labeled on the objects and filling the spaces with polarizations instead. The polarization altogether replaces what had been the "charge." Every charge is inherently dual: every induced positive charge necessarily entails an equal induced negative charge as well. Note that this means that the *earth* has a crucial role in many of these figures: if no body is ever really "charged" except *with respect to other bodies,* then a seemingly isolated charge is really polarized with respect to some object in its surroundings, such as the floor beneath it, the walls, or the earth itself:

> Another ever-present question on my mind has been, whether electricity has an actual and independent existence as a fluid or fluids, or was a mere power of matter, like what we conceive of the attraction of gravitation. . . . It was in attempts to prove the existence of electricity separate from matter, by giving an independent charge of either positive or negative power only, to some one substance, and the utter failure of all such attempts . . . that first drove me to look upon induction as an action of the particles of matter, each having *both* forces developed in it in exactly equal amounts. . . . Charge always implies *induction.*

Faraday made immense efforts to test his hypothesis by charging a body "absolutely"—*really* to electrify one single body in isolation and hence not merely relatively. This effort led him to one of the memorable experiments of the nineteenth century. He built a sizable chamber—a 12-ft cube of frame and copper wire—in his lecture hall at the Royal Institution and he charged it with all the combined power of his electrical machines. Sparks must have been flying in all directions and the atmosphere redolent of ozone!

> I went into the cube and lived in it, and using lighted candles, electrometers, and all other tests of electrical states, I could not find the least influence upon them . . . though all the time the outside of the cube was powerfully charged, and large sparks and brushes were darting off from every part of its outer surface. The conclusion I have come to is, that non-conductors, as well as conductors, have never yet had an absolute and independent charge of one electricity communicated to them, and that to all appearance such a state of matter is impossible.

There was no trace of an electric fluid *in* the chamber, rather only a tense polarization of the space *surrounding* his cube. To this polarized medium, which, it seems, was all that electricity had ever been, Faraday at times gives a name of momentous significance for the future of electrical science: he sometimes calls it the "field."

## Magnetism

Magnetism, as frictional electricity, is a phenomenon which has long been known. Magnets are in some sense "polar," with the end of the bar magnet which points toward the geographical north being known as the "north" pole. By contrast with charged bodies, which *appear* to act alone, it is always very evident in the case of magnetism that for each pole, there is a counterpart, opposite pole. Each body which has a north pole has also an equal and opposite south pole. The simple governing rule for the force between charges has its precise counterpart for magnets: "like *poles* repel; unlike *poles* attract." Again, quantitative investigation leads to a law of force between poles—and again the force law takes the form of the inverse square:

$$f \propto \frac{m_1 m_2}{r^2}$$

In this case, however, there is something disturbingly artificial about experiments carried out with "isolated" poles, for we have observed from the outset in the case of magnetism that no pole is in fact ever "isolated"—no north without its corresponding south. While it seemed reasonable, at least prior to Faraday's critique, to believe that isolated *charges* had been produced, it was manifest that poles always occurred in pairs. At most, to produce the effect of an *isolated* pole we can only hide the opposite pole from the vicinity of an experiment, removing it as far as possible by contriving a very long, slender magnet.

Faraday, on the other hand—never one to be interested in "poles" or a "law of force" based upon them—has carried out extensive experiments with magnets by tracing "lines of force," using small compasses, or iron filings, to produce patterns of the sort we have seen earlier (Fig. 1.11). Here, as in the case of electrostatics, William Thomson has found powerful ways, following the French theorist Poisson, to incorporate Faraday's findings into formal theory.[15] But again, Thomson has stopped short of constructing a single, coherent formal theory and thus Faraday's point of view remains very much his own. Maxwell, we know, has been studying Thomson's electrical papers, but they imply that the major work of following Faraday's insights to a single vision of electromagnetism still lies ahead. In magnetism as in electrostatics, for Faraday everything points toward a crucial role for an intervening medium.

Why does a lodestone pick up pieces of soft iron that are themselves initially unmagnetized? The conventional answer, couched in terms of "poles," is that the soft iron, initially unmagnetized, becomes polarized *by induction.* As it thus becomes magnetized, its particles align under the action of a magnet in such a way that, for example, a new south *pole* is induced in the soft iron by the north *pole* of an approaching magnet and the two poles in turn "attract" (Fig. 1.28). With these supposed "poles" taken as centers of force, such a "polar" theory fits well with the point of view which bases all analyses on a Newtonian force law between such force centers, only now with the complication of multiple or distributed centers. For Faraday, however, all this seems a mathematician's artifice and altogether unpersuasive.

About a decade before Maxwell enters the discussion with "Faraday's Lines," Faraday has experienced what has evidently been one of the most exciting moments of his dramatic scientific life. Using a new, powerful electromagnet which had just then become available to him, he discovered an astonishing magnetic effect: "magnetism" is not limited to iron and the very few closely related substances to which it has always seemed confined. Suddenly, other substances have revealed a slight but distinct magnetic

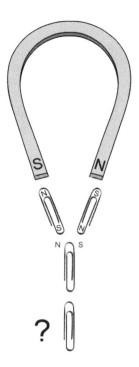

Fig. 1.28. Magnetism by induction.

property, in a sense the opposite of that of iron. They do not align as iron does to point *with* the magnetic axis between "north" and "south" poles, but rather—*repelled* under induction—point *across* the axis: when shaped as needles, pointing *east and west* rather than north and south (Fig. 1.30).

These Faraday has distinguished as "diamagnetic" substances, while iron and other materials which act as it does he has called "paramagnetic." In the course of his investigation of this new effect, Faraday has tried a wide variety of readily available substances, ranging from rock crystal through caffeine, sealing wax, mutton, beef, and blood—each "fresh" and "dried"—to leather, apple, and bread. As always, the candid humanity of Faraday dispels without effort what has since set as the hard shell of scientific reporting. He tells us:

> It is curious to see such a list as this of bodies presenting on a sudden this remarkable property, and it is strange to find a piece of wood, or beef, or apple obedient to or repelled by a magnet. If a man could be suspended, with sufficient

Fig. 1.29. The great Royal Institution electromagnet.

delicacy . . . and placed in the magnetic field, he would point equatorially; for all the substances of which he is formed, including the blood, possess this property.

What he has found—"on a sudden"—is, he says, "the existence of a magnetic property in matter, new to our knowledge."

This discovery came in 1845.[16] Five years later, he is proposing a way of understanding this as well as ordinary paramagnetic induction in terms of the role of a *medium:*

The remarkable results given in a former series of these *Researches* . . . led me to the idea, that if bodies possess different degrees of *conducting power* for magnetism, that difference may account for the phenomena; and, further, that if such an idea be considered, it may assist in developing the nature of magnetic force. . . .

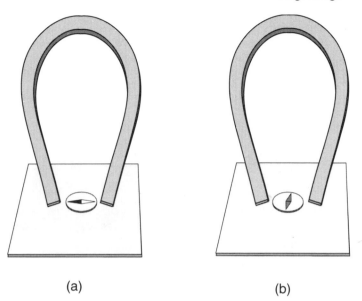

(a)                                                        (b)

Fig. 1.30. Paramagnetism and diamagnetism.

If a medium having a certain conducting power occupy the magnetic field, and then a portion of another medium or substance be placed in the field having a greater conducting power, the latter will tend to draw up towards the place of greatest force, displacing the former. Such at least is the case with bodies that are freely magnetic, as iron . . . and such a result is in analogy with the phaenomena produced by electric induction. If a portion of still higher conducting power be brought into play, it will approach the axial line and displace that which had just gone there; so that a body having a certain amount of conducting power, will appear as if attracted in a medium of weaker power, and as if repelled in a medium of stronger power by this differential kind of action. . . .

In Faraday's mind, revealed here at work, we see "attraction" and "repulsion" unveiled once again as mere *appearances,* the consequences of something perhaps really happening in the medium itself. He cannot avoid the conclusion that "mere space is magnetic," and that two figures incorporating the lines of force will represent the magnetic properties of all matter (Fig. 1.31). He explains his two diagrams in this way:

When a paramagnetic conductor, as for instance, a sphere of oxygen, is introduced into such a magnetic field, considered previously as free from matter, it will cause a concentration of the lines of force on and through it, so that the space occupied

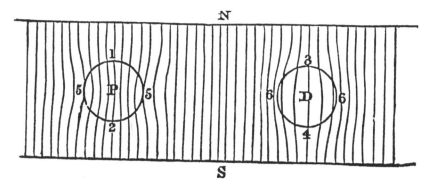

Fig. 1.31. Faraday's interpretation of diamagnetic and paramagnetic action in terms of conduction. The left and right configurations correspond, respectively, to fig. 1 and fig. 2 of Faraday's account.

by it transmits more magnetic power than before (fig. 1). If, on the other hand, a sphere of diamagnetic matter be placed in a similar field, it will cause a divergence or opening out of the lines (fig. 2); and less magnetic power will be transmitted through the space it occupies than if it were away.

It is as "good conductors" that paramagnetic bodies align *with* the field; while diamagnetics, as "poor conductors," move *out* of the field (and, hence, when given an appropriate needle shape, point "equatorially," *across* the field, as had been observed.)

The lines traced by the filings around a simple bar magnet inexorably advance in his mind from "mere representations" to an insistent suggestion of the presence of substantive beings. The body of lines about the isolated bar magnet of Fig. 1.10 is not for him merely a diagram, but an entity of some sort operative in the world. It looks to him beetle-like in shape. Thus when he asks a learned friend for an appropriate Greek word, as is his custom, the answer comes back "sphondyloid" (beetle-form, in Greek), and in his usage becomes the *sphondyloid of power.*[17] Forces between magnets begin to look like interactions between their sphondyloids, which may align so that their lines of force "coalesce," and the sphondyloids converge, or be compressed and thus repel—his reading of the patterns of interaction in Fig. 1.10.

Ongoing speculations of this sort on Faraday's part have continued to the very point of publication of the *Researches.* In "On Some Points of Magnetic Philosophy," in the *Philosophical Magazine* for February

1855, he writes: "Within the last three years I have been bold enough, though only as an experimentalist, to put forth new views of magnetic action." Here, he reviews alternative "general hypotheses of the physical nature of magnetic action," and locates among them his own:

> My physico-hypothetical notion does . . . not profess to say how the magnetic force is orignated or sustained in a magnet. . . . Accepting the magnet as a centre of power surrounded by lines of force, which, as representants of the power, are now justified by mathematical analysis . . . it views these lines as *physical* lines of power, essential both to the existence of the force within the magnet, and to its conveyance to, and exertion upon, magnetic bodies at a distance.[18]

This is the point, then, of a high pitch of Faraday's thought, at which Maxwell takes up the work. It seems likely that these intense speculations about magnetism—fueled in part by debates in the current literature—more than anything else may have led Maxwell to work on his own task in "Faraday's Lines."

## Current Electricity

In the absence of a truly practical electrical dynamo or "generator," virtually the sole source of an electric current is the chemical cell. This is commonly referred to as the "galvanic" cell, and the current spoken of as "galvanic electricity." Especially with the new demands of the telegraph, a considerable technology of galvanic cells has sprung up in a great variety of types involving different electrolytes and electrodes and a range of configurations.

The presence of an electric current in a wire can be shown electrolytically (by causing a chemical action in a conducting solution, or electrolyte), or (as we shall see) by the magnetic action produced by the current. By such means, the intensity of the current can be indicated, though definitions and standards are lacking by which to express results in generally comparable units. It is also possible to use delicate electrostatic means to determine the potential difference between the terminals of the cell that is driving the current and, with this known, to look for a relation between the driving potential and the resulting current intensity. This relation depends as well, of course, on the nature of the material of the wire; some substances such as copper are good "conductors" of current and others either insulators or poor conductors.

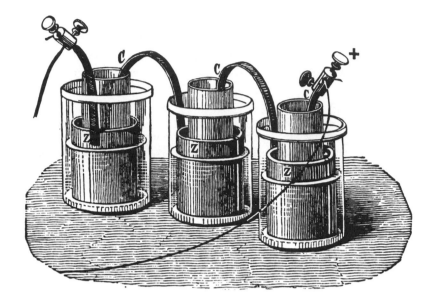

Fig. 1.32. A battery of Daniell cells.

Ohm showed that for a given material the intensity of the current through a conductor was directly proportional to the potential difference across it.[19] If we use the symbol $I$ to represent the intensity of the current, and E to represent the potential or "electromotive force" driving the current, we can express this in symbols as a proportionality:

$$I \propto E$$

We may now introduce a parameter to indicate the resistance of the conductor, $R$. This will appear as a constant factor, or "constant of proportionality," relating current to potential in "Ohm's law." Since high resistance means small current for a given potential, $R$ appears in the denominator:

$$I = \frac{E}{R}$$

To write such an equation, however, we must have established some agreed and consistent system of units; as we shall see, Maxwell will become very much involved in the establishment of such a system, which in fact will in a surprising way lie at the center of his own theory. At the outset of his work, however, such a system of practical units is still

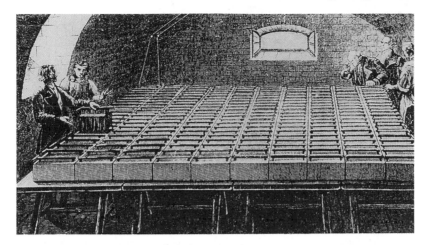

Fig. 1.33. The great battery of the Royal Institution. The "series" connection will produce very high voltages.

lacking, so that we are anticipating a bit in writing Ohm's law in this simple equation form. Today, what are termed "practical units" for the measurement of these quantities are respectively, the ampere, the volt, and the ohm.

A very simple analogy to water flow through a narrow pipe is very helpful in thinking about electric currents. Though we would not want to take from it any suggestion of an actual electric fluid in the wire corresponding to water in the pipe, there will be no harm in pointing out that electrical potential difference corresponds to water pressure in the pipe. As the pipe in turn offers "resistance" to the flow of water, so a wire offers resistence to the flow of the electric current. We can use an electrometer such as shown in Fig. 1.7 to measure potential; it becomes our electrical "pressure" meter. If we presuppose a corresponding meter for measuring current—an assumption we will justify in the next section—we may represent in Fig. 1.34 a circuit to which Ohm's equation applies. In this figure—in which, admittedly, for the sake of clarity for the reader new to electricity we have introduced certain more modern simplifications—V denotes a meter (now termed a "voltmeter") for the measurement of potential; A denotes a meter (an "ammeter") for the measurement of current flow, and R denotes a device (a "resistor") deliberately introduced to control current flow. The battery, which might be the Daniell cells of Fig. 1.32, with positive and negative plates, is

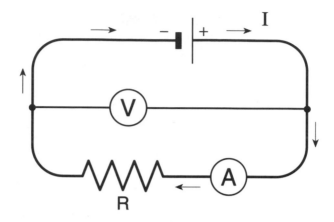

Fig. 1.34. Modern conventional representation of a battery passing current through a resistor. Current passes *through* the ammeter A, which measures flow in amperes. The voltmeter V is connected *across* the circuit to measure "pressure," the electromotive force *E* in volts.

drawn at the top. Current, indicated in the diagram by an arrow, is conventionally thought of as flowing out of the positive terminal of the battery and *through* the ammeter and the resistance. The voltmeter, whose purpose is merely to test the potential or electric "pressure" supplied by the battery, is placed *across* its terminals; the circuit's current does not flow through it. Battery, ammeter, and resistor constitute a *series* circuit (connected like elephants trunk-to-tail in a parade); a single current flows through these series elements. The voltmeter, through which the current does not flow, is attached in *parallel* to this series circuit.

In all of this, it is important to keep in mind that with neither the terminology nor the units and standards of measurement agreed upon, discussions which later seem quite elementary in fact in 1855 require massive efforts of translation and interpretation.

## The Oersted Effect

A phenomenon, altogether unanticipated at the time, had been discovered in 1820 by Hans Christian Oersted.[20] He had noted that an electric current has an effect at a distance upon a magnetic needle. Careful study showed that in the case of a long straight conductor, the orientation of the needle would be perpendicular to the radius joining the center of the needle with the wire (Fig. 1.35).

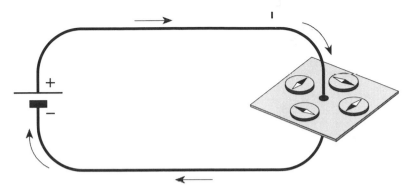

Fig. 1.35. The Oersted effect.

The force reversed with reversal of the direction of the current. By balancing the force of the current on the needle against that of the earth's magnetic field as a reference, it was possible to show that the force was directly proportional to the magnitude of the current in the wire (as measured, for example, by an electrolytic cell) and inversely proportional to the distance from the wire. If the wire is wound in a circular coil of many turns, the effect in the center combines and the result becomes a sensitive instrument—the *galvanometer*—of immense utility by which electric currents can be conveniently measured. Thomson's practical form of galvanometer, used in connection with the cable-laying enterprise, was illustrated in Fig. 1.7. Figure 1.36 illustrates another, designed by Weber with radical simplicity of form to achieve precise measurements of electrical quantities. Weber's instrument illustrates the principle of the tangent galvanometer. Here the magnetic field of the current, which lies along the axis of the coil, is oriented at right angles to that of the earth so that the field of the current is balanced against that of the earth. As a result, the compass needle deflects through an angle whose tangent measures the magnitude of the current.

The Oersted effect of many turns of wire, when wound on an iron core and driven by a strong battery, can be used to produce very powerful magnetic fields, as in the case of the great Royal Institution electromagnet used by Faraday (Fig. 1.29). Efforts to derive practical mechanical power from the Oersted effect suggest motors of the form of Fig. 1.17.

For mathematicians who have really no ground to stand on other than force laws of Newton's type, this phenomenon is deeply disturbing. Forces ought to act in direct lines *between centers.* Newton's laws of

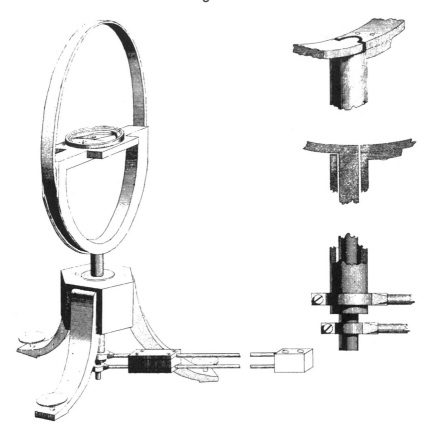

Fig. 1.36. Tangent galvanometer designed by Weber, for making absolute measurements.

motion assumed this, for how otherwise could there be a balance of action and reaction—if a ponderable body were to be moved by the Oersted force, *what* would be pushing against *what* in reaction (against what would the force on the body get any foothold or leverage?). For Faraday, however, conveniently unconcerned with Newtonian problems, a force moving in curved lines presented no difficulty, but rather evidence for the role of a medium. Working with this "curved" force, he has been able to devise an ingenious apparatus for extracting from the circular "curves of magnetic power" a *continuous circular motion.*

The question of the possibility of producing continuous rotations from the magnetic field of a current thus had become an issue under serious contention and a broad challenge to both inventive genius and mathematical theory in force in 1821, immediately following Oersted's obscure announcement of his discovery. It was in pursuit of this question that Faraday began his researches on electricity at the Royal Institution in earnest. He soon produced a device which solved the problem, drawing on ingenuity which never ceased to impress Maxwell. On the one hand, Faraday had shown that the way was, in principle, open to the electric motor, continuously extracting energy from the electric field in the form of mechanical motion (energy which must of course be supplied, equally continuously, from some electric source). On the other, he had given striking evidence—to himself decisive, though as we shall see, never ultimately convincing to the mathematicians—of the real existence of *curved lines* of the magnetic force.

## Electromagnetic Induction

Once the Oersted effect had become known, it was felt by many that there ought in principle to be a second, symmetric effect: if the electric current gave rise to magnetism, then in some way, magnetism should give rise to an electric current. Others had labored to demonstrate the effect, but Faraday hit upon the answer: it was not magnetism *per se* but the *moving* magnet, or *changing* magnetism, which would produce the result.[21] Moving a magnet into or out of a coil of wire produces an "induced" current in the coil, as evidenced by a galvanometer (Fig. 1.37).

Further, we know that by the Oersted effect, a coil of wire carrying current becomes itself a magnet. If one coil of the sort we have just discussed is used in this way to act as a magnet (the "primary"), then another (the "secondary") connected to a galvanometer can be used to reveal another version of the effect Faraday discovered. The effect is strongest if they are linked by a common iron core (Fig. 1.38). *Interrupting* the current in the first will produce a sudden *change* in the magnetic force at the second, resulting in a surge of current through the galvanometer to which it is connected. Interruption of the current in the primary of Fig. 1.38 is equivalent to removing the magnet of Fig. 1.37; initiation of the current corresponds to inserting the magnet. The current is again said to be "induced" in the secondary, and it is found that the induced current is proportional to the *rate of change* of the current in the primary.

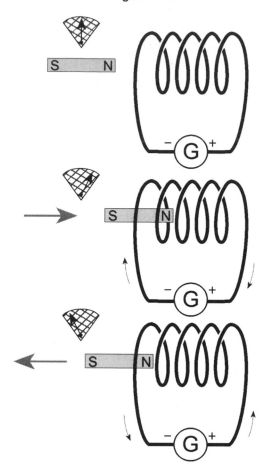

Fig. 1.37. Electromagnetic induction.

## Electrodynamics

Since the Oersted effect shows that a current-carrying conductor acts like a magnet in exerting a force on a magnetic needle at a distance, and hence is affected *by* a magnet as well, it may not seem surprising that two current-carrying conductors exert forces on one another. The general realm of mechanical actions of one current on another has become known as "electrodynamics." This seems yet another case of action of a force at a distance but, again, it defies analysis in the Newtonian manner. Newton assumes that two bodies will be joined by a force, attractive or

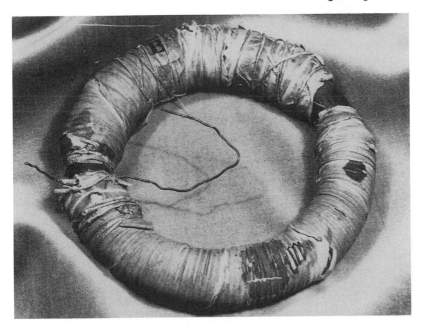

Fig. 1.38. Coil used by Faraday at the Royal Institution: One winding serves as the primary and the other as the secondary.

repulsive, acting along the line of their centers. The electrodynamic force, by contrast, seems to be neither one of attraction nor repulsion, but in this case to act in a strangely *sidewise* manner. In Fig. 1.39, current $i$ in the conductor $AB$ demonstrably causes a force $f$ on a second conductor $CD$ in the direction *perpendicular* to the line between $AB$ and $CD$! As in the case of the Oersted effect, there seems no way to apply Newton's crucial third law, requiring an equal and opposite reaction to every action. There is apparently no way to bring these effects within the compass of classical force laws.

It is at this critical point that André-Marie Ampère intervenes with a general force law for electrodynamics which in principle saves the science for Newton.[22] Basing his work on a series of ingenious experiments with wire frames carrying currents, he interprets the results in terms of forces between *current-elements*. (Perhaps we should speak of these as "putative" current-elements, since only by an act of heroic abstraction can such "elements" be removed from the continuous circuits in which the currents are in fact flowing.) Ampère contrives a complex,

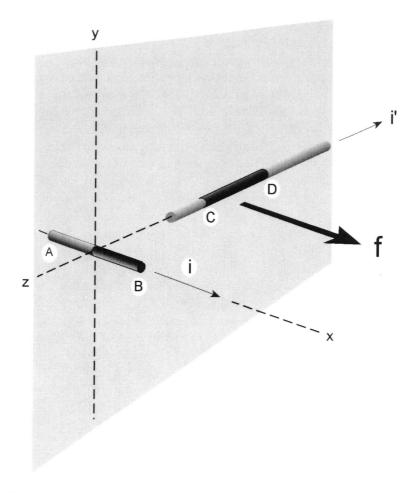

Fig. 1.39. The electrodynamic force.

rather tricky force law which indeed does preserve the principle of action directly along the line of centers of the interacting elements. *Pairs* of forces are shown to exert turning moments, and thereby to generate the mysterious rotational effect in Oersted's experiment. Further, supposing that magnetism in permanent magnets is due to permanent molecular currents in the material, Ampère is able to extend his theory to give a theoretical account of permanent magnetism and thus to link these two major oases of electrical theory. All this seems a great victory for theoretical analysis—but perhaps, a Pyrrhic victory. The complexity of

Ampère's formulas and the difficulty of applying them in practice make his theory—for many others, as for Maxwell—altogether unpersuasive.

We should also note that at the time he writes "Faraday's Lines," Maxwell is aware that the German theorists he is reading have proposed formal equations which attempt to gather these effects together, and these are important thrusts in the direction of unifying the principal phenomena of electricity and magnetism into a single science. Perhaps the most impressive of these has been an equation of Weber's. It is conceived in terms of moving charges rather than current elements or portions of a current-carrying circuit in the manner of Ampère. It consists of several terms, one of which gives the Coulomb force between static charges, while others yield Faraday's electromagnetic induction and Ampère's forces between conductors. By assuming induced molecular currents as well as Ampère's permanent currents in iron, it appears able to account for both paramagnetism and diamagnetism.

At this point, it would appear that nothing is omitted. The entire account is in terms of direct action upon one another of charges either resting or in uniform or accelerated relative motions, with no assumptions required concerning any intervening medium. It departs fundamentally from Newtonian theory in that the forces are dependent on the velocities and accelerations of the bodies and not on their positions alone. Certain important complications do arise from this, yet Maxwell is clearly impressed with Weber's theory. At the time "Faraday's Lines" is being written, it is at least possible that Weber's account will prove complete and, in its own way, faultless.

Why, then, does Maxwell describe "the present state of electrical science" as so unsatisfactory? Has not Thomson given all the clues that are needed and has not Weber found the way to full unification and understanding? The answer must be, as we have seen, that however ingenious Thomson's devices and however elegant and complete the equations of Ampère or Weber, they are all in the realm of mathematics alone. Neither explains electromagnetism to the satisfaction of the *speculative* mind. Maxwell is seeking something more than just numbers and equations.

With Maxwell, then, we are embarking on a precarious enterprise—the effort to devise an account that will satisfy the strict mathematician but at the same time go further, to meet the demand for intelligible explanation by incorporating in the very mathematics Faraday's lines of force. To achieve this would be to comprehend the world in a new way

and in the process perhaps to push the possibility of our human understanding of nature to its limits, or beyond. We shall see, as we go through
these successive papers, how the effort fares and what the limits may
prove to be of the power of our sciences to speak of a reality underlying
the numbers which move in the mathematician's equations.

# Chapter 2

# *On Faraday's Lines of Force*

We turn now without further ado to the reading of the first of the three papers of Maxwell's, "On Faraday's Lines of Force." As an appropriate text for our present study, the first half of "Faraday's Lines" is reproduced here, with the exception of certain portions whose omission is indicated by asterisks. In this and the succeeding papers, selections have been made which seem likely to yield a manageable and coherent reading, while the reader is of course encouraged to consult the original in Maxwell's *Scientific Papers.*

Following Maxwell's text are to be found our "notes" as well as a number of interpretive diagrams; these are keyed to Maxwell's text by line numbers. We might question whether it is a good idea to provide diagrams, as is done here. Maxwell published the text without illustrations and he may have done so because he wished to make the fullest demand on the reader's unassisted imagination. Most of us, however, find such figures useful and Maxwell's own support for the full use of the power of the visual imagination is evidenced in many ways. The lines of force themselves constitute a powerful appeal to the mind's eye; some of our figures are taken from his own later *Treatise on Electricity and Magnetism.* Just one example of his own special interest in diagraming and scientific visualization is a remarkable article he provided for the eighth edition of the *Encyclopedia Britannica,* of which he was the first scientific editor, on "Diagrams."

In addition to the "notes" just referred to, there are a few longer reflections on particular topics as they come along. These are referred to as "discussions" and will be found following the notes. They are numbered in sequence and referenced in the text with a notation in the form [D1], for "Discussion 1." A corresponding entry on the notes page will indicate the topic of each discussion.

On top of all this—or, rather, at the bottom—are the endnotes; for our purposes, these have been kept to a minimum. They are indicated in the usual way by superscripts in the text of the notes and are to be found gathered by chapters at the end of the book. By contrast, the footnotes that appear in the course of Maxwell's text are his own. A selective bibliography is included for each chapter; these, too, are gathered at the end of the work.

It is to be hoped that each reader will find a way to use this panoply of devices in a way that is personally most comfortable. They are meant to enable, and never to constrain. *Bon voyage!*

[From the *Transactions of the Cambridge Philosophical Society,* Vol x.
Part I.]
VIII. On Faraday's Lines of Force
[Read Dec. 10. 1855, and Feb. 11, I856]

The present state of electrical science seems peculiarly unfavourable
to speculation. The laws of the distribution of electricity on the surface
of conductors have been analytically deduced from experiment; some
parts of the mathematical theory of magnetism are established, while
5   in other parts the experimental data are wanting; the theory of the conduction of galvanism and that of the mutual attraction of conductors
have been reduced to mathematical formulae, but have not fallen into
relation with the other parts of the science. No electrical theory can
now be put forth, unless it shews the connexion not only between elec-
10   tricity at rest and current electricity, but between the attractions and inductive effects of electricity in both states. Such a theory must
accurately satisfy those laws, the mathematical form of which is
known, and must afford the means of calculating the effects in the limiting cases where the known formulae are inapplicable. In order there-
15   fore to appreciate the requirements of the science, the student must
make himself familiar with a considerable body of most intricate
mathematics, the mere retention of which in the memory materially in-
terferes with further progress. The first process therefore in the effec-
tual study of the science, must be one of simplification and reduction
20   of the results of previous investigation to a form in which the mind can
grasp them. The results of this simplification may take the form of a
purely mathematical formula or of a physical hypothesis. In the first
case we entirely lose sight of the phenomena to be explained; and
though we may trace out the consequences of given laws, we can
25   never obtain more extended views of the connexions of the subject. If,
on the other hand, we adopt a physical hypothesis, we see the phenomena only through a medium, and are liable to that blindness to
facts and rashness in assumption which a partial explanation encour-
ages. We must therefore discover some method of investigation
30   which allows the mind at every step to lay hold of a clear physical conception, without being committed to any theory founded on the physical science from which that conception is borrowed, so that it is
neither drawn aside from the subject in pursuit of analytical subtleties,
nor carried beyond the truth by a favourite hypothesis.

35   In order to obtain physical ideas without adopting a physical theory we must make ourselves familiar with the existence of physical analogies. By a physical analogy I mean that partial similarity between the laws of one science and those of another which makes each of them illustrate the other. Thus all the mathematical sciences are founded

40   on relations between physical laws and laws of numbers, so that the aim of exact science is to reduce the problems of nature to the determination of quantities by operations with numbers. Passing from the most universal of all analogies to a very partial one, we find the same resemblance in mathematical form between two different phenomena

45   giving rise to a physical theory of light.

The changes of direction which light undergoes in passing from one medium to another, are identical with the deviations of the path of a particle in moving through a narrow space in which intense forces act [D1] . This analogy, which extends only to the direction, and not to the

50   velocity of motion, was long believed to be the true explanation of the refraction of light; and we still find it useful in the solution of certain problems, in which we employ it without danger, as an artificial method. The other analogy, between light and the vibrations of an elastic medium [D2], extends much farther, but, though its importance

55   and fruitfulness cannot be over-estimated, we must recollect that it is founded only on a resemblance *in form* between the laws of light and those of vibrations. By stripping it of its physical dress and reducing it to a theory of "transverse alternations," we might obtain a system of truth strictly founded on observation, but probably deficient both in the

60   vividness of its conceptions and the fertility of its method [D3]. I have said thus much on the disputed questions of Optics, as a preparation for the discussion of the almost universally admitted theory of attraction at a distance.

We have all acquired the mathematical conception of these attrac-

65   tions. We can reason about them and determine their appropriate forms or formulae. These formulae have a distinct mathematical significance, and their results are found to be in accordance with natural phenomena. There is no formula in applied mathematics more consistent with nature than the formula of attractions, and no theory better

70   established in the minds of men than that of the action of bodies on one another at a distance. The laws of the conduction of heat in uniform media appear at first sight among the most different in their physical relations from those relating to attractions. The quantities

which enter into them are *temperature, flow of heat, conductivity.* The
75  word *force* is foreign to the subject. Yet we find that the mathematical
laws of the uniform motion of heat in homogeneous media are identical in form with those of attractions varying inversely as the square of
the distance [D4]. We have only to substitute *source of heat* for *centre
of attraction, flow of heat* for *accelerating effect of attraction* at any
80  point, and *temperature* for *potential,* and the solution of a problem in
attractions is transformed into that of a problem in heat.

    This analogy between the formulae of heat and attraction was, I be-
lieve, first pointed out by Professor William Thomson in the *Camb.
Math. Journal,* Vol. III.

85      Now the conduction of heat is supposed to proceed by an action between contiguous parts of a medium, while the force of attraction is a
relation between distant bodies, and yet, if we knew nothing more
than is expressed in the mathematical formulae, there would be nothing to distinguish between the one set of phenomena and the other.

90      It is true, that if we introduce other considerations and observe additional facts, the two subjects will assume very different aspects, but
the mathematical resemblance of some of their laws will remain, and
may still be made useful in exciting appropriate mathematical ideas.

    It is by the use of analogies of this kind that I have attempted to
95  bring before the mind, in a convenient and manageable form, those
mathematical ideas which are necessary to the study of the phenomena of electricity. The methods are generally those suggested by the
processes of reasoning which are found in the researches of Faraday,* and which, though they have been interpreted mathematically
100  by Prof. Thomson and others, are very generally supposed to be of an
indefinite and unmathematical character, when compared with those
employed by the professed mathematicians. By the method which I
adopt, I hope to render it evident that I am not attempting to establish
any physical theory of a science in which I have hardly made a single
105  experiment, and that the limit of my design is to shew how, by a strict
application of the ideas and methods of Faraday, the connexion of the
very different orders of phenomena which he has discovered may be
clearly placed before the mathematical mind. I shall therefore avoid as
much as I can the introduction of anything which does not serve as a

---

*See especially Series XXXVIII. [XXVIII (ed.)] of the *Experimental Researches,*
and *Phil. Mag.* 1852.

110  direct illustration of Faraday's methods, or of the mathematical deductions which may be made from them. In treating the simpler parts of
the subject I shall use Faraday's mathematical methods as well as his
ideas. When the complexity of the subject requires it, I shall use
analytical notation, still confining myself to the development of ideas
115  originated by the same philosopher.

I have in the first place to explain and illustrate the idea of "lines of
force."

When a body is electrified in any manner, a small body charged with
positive electricity, and placed in any given position, will experience a
120  force urging it in a certain direction. If the small body be now negatively
electrified, it will be urged by an equal force in a direction exactly opposite.

The same relations hold between a magnetic body and the north or
south poles of a small magnet. If the north pole is urged in one direction, the south pole is urged in the opposite direction.
125  In this way we might find a line passing through any point of space,
such that it represents the direction of the force acting on a positively
electrified particle, or on an elementary north pole, and the reverse direction of the force on a negatively electrified particle or an elementary south pole. Since at every point of space such a direction may be
130  found, if we commence at any point and draw a line so that, as we go
along it, its direction at any point shall always coincide with that of the
resultant force at that point, this curve will indicate the direction of that
force for every point through which it passes, and might be called on
that account a *line of force*. We might in the same way draw other
135  lines of force, till we had filled all space with curves indicating by their
direction that of the force at any assigned point.

We should thus obtain a geometrical model of the physical phenomena, which would tell us the *direction* of the force, but we should still
require some method of indicating the *intensity* of the force at any
140  point. If we consider these curves not as mere lines, but as fine tubes
of variable section carrying an incompressible fluid, then, since the velocity of the fluid is inversely as the section of the tube, we may make
the velocity vary according to any given law, by regulating the section
of the tube, and in this way we might represent the intensity of the
145  force as well as its direction by the motion of the fluid in these tubes.
This method of representing the intensity of a force by the velocity of
an imaginary fluid in a tube is applicable to any conceivable system of
forces, but it is capable of great simplification in the case in which the

forces are such as can be explained by the hypothesis of attractions
150 varying inversely as the square of the distance, such as those observed in electrical and magnetic phenomena. In the case of a perfectly arbitrary system of forces, there will generally be interstices
between the tubes; but in the case of electric and magnetic forces it is
possible to arrange the tubes so as to leave no interstices [D5]. The
155 tubes will then be mere surfaces, directing the motion of a fluid filling
up the whole space. It has been usual to commence the investigation
of the laws of these forces by at once assuming that the phenomena
are due to attractive or repulsive forces acting between certain points.
We may however obtain a different view of the subject, and one more
160 suited to our more difficult inquiries, by adopting for the definition of
the forces of which we treat, that they may be represented in magnitude and direction by the uniform motion of an incompressible fluid.

I propose, then, first to describe a method by which the motion of
such a fluid can be clearly conceived; secondly to trace the conse-
165 quences of assuming certain conditions of motion, and to point out the
application of the method to some of the less complicated phenomena
of electricity, magnetism, and galvanism; and lastly to shew how by
an extension of these methods, and the introduction of another idea
due to Faraday, the laws of the attractions and inductive actions of
170 magnets and currents may be clearly conceived, without making any
assumptions as to the physical nature of electricity, or adding anything to that which has been already proved by experiment [D6].

By referring everything to the purely geometrical idea of the motion
of an imaginary fluid, I hope to attain generality and precision, and to
175 avoid the dangers arising from a premature theory professing to explain the cause of the phenomena. If the results of mere speculation
which I have collected are found to be of any use to experimental philosophers, in arranging and interpreting their results, they will have
served their purpose, and a mature theory, in which physical facts will
180 be physically explained, will be formed by those who by interrogating
Nature herself can obtain the only true solution of the questions which
the mathematical theory suggests.

## I. Theory of the Motion of an incompressible Fluid

(1) The substance here treated of must not be assumed to possess
185 any of the properties of ordinary fluids except those of freedom of motion and resistance to compression. It is not even a hypothetical fluid

which is introduced to explain actual phenomena. It is merely a collection of imaginary properties which may be employed for establishing
certain theorems in pure mathematics in a way more intelligible to
190    many minds and more applicable to physical problems than that in
which algebraic symbols alone are used. The use of the word "Fluid"
will not lead us into error, if we remember that it denotes a purely
imaginary substance with the following property:
*The portion of fluid which at any instant occupied a given volume,*
195    *will at any succeeding instant occupy an equal volume.*
This law expresses the incompressibility of the fluid, and furnishes
us with a convenient measure of its quantity, namely its volume. The
unit of quantity of the fluid will therefore be the unit of volume [D7].
(2) The direction of motion of the fluid will in general be different at
200    different points of the space which it occupies, but since the direction
is determinate for every such point, we may conceive a line to begin
at any point and to be continued so that every element of the line indicates by its direction the direction of motion at that point of space.
Lines drawn in such a manner that their direction always indicates the
205    direction of fluid motion are called *lines of fluid motion.*
If the motion of the fluid be what is called *steady motion,* that is, if
the direction and velocity of the motion at any fixed point be independent of the time, these curves will represent the paths of individual
particles of the fluid, but if the motion be variable this will not generally
210    be the case. The cases of motion which will come under our notice
will be those of steady motion [D8].
(3) If upon any surface which cuts the lines of fluid motion we draw
a closed curve, and if from every point of this curve we draw a line of
motion, these lines of motion will generate a tubular surface which we
215    may call a *tube of fluid motion.* Since this surface is generated by
lines in the direction of fluid motion no part of the fluid can flow across
it, so that this imaginary surface is as impermeable to the fluid as a
real tube.
(4) The quantity of fluid which in unit of time crosses any fixed sec-
220    tion of the tube is the same at whatever part of the tube the section be
taken. For the fluid is incompressible, and no part runs through the
sides of the tube, therefore the quantity which escapes from the second section is equal to that which enters through the first.
If the tube be such that unit of volume passes through any section
225    in unit of time it is called a *unit tube of fluid motion* [D9].

(5) In what follows, various units will be referred to, and a finite number of lines or surfaces will be drawn, representing in terms of those units the motion of the fluid. Now in order to define the motion in every part of the fluid, an infinite number of lines would have to be
230 drawn at indefinitely small intervals; but since the description of such a system of lines would involve continual reference to the theory of limits, it has been thought better to suppose the lines drawn at intervals depending on the assumed unit, and afterwards to assume the unit as small as we please by taking a small submultiple of the stand-
235 ard unit.

(6) To define the motion of the whole fluid by means of a system of unit tubes.

Take any fixed surface which cuts all the lines of fluid motion, and draw upon it any system of curves not intersecting one another. On
240 the same surface draw a second system of curves intersecting the first system, and so arranged that the quantity of fluid which crosses the surface within each of the quadrilaterals formed by the intersection of the two systems of curves shall be unity in unit of time. From every point in a curve of the first system let a line of fluid motion be
245 drawn. These lines will form a surface through which no fluid passes. Similar impermeable surfaces may be drawn for all the curves of the first system. The curves of the second system will give rise to a second system of impermeable surfaces, which, by their intersection with the first system, will form quadrilateral tubes, which will be tubes of
250 fluid motion. Since each quadrilateral of the cutting surface transmits unity of fluid in unity of time, every tube in the system will transmit unity of fluid through any of its sections in unit of time. The motion of the fluid at every part of the space it occupies is determined by this system of unit tubes; for the direction of motion is that of the tube
255 through the point in question, and the velocity is the reciprocal of the area of the section of the unit tube at that point.

(7) We have now obtained a geometrical construction which completely defines the motion of the fluid by dividing the space it occupies into a system of unit tubes. We have next to shew how by means of
260 these tubes we may ascertain various points relating to the motion of the fluid.

A unit tube may either return into itself, or may begin and end at different points, and these may be either in the boundary of the space in which we investigate the motion, or within that space. In the first case

265 there is a continual circulation of fluid in the tube, in the second the
fluid enters at one end and flows out at the other. If the extremities of
the tube are in the bounding surface, the fluid may be supposed to be
continually supplied from without from an unknown source, and to
flow out at the other into an unknown reservoir; but if the origin of the
270 tube or its termination be within the space under consideration, then
we must conceive the fluid to be supplied by a *source* within that
space, capable of creating and emitting unity of fluid in unity of time,
and to be afterwards swallowed up by a *sink* capable of receiving and
destroying the same amount continually.

275     There is nothing self-contradictory in the conception of these
sources where the fluid is created, and sinks where it is annihilated.
The properties of the fluid are at our disposal, we have made it incompressible, and now we suppose it produced from nothing at certain
points and reduced to nothing at others. The places of production will
280 be called *sources,* and their numerical value will be the number of
units of fluid which they produce in unit of time. The places of reduction will, for want of a better name, be called *sinks,* and will be estimated by the number of units of fluid absorbed in unit of time. Both
places will sometimes be called sources, a source being understood
285 to be a sink when its sign is negative.

    (8) It is evident that the amount of fluid which passes any fixed sur-
face is measured by the number of unit tubes which cut it, and the direction in which the fluid passes is determined by that of its motion in
the tubes. If the surface be a closed one, then any tube whose termi-
290 nations lie on the same side of the surface must cross the surface as
many times in the one direction as in the other, and therefore must
carry as much fluid out of the surface as it carries in. A tube which begins within the surface and ends without it will carry out unity of fluid;
and one which enters the surface and terminates within it will carry in
295 the same quantity. In order therefore to estimate the amount of fluid
which flows out of the closed surface, we must subtract the number of
tubes which end within the surface from the number of tubes which begin
there. If the result is negative the fluid will on the whole flow inwards.

    If we call the beginning of a unit tube a unit source, and its termina-
300 tion a unit sink, then the quantity of fluid produced within the surface is
estimated by the number of unit sources minus the number of unit
sinks, and this must flow out of the surface on account of the incompressibility of the fluid.

In speaking of these unit tubes, sources and sinks, we must remem-
305 ber what was stated in (5) as to the magnitude of the unit, and how by
diminishing their size and increasing their number we may distribute
them according to any law however complicated.

(9) If we know the direction and velocity of the fluid at any point in
two different cases, and if we conceive a third case in which the direc-
310 tion and velocity of the fluid at any point is the resultant of the velocities in the two former cases at corresponding points, then the amount
of fluid which passes a given fixed surface in the third case will be the
algebraic sum of the quantities which pass the same surface in the
two former cases. For the rate at which the fluid crosses any surface
315 is the resolved part of the velocity normal to the surface, and the resolved part of the resultant is equal to the sum of the resolved parts of
the components.

Hence the number of unit tubes which cross the surface outwards
in the third case must be the algebraical sum of the numbers which
320 cross it in the two former cases, and the number of sources within any
closed surface will be the sum of the numbers in the two former
cases. Since the closed surface may be taken as small as we please,
it is evident that the distribution of sources and sinks in the third case
arises from the simple superposition of the distributions in the two for-
325 mer cases.

## II. Theory of the uniform motion of an imponderable incompressible fluid through a resisting medium

(10) The fluid is here supposed to have no inertia, and its motion is
opposed by the action of a force which we may conceive to be due to
330 the resistance of a medium through which the fluid is supposed to
flow. This resistance depends on the nature of the medium, and will in
general depend on the direction in which the fluid moves, as well as
on its velocity. For the present we may restrict ourselves to the case
of a uniform medium, whose resistance is the same in all directions.
335 The law which we assume is as follows.

*Any portion of the fluid moving through the resisting medium is di-
rectly opposed by a retarding force proportional to its velocity.*

If the velocity be represented by *v*, then the resistance will be a
force equal to *kv* acting on unit of volume of the fluid in a direction con-
340 trary to that of motion [D10]. In order, therefore, that the velocity may
be kept up, there must be a greater pressure behind any portion of

the fluid than there is in front of it, so that the difference of pressures may neutralise the effect of the resistance. Conceive a cubical unit of fluid [which we may make as small as we please, by (5)], and
345    let it move in a direction perpendicular to two of its faces. Then the resistance will be *kv,* and therefore the difference of pressures on the first and second faces is *kv,* so that the pressure diminishes in the direction of motion at the rate of *kv* for every unit of length measured along the line of motion; so that if we measure a length equal to *h*
350    units, the difference of pressure at its extremities will be *kvh.*

(11) Since the pressure is supposed to vary continuously in the fluid, all the points at which the pressure is equal to a given pressure *p* will lie on a certain surface which we may call the *surface (p) of equal pressure.* If a series of these surfaces be constructed
355    in the fluid corresponding to the pressures 0,1,2,3 &c., then the number of the surface will indicate the pressure belonging to it, and the surface may be referred to as the surface 0, 1, 2, or 3. The unit of pressure is that pressure which is produced by unit of force acting on unit of surface. In order therefore to diminish the unit of
360    pressure as in (5) we must diminish the unit of force in the same proportion.

(12) It is easy to see that these surfaces of equal pressure must be perpendicular to the lines of fluid motion; for if the fluid were to move in any other direction, there would be a resistance to its mo-
365    tion which could not be balanced by any difference of pressures [D11]. (We must remember that the fluid here considered has no inertia or mass, and that its properties are those only which are formally assigned to it, so that the resistances and pressures are the only things to be considered.) There are therefore two sets of sur-
370    faces which by their intersection form the system of unit tubes, and the system of surfaces of equal pressure cuts both the others at right angles. Let h be the distance between two consecutive surfaces of equal pressure measured along a line of motion, then since the difference of pressures = 1,

375

$$kvh = 1,$$

which determines the relation of *v* to *h,* so that one can be found when the other is known. Let *s* be the sectional area of a unit tube measured on a surface of equal pressure, then since by the definition of a unit tube

380                        $$vs = 1,$$

we find by the last equation

$$s = kh.$$

(13) The surfaces of equal pressure cut the unit tubes into portions whose length is *h* and section *s*. These elementary portions of unit
385   tubes will be called *unit cells* [D12]. In each of them unity of volume of fluid passes from a pressure *p* to a pressure (*p* – 1) in unit of time, and therefore overcomes unity of resistance in that time. The work spent in overcoming resistance is therefore unity in every cell in every unit of time [D13].

(14) If the surfaces of equal pressure are known, the direction and
390   magnitude of the velocity of the fluid at any point may be found, after which the complete system of unit tubes may be constructed, and the beginnings and endings of these tubes ascertained and marked out as the sources whence the fluid is derived, and the sinks where it disappears. In order to prove the converse of this, that if the distribution
395   of sources be given, the pressure at every point may be found, we must lay down certain preliminary propositions.

(15) If we know the pressures at every point in the fluid in two different cases, and if we take a third case in which the pressure at any point is the sum of the pressures at corresponding points in the two
400   former cases, then the velocity at any point in the third case is the resultant of the velocities in the other two, and the distribution of sources is that due to the simple superposition of the sources in the two former cases [D14].

For the velocity in any direction is proportional to the rate of decrease of
405   the pressure in that direction; so that if two systems of pressures be added together, since the rate of decrease of pressure along any line will be the sum of the combined rates, the velocity in the new system resolved in the same direction will be the sum of the resolved parts in the two original systems. The velocity in the new system will therefore be the resultant of the
410   velocities at corresponding points in the two former systems.

It follows from this, by (9), that the quantity of fluid which crosses any fixed surface is, in the new system, the sum of the corresponding quantities in the old ones, and that the sources of the two original systems are simply combined to form the third.
415      It is evident that in the system in which the pressure is the difference of pressure in the two given systems the distribution of sources

will be got by changing the sign of all the sources in the second system and adding them to those in the first.

(16) If the pressure at every point of a closed surface be the same
420    and equal to $p$, and if there be no sources or sinks within the surface, then there will be no motion of the fluid within the surface, and the pressure within it will be uniform and equal to $p$.

For if there be motion of the fluid within the surface there will be tubes of fluid motion, and these tubes must either return into them-
425    selves or be terminated either within the surface or at its boundary. Now since the fluid always flows from places of greater pressure to places of less pressure, it cannot flow in a re-entering curve; since there are no sources or sinks within the surface, the tubes cannot begin or end except on the surface; and since the pressure at all points
430    of the surface is the same, there can be no motion in tubes having both extremities on the surface. Hence there is no motion within the surface, and therefore no difference of pressure which would cause motion, and since the pressure at the bounding surface is $p$, the pressure at any point within it is also $p$.

435    (17) If the pressure at every point of a given closed surface be known, and the distribution of sources within the surface be also known, then only one distribution of pressures can exist within the surface.

For if two different distributions of pressures satisfying these conditions could be found, a third distribution could be formed in which the
440    pressure at any point should be the difference of the pressures in the two former distributions. In this case, since the pressures at the surface and the sources within it are the same in both distributions, the pressure at the surface in the third distribution would be zero, and all the sources within the surface would vanish, by (15).

445    Then by (16) the pressure at every point in the third distribution must be zero; but this is the difference of the pressures in the two former cases, and therefore these cases are the same, and there is only one distribution of pressure possible.

(18) Let us next determine the pressure at any point of an infinite
450    body of fluid in the centre of which a unit source is placed, the pressure at an infinite distance from the source being supposed to be zero [D15].

The fluid will flow out from the centre symmetrically, and since unity of volume flows out of every spherical surface surrounding the point in
455    unit of time, the velocity at a distance $r$ from the source will be

$$v = \frac{1}{4\pi r^2}.$$

The rate of decrease of pressure is therefore $kv$ or

$$\frac{k}{4\pi r^2},$$

and since the pressure $= 0$ when $r$ is infinite, the actual pressure at
460   any point will be

$$p = \frac{k}{4\pi r}.$$

The pressure is therefore inversely proportional to the distance
from the source.

It is evident that the pressure due to a unit sink will be negative and
465   equal to

$$-\frac{k}{4\pi r}.$$

If we have a source formed by the coalition of $S$ unit sources, then
the resulting pressure will be

$$p = \frac{kS}{4\pi r},$$

470   so that the pressure at a given distance varies as the resistance and
number of sources conjointly.

(19) If a number of sources and sinks coexist in the fluid, then in order to determine the resultant pressure we have only to add the pressures which each source or sink produces. For by (15) this will be a
475   solution of the problem, and by (17) it will be the only one. By this
method we can determine the pressures due to any distribution of
sources, as by the method of (14) we can determine the distribution of
sources to which a given distribution of pressures is due.

(20) We have next to shew that if we conceive any imaginary sur-
480   face as fixed in space and intersecting the lines of motion of the fluid,
we may substitute for the fluid on one side of this surface a distribution of sources upon the surface itself without altering in any way the
motion of the fluid on the other side of the surface.

For if we describe the system of unit tubes which defines the mo-
485   tion of the fluid, and wherever a tube enters through the surface place
a unit source, and wherever a tube goes out through the surface

place a unit sink, and at the same time render the surface impermeable to the fluid, the motion of the fluid in the tubes will go on as
before.

490    (21) If the system of pressures and the distribution of sources
which produce them be known in a medium whose resistance is measured by $k$, then in order to produce the same system of pressures in a
medium whose resistance is unity, the rate of production at each
source must be multiplied by $k$. For the pressure at any point due to a
495    given source varies as the rate of production and the resistance conjointly; therefore if the pressure be constant, the rate of production
must vary inversely as the resistance.

* * *

## Application of the Idea of Lines of Force

500    I have now to shew how the idea of lines of fluid motion as described above may be modified so as to be applicable to the sciences
of statical electricity, permanent magnetism, magnetism of induction,
and uniform galvanic currents, reserving the laws of electro-magnetism for special consideration.
505    I shall assume that the phenomena of statical electricity have been
already explained by the mutual action of two opposite kinds of matter. If we consider one of these as positive electricity and the other as
negative, then any two particles of electricity repel one another with a
force which is measured by the product of the masses of the particles
510    divided by the square of their distance.
Now we found in (18) that the velocity of our imaginary fluid due to a
source $S$ at a distance $r$ varies inversely as $r^2$. Let us see what will be the
effect of substituting such a source for every particle of positive electricity
[D16]. The velocity due to each source would be proportional to the attrac-
515    tion due to the corresponding particle, and the resultant velocity due to all
the sources would be proportional to the resultant attraction of all the particles. Now we may find the resultant pressure at any point by adding the
pressures due to the given sources, and therefore we may find the resultant velocity in a given direction from the rate of decrease of pressure in
520    that direction, and this will be proportional to the resultant attraction of the
particles resolved in that direction.
Since the resultant attraction in the electrical problem is proportional to the decrease of pressure in the imaginary problem, and since

we may select any values for the constants in the imaginary problem,
525 we may assume that the resultant attraction in any direction is numerically equal to the decrease of pressure in that direction, or

$$X = -\frac{dp}{dx}.$$

By this assumption we find that if $V$ be the potential,

$$dV = X\,dx + Y\,dy + Z\,dz = -\,dp,$$

530 or since at an infinite distance $V = 0$ and $p = 0$, $V = -p$. [D17]
In the electrical problem we have

$$v = -\Sigma\left(\frac{dm}{r}\right).$$

In the fluid

$$p = \Sigma\left(\frac{k}{4\pi}\,\frac{S}{r}\right);$$

535
$$\therefore S = \frac{4\pi}{k}dm.$$

If $k$ be supposed very great, the amount of fluid produced by each source in order to keep up the pressures will be very small.
The potential of any system of electricity on itself will be

$$\Sigma(p\,dm) = \frac{k}{4\pi}W, \qquad \Sigma(pS) = \frac{k}{4\pi}W$$

540 If $\Sigma\,(dm)$, $\Sigma\,(dm')$ be two systems of electrical particles and $p$, $p'$ the potentials due to them respectively, then by (32)

$$\Sigma(p\,dm') = \frac{k}{4\pi}\Sigma(pS') = \frac{k}{4\pi}\Sigma(p'S) = \Sigma(p'\,dm),$$

or the potential of the first system on the second is equal to that of the second system on the first.
545 So that in the ordinary electrical problems the analogy in fluid motion is of this kind:

$$V = -\,p,$$

$$X = -\frac{dp}{dx} = ku,$$

$$dm = \frac{k}{4\pi}S,$$

550  whole potential of a system $= -\Sigma V\,dm = \dfrac{k}{4\pi}W,$

where $W$ is the work done by the fluid in overcoming resistance.

The lines of force are the unit tubes of fluid motion, and they may be estimated numerically by those tubes.

## Theory of Dielectrics

555   The electrical induction exercised on a body at a distance depends not only on the distribution of electricity in the inductric, and the form and position of the inducteous body, but on the nature of the inter-posed medium, or dielectric. Faraday* expresses this by the concep-tion of one substance having a *greater inductive capacity,* or
560  conducting the lines of inductive action more freely than another. If we suppose that in our analogy of a fluid in a resisting medium the resis-tance is different in different media, then by making the resistance less we obtain the analogue to a dielectric which more easily con-ducts Faraday's lines.

565   It is evident from (23) that in this case there will always be an appar-ent distribution of electricity on the surface of the dielectric, there be-ing negative electricity where the lines enter and positive electricity where they emerge. In the case of the fluid there are no real sources on the surface, but we use them merely for purposes of calculation. In
570  the dielectric there may be no real charge of electricity, but only an ap-parent electric action due to the surface.

If the dielectric had been of less conductivity than the surround-ing medium, we should have had precisely opposite effects, namely, positive electricity where lines enter, and negative where
575  they emerge.

If the conduction of the dielectric is perfect or nearly so for the small quantities of electricity with which we have to do, then we have the case of (24). The dielectric is then considered as a conductor, its sur-face is a surface of equal potential, and the resultant attraction near
580  the surface itself is perpendicular to it.

## Theory of Permanent Magnets

A magnet is conceived to be made up of elementary magnetized particles, each of which has its own north and south poles, the action

*Series XI.

585 of which upon other north and south poles is governed by laws mathematically identical with those of electricity. Hence the same application of the idea of lines of force can be made to this subject, and the same analogy of fluid motion can be employed to illustrate it.

But it may be useful to examine the way in which the polarity of the 590 elements of a magnet may be represented by the unit cells in fluid motion. In each unit cell unity of fluid enters by one face and flows out by the opposite face, so that the first face becomes a unit sink and the second a unit source with respect to the rest of the fluid. It may therefore be compared to an elementary magnet, having an equal quantity 595 of north and south magnetic matter distributed over two of its faces. If we now consider the cell as forming part of a system, the fluid flowing out of one cell will flow into the next, and so on, so that the source will be distributed entirely on that surface, and in the case of a magnet which has what has been called a solenoidal or tubular distribution of 600 magnetism, all the imaginary magnetic matter will be on the surface.*

## Theory of Paramagnetic and Diamagnetic Induction

Faraday** has shewn that the effects of paramagnetic and diamagnetic bodies in the magnetic field may be explained by supposing paramagnetic bodies to conduct the lines of force better, and diamag- 605 netic bodies worse, than the surrounding medium. By referring to (23) and (26), and supposing sources to represent north magnetic matter, and sinks south magnetic matter, then if a paramagnetic body be in the neighbourhood of a north pole, the lines of force on entering it will produce south magnetic matter, and on leaving it they will produce an 610 equal amount of north magnetic matter. Since the quantities of magnetic matter on the whole are equal, but the southern matter is nearest to the north pole, the result will be attraction. If on the other hand the body be diamagnetic, or a worse conductor of lines of force than the surrounding medium, there will be an imaginary distribution of 615 northern magnetic matter where the lines pass into the worse conductor, and of southern where they pass out, so that on the whole there will be repulsion.

We may obtain a more general law from the consideration that the potential of the whole system is proportional to the amount of work

*See Professor Thomson *On the Mathematical Theory of Magnetism,* Chapters III and V. *Phil. Trans.* 1851.
** *Experimental Researches* (3292).

620 done by the fluid in overcoming resistance. The introduction of a second medium increases or diminishes the work done according as the resistance is greater or less than that of the first medium. The amount of this increase or diminution will vary as the square of the velocity of the fluid.

625    Now, by the theory of potentials, the moving force in any direction is measured by the rate of decrease of the potential of the system in passing along that direction, therefore when $k'$, the resistance within the second medium, is greater than $k$, the resistance in the surrounding medium, there is a force tending from places where the resultant

630 force v is greater to where it is less, so that a diamagnetic body moves from greater to less values of the resultant force.*

In paramagnetic bodies $k'$ is less than $k$, so that the force is now from points of less to points of greater resultant magnetic force. Since these results depend only on the relative values of $k$ and $k'$, it is evi-

635 dent that by changing the surrounding medium, the behaviour of a body may be changed from paramagnetic to diamagnetic at pleasure.

It is evident that we should obtain the same mathematical results if we had supposed that the magnetic force had a power of exciting a polarity in bodies which is in the *same* direction as the lines in param-

640 agnetic bodies, and in the *reverse* direction in diamagnetic bodies.**
In fact we have not as yet come to any facts which would lead us to choose any one out of these three theories, that of lines of force, that of imaginary magnetic matter, and that of induced polarity. As the theory of lines of force admits of the most precise, and at the same time

645 least theoretic statement, we shall allow it to stand for the present.

* * *

## Theory of the Conduction of Current Electricity

It is in the calculation of the laws of constant electric currents that the theory of fluid motion which we have laid down admits of the most

650 direct application. In addition to the researches of Ohm on this subject, we have those of M. Kirchhoff, *Ann. de Chim.* XLI. 496, and of M. Quincke, XLVII. 203, on the Conduction of Electric Currents in Plates.

*Experimental Researches* (2797), (2798). See Thomson, *Cambridge and Dublin Mathematical Journal,* May, 1847.
***Exp. Res.* (2429), (3320). See Weber, Poggendorff, LXXXVII. p. 145. Prof. Tyndall, *Phil. Trans.* 1856, p. 237.

According to the received opinions we have here a current of fluid
moving uniformly in conducting circuits, which oppose a resistance to
655 the current which has to be overcome by the application of an electromotive force at some part of the circuit. On account of this resistance
to the motion of the fluid the pressure must be different at different
points in the circuit. This pressure, which is commonly called electrical
tension, is found to be physically identical with the *potential* in statical
660 electricity, and thus we have the means of connecting the two sets of
phenomena. If we knew what amount of electricity, measured stati-
cally, passes along that current which we assume as our unit of current, then the connexion of electricity of tension with current electricity
would be completed.* This has as yet been done only approximately,
665 but we know enough to be certain that the conducting powers of different substances differ only in degree, and that the difference between
glass and metal is, that the resistance is a great but finite quantity in
glass, and a small but finite quantity in metal. Thus the analogy between statical electricity and fluid motion turns out more perfect than
670 we might have supposed, for there the induction goes on by conduction just as in current electricity, but the quantity conducted is insensi-
ble owing to the great resistance of the dielectrics.**

## On Electro-motive Forces

When a uniform current exists in a closed circuit it is evident that
675 some other forces must act on the fluid besides the pressures. For if
the current were due to difference of pressures, then it would flow
from the point of greatest pressure in both directions to the point of
least pressure, whereas in reality it circulates in one direction constantly. We must therefore admit the existence of certain forces capable of keeping up a constant current in a closed circuit. Of these the
680 most remarkable is that which is produced by chemical action. A cell
of a voltaic battery, or rather the surface of separation of the fluid of
the cell and the zinc, is the seat of an electro-motive force which can
maintain a current in opposition to the resistance of the circuit. If we
adopt the usual convention in speaking of electric currents, the posi-
685 tive current is from the fluid through the platinum, the conducting circuit, and the zinc, back to the fluid again. If the electro-motive force

*See *Exp. Res.* (371).
**Exp. Res.* Vol. III. p. 513.

act only in the surface of separation of the fluid and zinc, then the tension of electricity in the fluid must exceed that in the zinc, by a quantity depending on the nature and length of the circuit and on the
690 strength of the current in the conductor. In order to keep up this difference of pressure there must be an electro-motive force whose intensity is measured by that difference of pressure. If $F$ be the
electro-motive force, $I$ the quantity of the current or the number of electrical units delivered in unit of time, and $K$ a quantity depending on the
695 length and resistance of the conducting circuit, then

$$F = IK = p - p',$$

where $p$ is the electric tension in the fluid and $p'$ in the zinc.

   If the circuit be broken at any point, then since there is no current
the tension of the part which remains attached to the platinum will be
700 $p$, and that of the other will be $p'$, $p - p'$ or $F$ affords a measure of the
intensity of the current. This distinction of quantity and intensity is very
useful,* but must be distinctly understood to mean nothing more than
this:-The quantity of a current is the amount of electricity which it transmits in unit of time, and is measured by $I$ the number of unit currents
705 which it contains. The intensity of a current is its power of overcoming
resistance, and is measured by $F$ or $IK$, where $K$ is the resistance of
the whole circuit.

* * *

## On the Action of closed Currents at a Distance

710    The mathematical laws of the attractions and repulsions of conductors have been most ably investigated by Ampère, and his results
have stood the test of subsequent experiments.

   From the single assumption, that the action of an element of one
current upon an element of another current is an attractive or repul-
715 sive force acting in the direction of the line joining the two elements,
he has determined by the simplest experiments the mathematical
form of the law of attraction, and has put this law into several most elegant and useful forms. We must recollect however that no experiments have been made on these elements of currents except under
720 the form of closed currents either in rigid conductors or in fluids, and
that the laws of closed currents can only be deduced from such ex-

*Exp. Res. Vol. III. p. 519.

periments. Hence if Ampere's formulae applied to closed currents give true results, their truth is not proved for *elements* of currents unless we assume that the action between two such elements must be
725    along the line which joins them. Although this assumption is most warrantable and philosophical in the present state of science, it will be more conducive to freedom of investigation if we endeavour to do without it, and to assume the laws of closed currents as the ultimate datum of experiment.

730                                    * * *

From these results it follows that the mutual action of two closed currents whose areas are very small is the same as that of two elementary magnetic bars magnetized perpendicularly to the plane of the currents.

The direction of magnetization of the equivalent magnet may be pre-
735    dicted by remembering that a current travelling round the earth from east to west as the sun appears to do, would be equivalent to that magnetization which the earth actually possesses, and therefore in the reverse direction to that of a magnetic needle when pointing freely.

If a number of closed unit currents in contact exist on a surface,
740    then at all points in which two currents are in contact there will be two equal and opposite currents which will produce no effect, but all round the boundary of the surface occupied by the currents there will be a residual current not neutralized by any other; and therefore the result will be the same as that of a single unit current round the boundary of
745    all the currents.

From this it appears that the external attractions of a shell uniformly magnetized perpendicular to its surface are the same as those due to a current round its edge; for each of the elementary currents in the former case has the same effect as an element of the magnetic shell.

750    If we examine the lines of magnetic force produced by a closed current, we shall find that they form closed curves passing round the current and *embracing* it, and that the total intensity of the magnetizing force all along the closed line of force depends on the quantity of the electric current only. The number of unit lines* of magnetic force due
755    to the closed current depends on the form as well as the quantity of the current, but the number of unit cells** in each complete line of force is measured simply by the number of unit currents which

*Exp. Res.* (3122) See Art. (6) of this paper.
**Art. (13).

embrace it. The unit cells in this case are portions of space in which unity of magnetic quantity is produced by unity of magnetizing force.

760 The length of a cell is therefore inversely as the intensity of the magnetizing force, and its section inversely as the quantity of magnetic induction at that point.

The whole number of cells due to a given current is therefore proportional to the strength of the current multiplied by the number of

765 lines of force which pass through it. If by any change of the form of the conductors the number of cells can be increased, there will be a force tending to produce that change, so that there is always a force urging a conductor transverse to the lines of magnetic force, so as to cause more lines of force to pass through the closed circuit of which the con-

770 ductor forms a part.

The number of cells due to two given currents is got by multiplying the number of lines of inductive magnetic action which pass through each by the quantity of the currents respectively. Now by (9) the number of lines which pass through the first current is the sum of its

775 own lines and those of the second current which would pass through the first if the second current alone were in action. Hence the whole number of cells will be increased by any motion which causes more lines of force to pass through either circuit, and therefore the resultant force will tend to produce such a motion, and the work done by this

780 force during the motion will be measured by the number of new cells produced. All the actions of closed conductors on each other may be deduced from this principle.

## On Electric Currents produced by Induction

Faraday has shewn* that when a conductor moves transversely to

785 the lines of magnetic force, an electro-motive force arises in the conductor, tending to produce a current in it. If the conductor is closed, there is a continuous current, if open, tension is the result. * * * In all these cases the electro-motive force depends on the *change* in the number of lines of inductive magnetic action which pass through the

790 circuit.

* * *

It is natural to suppose that a force of this kind, which depends on a change in the number of lines, is due to a change of state which is

*Exp. Res. (3077), &c.

measured by the number of these lines. A closed conductor in a mag-
795 netic field may be supposed to be in a certain state arising from the
magnetic action. As long as this state remains unchanged no effect
takes place, but, when the state changes, electro-motive forces arise,
depending as to their intensity and direction on this change of state. I
cannot do better here than quote a passage from the first series of
800 Faraday's *Experimental Researches,* Art. (60).

"While the wire is subject to either volta-electric or magno-electric induction it
appears to be in a peculiar state, for it resists the formation of an electrical
current in it; whereas, if in its common condition, such a current would be
produced; and when left uninfluenced it has the power of originating a current,
805 a power which the wire does not possess under ordinary circumstances. This
electrical condition of matter has not hitherto been recognized, but it probably
exerts a very important influence in many if not most of the phenomena
produced by currents of electricity. For reasons which will immediately appear
(7) I have, after advising with several learned friends, ventured to designate it
810 as the *electro-tonic* state."

Finding that all the phenomena could be otherwise explained without reference to the electro-tonic state, Faraday in his second series
rejected it as not necessary; but in his recent researches* he seems
still to think that there may be some physical truth in his conjecture
815 about this new state of bodies.

The conjecture of a philosopher so familiar with nature may sometimes be more pregnant with truth than the best established experimental law discovered by empirical inquirers, and though not bound to
admit it as a physical truth, we may accept it as a new idea by which
820 our mathematical conceptions may be rendered clearer.

In this outline of Faraday's electrical theories, as they appear from
a mathematical point of view, I can do no more than simply state the
mathematical methods by which I believe that electrical phenomena
can be best comprehended and reduced to calculation, and my aim
825 has been to present the mathematical ideas to the mind in an embodied form, as systems of lines or surfaces, and not as mere symbols,
which neither convey the same ideas, nor readily adapt themselves to
the phenomena to be explained. The idea of the electro-tonic state,
however, has not yet presented itself to my mind in such a form that
830 its nature and properties may be clearly explained without reference
to mere symbols, and therefore I propose in the following investigation

*(3172) (3269).

to use symbols freely, and to take for granted the ordinary mathematical operations. By a careful study of the laws of elastic solids and of the motions of viscous fluids, I hope to discover a method of forming a
835 mechanical conception of this electro-tonic state adapted to general reasoning.*

*See Prof. W. Thomson *On a Mechanical Representation of Electric, Magnetic and Galvanic Forces. Camb. and Dub. Math. Jour.* Jan. 1847.

## Interpretive Notes

2    "Speculation"
This is a suggestive word, one which Faraday uses often as well. It hints at the fact that Maxwell is embarking on a thought process of a kind which, as he asserts below, is not intended to result in a theory. A *speculum* is a mirror: perhaps, as he goes on to list the successive fragments of the "present state" of his science, he hopes through reflections to catch them in a single image, as a whole. He speaks in terms of *analogy;* the "lines of force" to which he devotes this essay may be more like a movable metaphor than a strictly logical structure, functioning rather to suggest a unifying point of view than to yield a connected argument.

2    "The laws. . . ."
We can correlate the disparate parts of the science as Maxwell recites them here to the headings under which they have been discussed in our Chapter 1:

> "laws of the distribution of electricity"—*Electrostatics*
> "the mathematical theory of magnetism"—*Magnetism*
> "the conduction of galvanisms"—*Current electricity*
> "the mutual attraction of conductors"—*Oersted Effect*
> "the inductive effects"—*Electromagnetic induction*

16    "intricate mathematics"
We are reminded especially of Ampère's elaborate theory, to which Maxwell devotes particular attention as exemplar of an approach opposite to Faraday's.

23    "the first case" (the "purely mathematical formula")
It will be interesting to keep in mind this fundamental reservation concerning mathematical physics. Once again, the metaphor is one of *seeing,* and we are reminded of the term *speculation.* In equations, we "lose sight" and fail to get "an extended view": the algebraic symbols get into the line of sight.

26    "on the other hand"
In the distinction Maxwell is making here, a mathematical formula does not entail a hypothesis, at least not a *physical* one. We are reminded that the full title of Newton's *Principia* is in translation, *The Mathematical Principles of Natural Philosophy;* he is able to claim, "I frame no hypotheses" precisely because his work is kept strictly mathematical as distinct from physical.
      Pursuing his metaphor, Maxwell now describes a physical hypothesis as another disturbance of vision, an intervening medium which may also block our seeing: we see more if we assume less, or assume less "rashly."

29    "some method of investigation"
For present purposes, this seems to be Maxwell's version of "scientific method." It does not apparently consist of first presenting a hypothesis and then testing it by experiment, as we are often told. Maxwell is setting forth something much more complex and subtle; we will have to watch to see whether it *ever* comes to the point of testing anything by doing an experiment. We are, it seems, going to borrow an *idea* from another science, an idea that we can see *clearly.* Is all metaphor or analogy, we might wonder, such a "borrowing" for the sake of clearer *seeing*?

37    "physical analogy"
Here we have Maxwell's formal definition of his art, one which we will do well to keep in mind if the procedure becomes confusing later on. Since a physical analogy is frankly only a *partial* similarity, we are not excessively deflected by it from our pursuit: it does not substitute itself for the object of our search, but rather *throws light.* That is, I believe, what "illustration" does. Maxwell would probably have in mind the Latin roots of what becomes one of his favorite words: *in + lustrare,* to "throw light (*lux*) in/on."

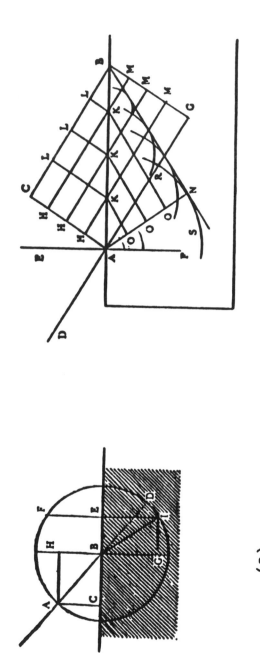

(a)

(b)

Fig. 2.1. Two theories of refraction: (a) Particle theory (Descartes); (b) Wave theory (Huygens) (see Discussions [D1] and [D2]).

43 "most universal of all analogies"
If the process of measurement thus leads the physical world and the world of numbers to illuminate one another, measurement must hint at a deeply underlying likeness of the two worlds. Is all measurement, then, metaphor?

49 [D1] On the Particulate Theory of Refraction

54 [D2] On the Wave Theory of Light

60 [D3] On a Theory of "Transverse Alternations"

62 "attraction at a distance"
We have spoken about this in the previous chapter as articulated by Newton in the *Principia,* where it is applied in the "System of the World" to gravity. As we saw then, the same concept of attraction at a distance had, by Maxwell's time, been extensively applied to electricity and magnetism. As a "mathematical conception," it is a mere formula, yet one which has proven widely "consistent with nature." In a sense, this formula is Maxwell's very target—it symbolizes the point of view which is the dialectical challenge to which Maxwell is responding by shifting the foundations of our reasoning about nature from such analytic formulas to the intuitive concept of the *field.*

78 [D4] On Fourier's Theory of the Uniform Motion of Heat

83 "Professor William Thomson"
Discovery of the analogy between heat flow and electrostatics had supplied the foundation of a series of Thomson's early mathematical successes. It becomes the mathematical clue on which this paper is based. Figure 2.2 suggests the way in which the analogy works. If an infinitely long rectangular metal plate is heated and cooled at once so that one end is held at 1°C while the sides and the remote end are held at 0°C, the temperature distribution will stabilize into the pattern shown in the figure. In the top images the vertical height is proportional to the temperature, while at the bottom the temperature is depicted by shadings of gray.

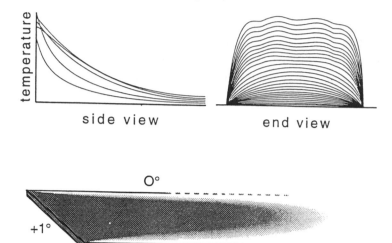

Fig. 2.2. Fourier's solution of the distribution of temperature over an infinite rectangular plate heated at one end and cooled on the sides. The same figure can be read by analogy as an electric potential distribution in a rectangular space. (See Discussion [D-4].)

If on the other hand for the bar we substitute a rectangular space the edges of which are held at the corresponding values of electric potential—1V at the end and 0 along the sides, the potential distribution over the space takes the very same form. Thus the thermal and the electrical realms mirror each other, and Fourier's solution of the thermal problem is at the same time the solution of the electrical. The figure is an approximation, computed using the first terms of Fourier's series solution, as explained in Discussion [D4].

94    "to bring before the mind"

To reflect on the analogy we have just been considering—how is it that it "brings before the mind" a "*mathematical* idea"? Evidently, in Maxwell's usage the term "mathematical" will not be identified with "algebraic" or "analytic"—musn't he mean something much broader? Perhaps heat flow, for example, is *physical* as far as it is concerned specifically with *heat,* but *mathematical* insofar as it bears the idea of conductivity and flow. Faraday had talked often in the latter terms, which, as Maxwell suggests below, were generally thought by an erudite scientific community to be merely "indefinite and unmathematical." By contrast, Maxwell is now proposing to show that such ideas can be "clearly placed before the mathematical mind." This clarity is not, however, to be achieved by recourse to analytic equations or a formal theory—such would definitely not be the "methods of Faraday."

We are asked to fix our minds on this concept of the motion of an ideal fluid—without the aid of equations, on the one hand, and without commitment to a physical system, on the other. Can "continuous flow" itself in abstraction from any physical fluid, as an *idea,* become an object for the mind, as is, for example, the idea of "triangle" or of "ellipse"? Maxwell seems to have in mind something like the Greek notion of "mathematics" as that domain of thought whose objects are fully knowable (*ta mathemata*). To return Faraday's ideas to him in this careful formulation, *mathematical* but not algebraic and thus accessible in a new way, may be a special gift of Maxwell's to Faraday. We might read this essay as Maxwell's "Letter to Faraday."[1]

113  "I shall use analytic notation"
In the second part of this paper, not reproduced here, Maxwell does have recourse to analytic methods; but this seems to come as a failure to achieve the grasp he would like to have, rather than as an advance.

125  "In this way"
A "way" seems to be defined here which ensures that a genuine geometrical entity has been determined (Fig. 2.3). We are reminded that since a force will exist at every point, no points are exempted from inclusion, and so the entity fills all space. It is pictured by the lines of force, as in Figs. 1.10 and 1.11, but they can be no more than partial representations of what is in concept a fully continuous, *space-filling* structure. Indeed, can such an extended continuum really be *seen,* even in the mind's eye? Unlike other simpler, better-behaved figures such as triangles, it

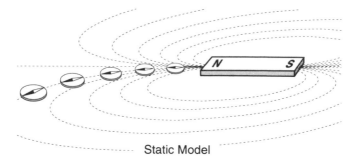

Static Model

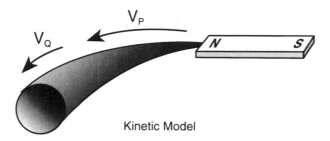

$V_P$

$V_Q$

Kinetic Model

Fig. 2.3. Defining the lines of force (Static Model), and tubes of flow (Kinetic Model).

seemingly has no boundaries, edges, or parts. Can a geometrical figure have *shape* without edges or parts? Maxwell's task is to show us that it can.

139 "intensity of the force"

The account we have been speaking of as *geometrical* is a mere "model" of the fluid we want to consider—a static model of the kinetic model. Ultimately, Maxwell's plan is apparently to refer, beyond models, to the objects of Faraday's interest, namely, the electrostatic and magnetic *forces*. A "force," however, as such, seems to give the mind's eye very little to contemplate. Two things now seem to be happening: (1) The geometrical model, which had been as static as a triangle, now springs into motion, as the lines of *direction* turn into lines of *flow*, and (2) the velocity of this flow varies from point to point and assumes significance. We are back, are we not, with the properties—though not the substance—of our thermal model, where *flow* (of heat) modeled *force* (in electricity).

Now specifically, the imagined flow will be modulated so that *velocity* of flow will represent *intensity* of force (Fig. 2.3). To give content to such a velocity, the lines must become fluid-conveying, as little pipes. Such a fluid-conveying "line" is to be called a *tube,* and space is no longer filled with *lines* but with these fluid-conveying *tubes.*

154   "leave no interstices"
      It is a special feature of forces which vary *inversely as the square* of the distance from the source, Maxwell says, that tubes constructed in this way fit neatly to fill the space, with no overlap and leaving no interstices. We shall see, as Maxwell leads us through the ensuing discussion, why this must be so. It seems that a special combination of intuition and good fortune has led Faraday and Maxwell, in invoking the image of a fluid, to just the analogy which will work to achieve his purposes. If the force law were otherwise, the analogy would not work, as Discussion [D5] shows.

154   [D5] On a Fluid-Flow Model That Does Not Work!

156   "It has been usual to commence"
      Here is the announcement of the revolution in thought which Maxwell has determined to pursue: to set aside the usual theories of action between points and to undertake a new "definition of the forces" in question. In the old view, they were called "inverse-square" forces. Under the new, they are to be defined, not by a formula, but *by the model—defined,* that is, by the tube of flow as a model. They *are* just those forces which *can* be modeled (in a connected, space-filling way) by the uniform motion of an incompressible fluid. Noninverse square forces *cannot* be so modeled, as exemplified in [D5]. By changing the mode of *definition* of force in this way, Maxwell has in effect challenged our notion of "force" itself—for a definition shapes our conception of the thing being defined. Aristotle calls a definition *"to ti ein einai"*—the-that-which-it-would-be-to-be the thing in question. What then will be the consequences of coming at the nature of *force* by way

of an image of *flow*? To trace the consequences of this new approach is really the work of our three papers.

163 "the motion of such a fluid can be clearly conceived"
This identifies our first task, which surprisingly has nothing to do with electricity *per se*—or, indeed, with anything in the world. We are simply asked to get into our heads *an altogether new idea*. It comes in the guise of a "perfect fluid" and its conceivable motions. But this is not a real "fluid," as we shall see, so the "fluid" is only the vehicle for the idea. And what then *is* the "idea"? It seems to be that of a fully connected system in which the whole is primary—in which a change anywhere can only be understood as a change everywhere. In a broader context, perhaps it is one approach to thinking about things which are whole, and not to be grasped as mere assemblages of parts—the sort of thing we might refer to as an organism or an ecology.

168 "another idea"
Maxwell is referring to the lurking idea which Faraday has called the "electrotonic state"—something like a state of tension in space, which seemed necessary to account for the surge of energy exhibited in induction. The fluid model is not very helpful to Maxwell in trying to track down that idea, the work of the second part of "Faraday's Lines," not reproduced here.

172 [D6] Reflections on Method

184 "The substance here treated of"
In Maxwell's first section, we embark with him on an adventure in pure thought. It is evidently important to him that we understand fully just *how* "speculative" this is to be. We are to *think into existence* an utterly nonphysical fluid. Note that his fluid has no inertia—*it is not made of matter* and hence presumably *could not* be thought of as existing in a real world.[2] It is a fluid that never was. It is not meant to exist, except as a thought: "it is not even a hypothetical fluid." Its purity (if that is the word for it) is guaranteed by the fact that it *has* no properties except those with which we endow it: total fluidity (zero viscosity—how fast then would it flow through a small hole?); total incompressibility (zero com-

pressibility, i.e., harder than any rock). Water, inept divers know, approaches the latter.

189  "in a way more intelligible"
The "theorems" which we are about to meet—established here by something more like intellectual intuition than by force of expressed logic—are also to be found systematically developed in standard texts of fluid mechanics, using analytic symbols and the full arsenal of algebraic arguments. Is the present method, which is "more intelligible to many minds," *inherently inferior* to such more formal methods? We already know, I suspect, Maxwell's answer from the first pages of this essay. Using the analytic method alone, however tight the argument, we inherently tend to "lose sight" and hence fail to get the "more extended views." If "sight" is what we want, it is not lower, but higher ground we are seeking by choosing this speculative method over the analytical.

196  "This law"
Maxwell calls this a "law"—but is it not rather a *definition* or perhaps part of a definition? The fact that volume and nothing else is preserved, i.e., *belongs to* this entity, seems to define very precisely an intellectual object called a "fluid." A perfect fluid has *exact* volume and *no* shape! (This is, perhaps, the very idea of "volume"—shapeless size.) But perhaps my quibble about "law" was misguided. In a realm of our own defining, "definition" may be exactly "law"!

197  "the unit of quantity"
When Maxwell tells us that the unit of quantity of our fluid will be the *volume,* he is placing—or rather perhaps displacing—our investigation. We will *not* be doing mechanics but strictly geometry. Contrast the quantity chosen by Newton as the foundation of the *Principia:* there, the unit is that of *mass,* by which Newton explains that he means the quantity of *matter.* Maxwell's fluid, it seems, has no "matter"!

198  [D7] On the Choice of the Unit of Quantity of the Fluid

205  "lines of fluid motion"
If this is to be a form of *geometry* which Maxwell is unfolding before our minds' eyes, it is interesting to compare it with other

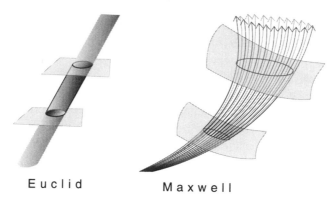

Euclid            Maxwell

Fig. 2.4. Alternative worlds: the "cylinder" in Euclid and in Maxwell.

geometries we may have known. Euclid begins with the point, followed by the line, as the foundations of his work; figures with volume come much later—"constructed," in Euclid's terms, of these "elements." Here by contrast we begin with the volume as our primary entity. What is last for Euclid is first for Maxwell. Further, we find ourselves in a realm in which *straightness* no longer has priority! Rather, fluid motion, flow, is a primary idea. If we pursue Euclid for a definition of "straightness," the concept seems to come down to a *line of sight.* Then in this new-configured, curvilinear world of Maxwell's, flow (whatever it is) assumes the role played by sight in Euclid's geometry of straightness.

211    [D8] On "Steady Motion"

215    "tube of fluid motion"
       The "tubes" are, it seems, the natural space-filling figures of the world of fluids. If lines of flow are like lines of sight in this fluid-configured world, would it be misleading to suggest that the lines which lie in the surface of a tube are the natural "parallels" here? In Euclid, a cylinder is a figure with a plane base and sides which are parallel. Then a tube of fluid motion is a cylinder of the fluid world (Fig. 2.4).
       It must be a clue to Maxwell's style that he should tease us with the notion of their "impenetrability." "Penetrate"—by *what*? If in

this world there is nothing but fluid, then to move with the fluid *is* . . . to be "rigid"!

225    "unit tube of fluid motion"
Despite its apparent simplicity, the definition we meet here of the "unit tube" may prove difficult to keep in mind: "A unit tube is one in which unit volume of fluid passes in unit time." More considerations of relationships in the unit tube are to be found in Discussion [D9]:

225    [D9] Further Reflections on the "Unit Tube"

229    "an infinite number of lines"
Our problem is, it seems, to get into our heads the idea of a continuum, configured throughout by ordered motion. What could it look like, to the mind's eye, if the number of lines to be "drawn" is infinite? The "lines," it would seem, must fuse and be lost to view! We seem to gain by stopping short while a finite tube—*however small*—still remains in our mental view. Since there is no lower limit to the "size" of these least-tubes that we draw, we may approach in this way the ultimate continuum as closely as we please.

236    "to define the motion of the whole fluid"
Here we fully confront our problem: if the fluid, with its motion *throughout* is our object, it is not enough to address it tube-by-tube or line-by-line. We must find ways to grasp the fluid in its motion *as the whole* that it essentially is. We are at this point on the doorstep of what is to become "field" theory.
     The first bold step in achieving this wholeness is to cut "all the lines" with one surface: this "cut" is the decisive mental act, which as a thought-cut is more of a survey than a slice (Fig. 2.5). Is it clear what "all" must include here? The reader is invited to look back at Fig. 1.10 to consider how surfaces of this sort might be passed in these cases. Will it matter if we happen to catch some of the lines more than once? Other figures, such as those of Fig. 1.11, present daunting challenges. We cannot expect the drawing to exhibit the whole space occupied by "all" the lines. In particular, we must imagine our ways from the two-dimensional slices represented in a plane drawing to the three-dimensional body filling space.

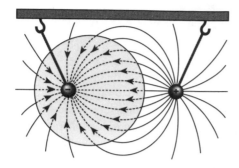

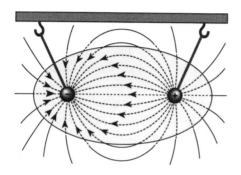

Fig. 2.5. Alternatives for drawing Maxwell's cutting surface in the case of two attracting pith balls. For the world inside the surface, the points of intersection with the lines are sources and sinks.

262    "may either return into itself, or may begin and end"
For the first case, we may look again at Fig. 1.10. Here the tubes "return into themselves" only if we imagine them completed through the body of the bar magnet, as Fig. 2.12 envisions. If we are not sure about that as an example of lines which make closed loops, we can think of the lines implicit in Fig. 1.35, which evidently make whole circles about the wire. By contrast in Fig. 2.5, the electrostatic case, the lines begin and end on the charged surfaces of the two conductors; in electrostatics there is never any closure. Lines here begin on the surface of the positive conductor; some end on the negative conductor, while others go on to terminate on negative charges induced on the walls, floor, and ceiling! It is an interesting challenge to go back to add such lines to Fig. 1.10. (We should recall that it is only by convention that lines are said to "begin" on the positive surface and "end" on the negative; it could equally be thought of the other way around.)

The bounding surfaces enclosing the spaces, drawn in thought, of which Maxwell speaks may be chosen so as to include one charged body, both charged bodies, or neither. The first two cases are drawn in Fig. 2.5; the third, with its implications, is left to the reader to envision.

275 "nothing self-contradictory"
Is self-consistency—or the avoidance of any lurking self-contradiction—the only criterion of *truth* in this mental world of our own construction?

308 "in two different cases"
The "different cases" are of our own choosing, as acts of the mind. This theorem thus secures our power to analyze and synthesize motions of every possible sort, and thus to relate simple components to whatever complex wholes we choose. Note that if we wish we may now take *components* of complex motions, with respect to convenient reference axes, deal separately with these simpler motions, and then synthesize them again to reconstruct the original complexity at will.

328 "no inertia"
In assigning the fluid "no inertia," electing instead to oppose its motion with a force proportional to its velocity, Maxwell takes us back to the physics of Aristotle, and leaves Newton with his laws of motion behind. Canceling inertia, we cancel as well the First Law of the *Principia,* and write instead: "A body not pushed forward by a continuing force will stop." The Second Law—by which, according to Newton, a body undergoes an acceleration proportional to the applied force—becomes rather: "A body will move with a velocity directly proportional to the force applied." The substitution here of *velocity* for Newton's *acceleration* is crucial. This is exactly what common sense, always allied with Aristotle, would have expected in the first place: push and the body goes; stop, and it stops. (What would the world be like if Aristotle had proved to be right and all physical bodies moved according to his law rather than Newton's?)

This is, in any case, just what "heat" will do according to the computations of Fourier, and thus is the law of motion of "caloric." If we think of caloric as a fluid being driven by the temperature gradient, the law is a simple proportionality: the greater the gradient, the greater the flow of heat. We see it at work every day in the cooling of a coffee cup: the coffee will cool at an ever-diminishing rate, always proportional to the difference between the temperature of the coffee and that of the air.

This means as well that the mathematical theory of Fourier can be imported to apply to our fluid: Maxwell's imaginary fluid will do just what "heat" will do, according to the computations of Fourier. We might conclude that the imaginary fluid of this essay is, in effect, "caloric."

338   "a force equal to $kv$"

Here $k$ is a constant proportional to the resistance of the "medium"—still utterly imaginary—through which our fluid is now understood to move. Our law of motion may now be written

$$f = kv$$

where $f$ is the net force of the medium opposing the motion, and $v$ is the velocity of a *unit volume* of the fluid: for a given velocity, to move greater volume will require proportionately greater force. (It is always a question of "net" force or pressure—the difference between that on the front and that on the rear surface.) We may call $f$ the "net force per unit volume."

For further discussion of this relation, which we may think of as the law of motion for the fluid, see [D10].

340   [D10] On Fluid Motion from the Point of View of Pressure

348   "at the rate"

When a force is distributed continuously over a surface, it is often useful to speak instead of "pressure," defined as force-per-unit-area. We have spoken of the "net pressure" on a portion of the fluid; this is the *difference* in pressure between the front and back faces. If we use the symbol "Δ" to denote "difference," we may write the law of motion as

$$\Delta p = kvh$$

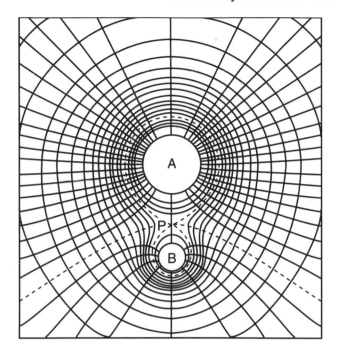

Fig. 2.6. Fluid motion, partitioned into unit tubes; "isobars" are also drawn. *A* = 20 units, *B* = 5 units. Flow is outward from both sources: what is the special role of point *P* and the dotted lines through it?

Further reasoning in these terms is developed in Discussion [D10].

366   [D11] On Isobars and Lines of Flow

385   .[D12] On Unit Cells

388   [D13] On Work and Energy in the Unit Cell

397   "in two different cases"
      Quantities may be classified as "scalar" or "vector." A scalar quantity such as temperature is specified with a single number. A vector, by contrast, has both magnitude and direction; to specify it numerically we may give, for example, its components—as in the case of a two-dimensional vector, which may be designated by

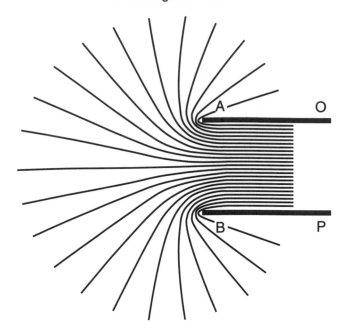

Fig. 2.7. Isobars between a long flat source *OA* (like a rectangular shower head) and a flat sink *PB* (the shower head in reverse) (see Discussion [D11]).

giving its east-west and its north-south components. Since it is essential to know which is east-west and which north-south, such a vector is said to be equivalent to an ordered pair of numbers. Similarly, a vector in three dimensions is equivalent to an ordered triple of numbers (as, east-west, north-south, up-down).

Happily, pressures are scalars and thus are simply additive. This gives us a very satisfying ability to work with pressure fields. Simple systems may be combined by simple addition. In Fig. 2.9 we look back at Fig. 2.6 to envision the same scene in terms of electric "pressure," or potential. Considered separately, *A* is a high-potential mountain symmetric about its peak, while *B* is a lower one. The combined figure is drawn by simply adding the two, the point *P* of Fig. 2.6 emerging as a saddle point between the two peaks. The vector field lines of Fig. 2.6 run directly "downhill" at every point of this mountainous terrain. We should note that the conductors *A* and *B* of Fig. 2.6, which are by definition equipotential planes, would cut off the mountain peaks of Fig. 2.10 as flat-topped mesas!

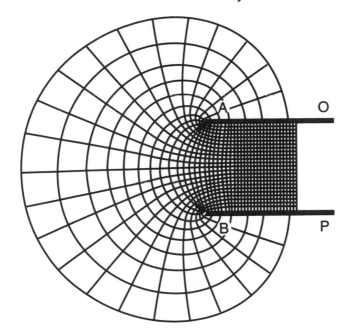

Fig. 2.8. The system of Fig. 2.7 divided into unit cells (see Discussions [D12] and [D13]).

403   [D14] Higher Ground: On the Scalar View of the Fluid Motion

419   "If the pressure at every point"
Maxwell's explanation seems convincing: this is the situation, for example, inside a charged sphere, within which there is no detectable electric force. A great moment at the Royal Institution was described in Chapter 1, in which Faraday entered a highly electrified cage, "and lived in it." This paragraph of Maxwell's accounts for Faraday's salvation in that experiment.

435   "every point of a given closed surface"
This is very elegant reasoning, which need be disturbed by no commentary. But we might reflect on the power of this conclusion. If we know the *pressure distribution over any surface*— though we may know nothing of the configuration of pressures and motions elsewhere in the space—we do know now that there is just one such spatial distribution possible. The information

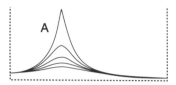

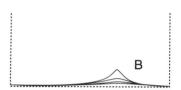

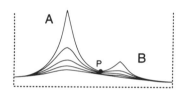

Fig. 2.9. Two scalar potential systems and the resultant of their combination. This corresponds to Fig. 2.6 seen from the side with potential plotted on the vertical axis.

associated with our one surface, we might say, is *controlling:* if our surface is defined, then all else is as well. This surely is a revealing clue to the connectedness of the system. To put it another way, the least change anywhere would be detected as a change, in some respect, somewhere over our surface.

449    "Let us next determine"

Here, a little analytic math intervenes—though we may hardly need it. We are asked to think of fluid emanating from a unit point source $Q$, that is, unit volume of fluid flowing out of $Q$ in unit time (Fig. 2.10).

The area of the sphere of radius $r$ is $4\pi r^2$. It may help to think of successive onion layers marking the advance of the fluid in unit intervals of time. Owing to the incompressibility of the fluid, each successive layer must hold the same volume. But now, thinking of that volume as the product of the area (as base) by the thickness of the layer (as height), we see that as the area of the expanding sphere increases, the thickness $\Delta r$ must decrease in proportion. However, the fluid advances by one "thickness" per unit time. Thus, the thickness measures the velocity, which similarly decreases with the area—that is, with the square of the radius:

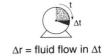

Δr = fluid flow in Δt

Fig. 2.10. Flow from a point source.

$$v = \frac{1}{4\pi r^2}$$

We now want to consider the pressure gradient as we move outward through this expanding volume of fluid. Recalling our law of force, we find the force *f per unit volume* to be:

$$f = kv$$

and we will get the actual force by multiplying by the volume:

$$F = kv(\text{vol}) = kv(4\pi r^2 \, \Delta r)$$

where $r$ and $\Delta r$ refer to the location and thickness of any layer, and the volume of the layer has been defined as the product of its surface area $4\pi r^2$ and its height $\Delta r$. (As a finite expression, this would be only an approximation but as we take it to the limit with a shrinking $\Delta r$, as a differential expression it will become strictly true. All such "approximate" expressions are to be understood as surrogates for limiting expressions, even if we do not go on to write out the latter explicitly. Hence we are not really cheating, as it might at first appear!)

Since we are interested in pressure, we divide by the area of the shell:

$$\Delta p = \frac{F}{4\pi r^2} = kv\Delta r = k\left(\frac{1}{4\pi r^2}\right)\Delta r$$

where we have substituted for $v$ the expression we derived for it just above.

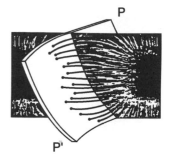

Fig. 2.11. Equivalence of surface and source.

This is really the *net pressure* on the layer, the difference $\Delta p$ between the pressure on the inside and outside surfaces. The "rate of decrease of pressure" (or *pressure gradient*) to which Maxwell refers will then be this expression divided by the thickness of the layer, $\Delta r$:

$$\frac{\Delta p}{\Delta r} = kv = \frac{k}{4\pi r^2}$$

452    [D15] Further Notes on the Region of a Point Source

467    "coalition of S unit sources"
       Do not confuse the $S$ introduced here (capital $S$) with the lower-case $s$ of the earlier equations!

472    "If a number of sources and sinks"
       Using the insight of an earlier section, we see that we now predict the result of flow from any number of sources or sinks simply by adding their pressure distributions. This was the genius of the "pressure" point of view, which takes us from the confusion of flow vectors to the calm of scalar addition. Fig. 2.9 was an example of such an easily computed result.

479    "We have next to shew"
       We may take the system of Fig. 2.11 as an example. If we draw, in our mind's eye, any surface $PP'$ and consider the region to the right of it, it is evident that the same tubes of force would arise if

we removed the source on the left—here the pole of a permanent magnet—and replaced it with a series of sources along the surface *PP'*—artfully chosen, of course, to generate exactly the same tubes of force.

The implication seems to be that a source is entirely equivalent to a surface distribution. If we in effect take everything to the left of the surface as a "black box," and know *only* the configuration of the system to the right, the source will be implicit. We are then free to interpret the configuration to mean either the existence of a single source or a distribution of sources over the surface. For that matter, we might have placed the "surface" anywhere. We see that *the source is becoming more implicit and dispensable,* yielding the possibility that the system with its distribution is *the only entity there is.* Little by little we are moving toward a view of the fluid motion suggesting Faraday's concept of the electric charge.

498   * * *

Maxwell begins now to think about alternative media represented by differing values of *k,* noting here that from the point of view of *pressure, k* and *v* play equivalent roles. We can get the same pressure distribution with low velocities in a resistant medium or high velocities in a more yielding one. He goes on to utilize this principle in a long section of the essay which we must omit for reasons of the economy of our study.

514   [D16] On Electrical Interpretation of Our Fluid Models

525   "numerically equal to the decrease"
This is a strange equation:

$$X = - \, dp/dx$$

Maxwell says, ". . . the resultant attraction in the electrical problem is proportional to the decrease of pressure in the imaginary problem . . ." and points out that since the imaginary problem is entirely of our own creation, we may choose what units we like and make the two quantities, for convenience, numerically equal. (We recall the conundrum of the fixing of a mental "unit," discussed in [D7].)

We get an equality but unlike any proper physical equation, this one draws its left side and its right side from two altogether disparate realms! *X* on the left-hand side stands for the "attraction in the electrical problem"—a physical quantity, then, measurable in the domain of electrostatics. However, *dp/dx*, denoting the limiting value of the change in pressure per change in distance, is the space-rate of change of *pressure* in a world of the imagination. Are we not violating a fundamental rule against "mixing apples with oranges"? Like good lawyers, we may argue that "the rule does not apply." For the rule assumes that the equation lies within a single universe. With our "equality" we are deliberately bridging universes, metaphorically balancing a physical world against an imaginary one.

528    "if *V* be the potential"

The concept of electric *potential* may be recalled from Chapter 1, where the potential associated with any point *P* was defined as the amount of work required to bring a unit charge from infinity to *P*. A potential difference between two points on the *x*-axis separated by distance $\Delta x$ will then be the work required to carry a unit charge over the distance $\Delta x$. We may express this a little more generally as the *work per unit charge* to carry any (small) test charge through that distance.

If *X* is the *x*-component of the electric field at this point, this will be by definition the *force per unit charge* there. But work is, by definition, just *force times distance*. Thus the potential difference between our two points on the *x*-axis will be $X\Delta x$. Since potentials, as scalars, are additive, if a displacement of the test charge has three components, $\Delta x$, $\Delta y$, and $\Delta z$, while the electric field has three corresponding components *X, Y,* and *Z,* the total potential difference will be

$$\Delta V = X\Delta x + Y\Delta y + Z\Delta z$$

Maxwell now begins to use what is known as *differential notation,* a kind of mathematical shorthand which will be of great help to us. Briefly, we may think of it in this way: when Maxwell writes

$$dV = X\,dx + Y\,dy + Z\,dz = -dp$$

each term of the type $dV$, $dx$, etc. represents a small increment in the quantity, of the sort we have been writing as $\Delta V$ and $\Delta x$. That is, we might substitute the equation:

$$\Delta V = X\Delta x + Y\Delta y + Z\Delta z = -\Delta p$$

except that this relationship would be only approximately true. It becomes strictly true only as a limiting expression, as the increments approach zero, that is, as

$$\Delta x \to 0, \qquad \Delta y \to 0, \qquad \Delta z \to 0, \qquad \Delta p \to 0, \qquad \Delta V \to 0$$

That is, the differential expression is not an approximation, but an equation strictly true *in the limit* as the incremental quantities all approach zero. As they vanish, they do so in that strict relationship. This is one way of looking at the fundamental idea of the calculus. The shorthand differential notation, then, which we shall follow Maxwell in using freely, is equivalent to the fuller statement:
   "in the limit, as

$$\Delta x \to 0, \Delta y \to 0, \Delta z \to 0, \Delta p \to 0, \Delta V \to 0$$
$$\Delta V = = X\Delta x + Y\Delta y + Z\Delta z = -\Delta p"$$

We always understand that full statement to be intended when we write the shorthand notation

$$dV = Xdx + Ydy + Zdz = -dp$$

So stated, this is a strict truth, not the approximation which it might at first seem!

530   [D17] On Calibrating the Model

531   "In the electrical problem we have"
Here Maxwell introduces the summation operator $\Sigma$, which will prove highly useful from now on. Mathematical operators are in effect instructions to carry out certain operations. The expression he has written

$$\Sigma\left(\frac{dm}{r}\right)$$

we may understand as an instruction to take all the sources $dm$, each with at its own distance $r$, divide the magnitude of each source by its distance, and then sum all the resulting quotients. This can be spelled out more formally by using a more complete notation. If we are understood to have $n$ sources, the individual sources being designated $dm_i$, with $i$ running from 1 to $n$, we may write

$$\Sigma_1^n \left( \frac{dm_i}{r_i} \right) = \frac{dm_1}{r_1} + \frac{dm_2}{r_2} + \frac{dm_3}{r_3} + \ldots \frac{dm_n}{r_n}$$

Here the notation is telling us explicitly to take all the instances of $dm$'s, each with its corresponding $r$, perform the division, and add up the results. Note that everything depends on the individual potentials being scalar quantities of the same kind, which *are* additive in the way we have already seen.

Maxwell, then, omitting the subscripts is using a shortcut notation which serves well whenever, as here, the meaning is clear.

Maxwell introduces a negative sign in front of this summation, evidently because he is thinking here of the force as one of attraction. If a given $dm$ is positive, then the test charge we are bringing in to measure the potential must be negative. In general, however, we will think of the unit test charge brought in to measure potential as *positive* and so not write the negative sign Maxwell uses here. If $dm$ is positive (positively charged source) and the test body by agreement is also positive, we will do work against a repelling force in bringing in the test body and the potential about the positive source will similary be positive. That yields the potential "mountain" depicted in Fig. 1.25.

538    "The potential of any system of electricity on itself"

Let us imagine that we are going to assemble a system of electrostatic charges. We start with one charge, then bring up another, and another. To bring the second charge into the vicinity of the first, we have had to do *work*, which is then said to be stored in the charged system as "potential energy," as in a compressed spring. If we call the potential due to the first charge at some point $V$, then by definition that will be the *work per unit charge* that we have to do in bringing up the second charge $dm$:

$$dW_{elec} = V \, dm$$

Now, thoroughly exploiting his metaphor, Maxwell writes $pS$ for the fluid to *correspond* to $Vdm$ electrically. This is not a derivation but rather an affirmation. The "equation" represents a quantitative comparison; the quantity $V$ is modeled by the quantity $p$, while $dm$, the electrical source, is modeled by $S$, the corresponding "coalition of unit sources" Maxwell spoke of earlier in the fluid case. Thus,

$$dW_{\text{fluid}} = pS$$

(We might wish to write $dS$, for consistency.)

In total, for all the charges of the system, the energy of the assembled system (like a system of compressed springs) is

$$W_{\text{elec}} = \Sigma \, V \, dm$$

and if we insert the factor $\frac{k}{4\pi}$ [D17], which serves to calibrate the model,

$$W_{\text{elec}} = \Sigma \, Vdm = \frac{k}{4\pi} \Sigma \, pdS = \frac{k}{4\pi} W_{\text{fluid}}$$

Finally, we recognize the significance of the mysterious "work" which we were told was being dissipated in the ideal fluid. It now represents the *total energy stored in the electrostatic field.* This is tricky, for while it is true that although both sides represent energy, on the electric side the energy is stored, while on the fluid side it is being dissipated. One side is timeless while the other is a process in time.

Furthermore, we can now imagine the electrostatic energy as not merely present in total but actually configured through space in accordance with the density of the unit cells in the fluid model. The cells were *dissipating* energy in time; but now we stop time. The electrostatic energy is thus not "dissipated," but merely *present.* Note that there is no physical evidence that "energy" is present in the electrostatic field in this way. We have merely the force of our metaphor to tell us so. Are we merely *inventing* this world of the field?

541  "then by (32)"
Section (32) is not included in our selections. What Maxwell demonstrates there, for fluid pressures, is just the symmetry he is beginning with here, as he interprets the fluid system in terms of

electric potentials. If a fluid system consists of just two sources, and we think of the pressure due to each at the position of the other and multiply each source by the pressure due to the other, the two products are the same. The pressure of the first times the strength of the second equals the pressure of the second times the strength of the first. The argument is extended in (32) to include two whole systems of sources.

The product (source   pressure) turns out to measure the energy being dissipated by pumping the source against the ambient pressure, so that in the fluid case—as we see now in the electrical analogy—we are measuring the energy of the whole assembled system. The symmetry result seems to reassure us that the total energy of a given assembled system is uniquely defined no matter how we go about assembling it. The total energy of a system is what Maxwell is calling "the potential of a system on itself."

555    "The electrical induction"
At line 498, we found it necessary to skip the sections Maxwell now alludes to. The line of thought there can be sketched as follows. Applying our fluid model to electrostatics, we may go on to correlate the *resistance* of the fluid medium (denoted by $k$ in our discussions) with Faraday's *specific inductive capacity*. Introduction of a block of glass, for example, in the region of a charged body alters the lines of force. It *reduces* the electrostatic attraction and hence the potential. This in turn corresponds to a lessening of fluid pressure or to a *reduction* of $k$ of the medium.

Now as Maxwell points out, this same effect can be produced in thought with a *new source on the boundary* between the two media. If we think in terms of the fluid model, the effect in the fluid is equivalent to an apparent distribution of sources or sinks on the boundary (compare Fig. 2.11). Shifting back to the electrostatic interpretation of the model, we see that any effect of a dielectric on the field will have a counterpart in terms of a distribution of charges, whether real or merely apparent, on this surface. The result which we meet here in the model—should we say "inevitably"?—also appears as a phenomenon in electrostatics.

Indeed, empirically we do seem to observe a charge on the *surface* of the glass, which is said to be "charged by induction." But is it really "charged" with anything or only apparently so?

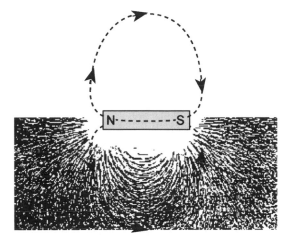

Fig. 2.12. Lines of force through a bar magnet.

Faraday certainly suggests that what we had thought of as "electricity" on a surface is no more than an epiphenomenon of a state of induction in the field. In his subsequent treatment of electrostatics, we will see Maxwell walking a delicate ontological tightrope on this fundamental question. Is "charge" really anything at all or *only* an epiphenomenon of the field?

581    "Theory of Permanent Magnets"

Magnetic lines of force seem to be "solenoidal," which means that unlike the electrostatic lines, they make closed curves. Even in the case of a permanent magnet, the lines may be understood to complete themselves through the body of the iron (Fig. 2.12).

Within the body of the magnet, we may imagine cells, each of which appears to be a source at one end and a sink at the other. Each cell will be in effect a little magnet, and we may think of the body of the magnetized material as "polarized" throughout in this way. In the larger view of the entire bar magnet, however, adjacent, juxtaposed poles will cancel one another, and the effect will be to leave the appearance of a polarization of the bar as a whole. We may represent this by imagining "sources" distributed over the north face of the bar magnet and "sinks" over the south. But from the point of view of the fluid metaphor, these "sources" and "sinks" of the traditional theory are merely apparent. Our solenoidal lines *have* no "sources."

590    "in fluid motion"
What evidence do we have that the "lines" really do continue through the body of the bar magnet? No test body can be introduced in the midst of the iron (except by carving out a cavity, which in turn confuses the issue). Faraday, however, had proved their existence to his satisfaction by the use of the "exploring wire," a loop of wire connected to a galvanometer. On the principle of electromagnetic induction, the exploring wire could detect the presence of the lines of force. Any change in the magnetic field through a loop would give rise to an impulse of current through the galvanometer. By wrapping the loop around the center of the bar magnet and then withdrawing the magnet completely, Faraday obtained a deflection of the galvanometer, which confirmed that there had been lines passing through it and thus that there were lines passing inside the bar magnet.

605    "By referring to 23 and 26"
As at line 555, Maxwell here alludes to sections we found it necessary to omit. Their content was summarized in the note to line 555.

618    "We may obtain a more general law"
Of the three interpretations of diamagnetism and paramagnetism that Maxwell offers, this, the second, seems the most interesting. In Fig. 1.31, Faraday depicted the paramagnetic and diamagnetic cases as, respectively, better and worse conductors of the imaginary fluid. Introduction of a paramagnetic material lowers the density of the unit cells and lowers, in turn, the energy of the whole system. Lowering the energy in this way, however, is like uncoiling a spring, for the system will tend to run in that direction. This tendency manifests itself as a "force" tending to orient the paramagnetic bar into alignment with the field, for when it is so aligned the paramagnetic bar will be in its lowest energy position. Just the opposite will be true of the diamagnet, which will orient crosswise to minimize its presence in the field. (It was these two

results that were illustrated in Fig. 1.30, depicting the diamagnetic effect that Faraday had discovered.)

From this "energy" point of view then, we see that *forces are no longer primary.* They *derive* from the configurations of the field, while considerations of *minimizing system energy* give direction to processes—that is, to "tendencies"—that begin to look almost purposeful, or teleological!

637   "the same mathematical results"
It is important to note that we are considering alternative modes of thinking about the same phenomena. Maxwell is interested in the fact that one can completely reverse the theory and yet end up with the same mathematical predictions. In the above accounts, bodies of both kinds are seen as polarized in the *same* sense as the lines. Now, in a third approach, diamagnetic bodies are supposed to polarize in the *reverse* sense. This was Tyndall's theory, to which Maxwell refers in his note.

646   * * *
We omit a brief section on "Magnecrystallic Induction."

647   "Theory of the Conduction"
The fluid model applies immediately—as if without translation—to electric current flow in conducting materials, whether solids, plates, or wire circuits. Here, motion in time really *is* in question and motion of "electricity" (whatever, if anything, it may be!) accords with that of the ideal fluid. "Electricity" does behave, then, like a perfect incompressible fluid and, interestingly, in the laws worked out by Ohm and Kirchoff, shows no sign of possessing inertia, a crucial property of Newtonian matter. Despite Maxwell's own later experiments, it is to be more than forty years before the first evidence of the association of charge with ponderable matter begins to emerge in the experiments of J. J. Thomson on "cathode rays" or, later still, those of Robert A. Millikan on the "electron."

661   "If we knew"
Determination of the amount of "electric charge" which passes when unit current flows becomes a highly critical measurement,

Fig. 2.13. The Leyden jar. One practical application of the Leyden jar is illustrated in Fig. 1.22.

linking as it does two otherwise distinct elements of electrical science. We will return to consider its importance and will be very much involved with its physical determination when we read the next paper "On Physical Lines."

666    "the difference between glass and metal"
This is a difficult passage, which seems, on the one hand, to come as close to a mistake as we get in our reading of this essay, yet, on the other, to contain the germ of what will be perhaps Maxwell's most creative and daring insights concerning electric science. It would seem that the polarization of an insulator is *not* rightly viewed as a case of poor conduction, as Maxwell claims here. If it were conduction, *however poor,* it would continue indefinitely at whatever low level—the current, however weak, would be steady. But actually this is not the case. The flow of current to a Leyden jar (diagramed in Fig. 2.13) is rapid at first, as the jar charges, but thereafter current flow virtually ceases and the glass reveals itself as an excellent insulator. The glass appears to store in its state of polarization the charge it has received. Long after, a good Leyden jar can be discharged with an impressive spark: it had been loaded, one might conclude, with "potential energy." This is very different, then, from merely a case of poor conduction.

In the absence of sensitive instruments and appropriate experiments, the phenomena were of course not so well-defined in Maxwell's time as they may seem today. In any case, it is fortunate that Maxwell holds to his seemingly perverse view, for reasons we will be able to see only when we meet their consequences in the next paper. As in the case of Faraday's conviction concerning "curved lines of force," this may be a brilliant instance of creative error in the sciences—one curious channel of deep insight.

701   "This distinction of quantity and intensity"
"Quantity" of current seems to be what is today called simply "current"; it is measured by some form of galvanometer or ammeter placed in "series" with the circuit (Fig. 1.34). Maxwell's "intensity of current" is confusing, as the term is now used in another sense. For Maxwell's "intensity of current," today one speaks of the "electromotive force" that drives the current, measured with a voltmeter. The fluid analogy now being applied strictly was already anticipated in our account of this in the first chapter, in which we likened "voltage" to fluid pressure and a battery or power source to a pump. As we saw, Ohm's law is written today as

$$I = E/R$$

where $I$ is the current (Maxwell's "quantity"); $E$ is the electromotive-force (Maxwell's "intensity" $F$); and $R$ is the resistance (Maxwell's $K$) of the conductor to the flow of current.

708   * * *
In the omitted section, Maxwell applies this same notion of flow to magnetism. I omit it reluctantly for reasons of economy in our study.

711   "investigated by Ampère"
We have referred to Ampère's investigations in our opening chapter and detected allusion to them in Maxwell's opening remarks concerning the "intricate mathematics" of other investigators.

Maxwell now draws upon Ampère's work but only with the ultimate intention of taking an altogether different approach.

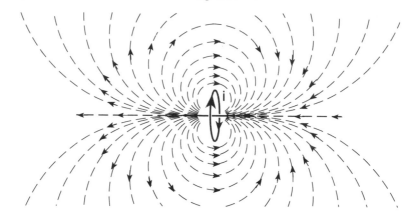

Fig. 2.14. Equivalence of a current and a magnetic dipole.

While Ampère has begun with *current elements*—little pieces of current—and has succeeded formally in building a mathematical theory upon them, there is no evidence for their existence and Maxwell feels entitled to "do without" such an assumption. The reversal of the foundations on which he embarks here is by now familiar to us: instead of starting with the small and building up, he will start with whole circuits. After all, if a current can flow only through a completed path, it is normally whole circuits with which we in fact experiment. In this case, Maxwell's approach through the "whole" seems less forced, closer to the actual phenomena. One might wonder, however, about those special circuits which seem instead to be "open," with current flow blocked, as in the case of the circuit which charges the Leyden jar (Fig. 2.13).

730    * * *

Once again for the sake of the economy of our own study, we will not attempt to track Ampère's argument here. A very useful review of his theory, together with a text of the essay, is available.[3] Rather, we turn directly to Maxwell's own approach, using the method of the "lines" in a remarkable and highly revealing way.

731    "From these results it follows"

This equivalence in the limit between a very small circuit and a very small magnetic dipole (Fig. 2.14), borrowed from Ampère's

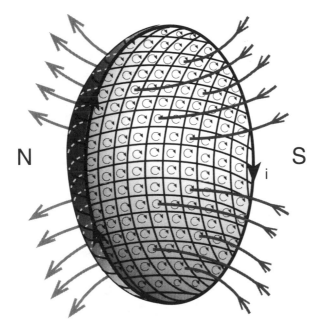

Fig. 2.15. Equivalence of a current and a magnetic shell.

theory, becomes the foundation of Maxwell's. A "right-hand" rule is useful in keeping track of this: if the fingers of the right hand indicate the direction of conventional current flow, the thumb points in the direction of the resulting magnetic lines.

739   "If a number"

The first step is to get from the elementary circuit element above to full-sized circuits. From the point of view of the currents, Maxwell points out that a large number of tiny current loops side by side over a surface have the effect of one large current around the periphery, since adjacent currents will cancel (Fig. 2.15). On the other hand, the *magnetic effects* of the elementary currents do *not* cancel, but combine to give the effect of a magnetized surface, which Maxwell speaks of as a "magnetic shell." Therefore, by a kind of physical syllogism, a single current around the periphery will have the effect of a magnetized surface. It is interesting that the *shape* of the surface does not matter—it need not be flat.

752    "the total intensity"

Since these lines "embrace" the current, they form complete loops, or are *solenoidal*. Since the line thus points in the same direction all the way around, if we think of our fluid and move opposite to that direction around the circuit, we will do work against a pressure gradient all the way (presumably we must insert a pump, representing the magnetizing force, to maintain the flow). That total work Maxwell calls here the "total intensity," and it will be proportional to the current flow.

763    "The whole number of cells"

The whole number of (unit) cells will be the total energy of the configuration. Since we have just seen that the total number of cells in *one* unit tube is a function of the quantity of the current only, the total number of cells will be proportional to the quantity of the current multiplied by the total number of lines *it* embraces (as through the shell of Fig. 2.15). The total number of lines embraced by the current has become known, plausibly, as the magnetic *flux* through the surface. Thus the *total energy in the whole configuration* will be proportional to the flux times the current. This relationship is illustrated by the field of a current-carrying loop (Fig. 2.16).

778    "the resultant force will tend"

What is still more striking is the realization that we now have a principle by which we can determine the mechanical forces between interacting magnets or currents from the configuration of the field alone. If the currents, and thus the fields due to them, are maintained constant, then the total energy of any combination of elements will be proportional to the total number of lines they produce. Since any system will run (like an uncoiling spring) to minimize its total energy, *mechanical forces will arise on the conductors tending to minimize the number of magnetic lines.* The magnetic field will tend to "uncoil" as does a complex spring.

As an example, think of the situation shown in Fig. 2.17. There, a wire carrying current vertically downward and has been placed in a uniform magnetic field, such as the field of the earth, running from right to left. Since the lines from the wire and the earth's

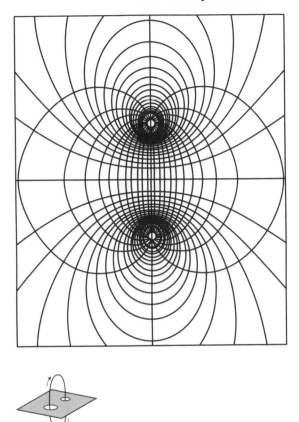

Fig. 2.16. Field lines delineating the energy of the magnetic field surrounding a circular current-carrying loop.

field *add* on the side toward *A,* increasing the density of the cells there, while a magnetic "vacuum" is created on the side toward *B,* the conductor will experience a mechanical force from *A* toward *B.* That is, it will move to relax the compressed field and thus reduce the total energy of the system. The same energy that had been stored in the field will now appear as the mechanical work done to move the wire. In field terms, we are seeing here the basic principle of every electric motor.

By the same principle, we can understand in field terms why it is that "like magnetic poles repel." In the image of repelling magnets in Fig. 1.10, let us suppose that these are equal, opposing

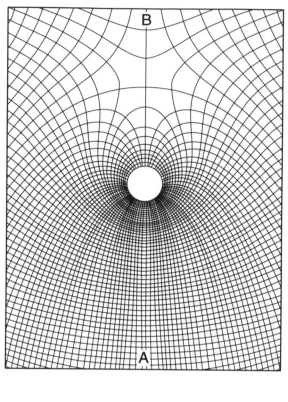

Fig. 2.17. Energy cells surrounding a wire carrying current vertically downward in a uniform field which runs from right to left. The wire tends to move toward B, an energy "vacuum" in the field.

north poles. The lines are then emerging symmetrically from both and are everywhere *adding* to increase the field intensity, hence, we now know the field energy density as well. As we bring the two magnets together, we feel that we are compressing the field as if it were a spring and storing energy. Released, the magnets will fly apart, relaxing the field and converting this energy into the mechanical energy of the accelerating magnets.

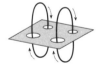

Fig. 2.18. The field of two parallel circular currents. The fields of two coils of the type shown in Fig. 2.16 have been allowed to coalesce. The approximation to a uniform field along the axis accounts for frequent laboratory use of this configuration.

The erstwhile "poles" have disappeared from this explanation. *The poles are gone* and with them the mystery of action-at-a-distance. By our new account, it was really the *fields* surrounding the magnets that were interacting—Faraday says they "coalesce"—when we told the old story about the "poles" "attracting" or "repelling." The reader is invited to work out the corresponding explanation for what we once called the "attraction" of "unlike poles," (Hint: take the *entire* fields into account, not just the portions drawn in our close-cropped figures!)

787    * * *

It has been necessary to omit a rather extensive discussion of this phenomenon.

828    "The idea of the electro-tonic state"

We leave Maxwell with this significant admission of temporary defeat for the method of the lines. I think we can identify the source of the difficulty, though not resolve it. The mental image of the lines which we so carefully built scrupulously excluded any reference to matter or inertia. What that meant was that *time* really had no place in the model we were making. It is true that we thought in terms of "flow", but without an inertial parameter there is no principle of connectedness between one "time" and the next.

What happens when a force is applied at one point in our thought-fluid? Is not the reaction *instantaneous* everywhere? Under our "Aristotelian" force law, velocity simply "is" proportional to the applied force—there is no time-value to the predication, no gradual acceleration. Similarly, the fluid is incompressible ("hard") and the *space is full*. Any impact at one point must be immediately felt throughout (contrast caloric, which *diffuses* into an "empty" space in heating a cold body). Here, it seems, is a fluid in which there can be no waves! For there is neither elasticity nor momentum, the two parameters fundamental in one form or another to the "swinging" character of any wave motion.

The "state" Faraday is naming "electro-tonic" would, it seems, be such a state of *momentum*—something like a "flywheel" effect in the space surrounding a magnet. Yet, as we see, momentum is just what Maxwell's fluid lacks. This seems to lie at the root of his abandonment of the project for the time being. We shall see what he decides to do about this situation in his next paper, "On Physical Lines of Force," the subject of our Chapter 3.

# Discussions

### D1. On the Particulate Theory of Refraction [line 49]

In the case of light, we have similarities to two altogether distinct sciences, so that it is evident that analogy to each can be only partial: the first is to the motion of a particle and the second to wave motion in an elastic medium. In the first, a ray of light is regarded as analogous to the path of a corpuscle, where motion is explained on mechanical principles. As a ray passes from air into an optically "denser" medium such as glass, it is bent toward the normal according to a simple law.

This is shown in Fig. 2.1a, taken from Descartes' *Optics*—a work appended to his *Discourse on the Method* to show the world how science should be done. If a particle of light moving from point *A* in one region enters a second, optically "denser" medium such as glass, Descartes assumes that it will be subjected to a sudden force *F* perpendicular to the surface *CE* as it enters the second region at point *B*. Thus it will be speeded up and its path correspondingly bent.

Descartes draws the circle *AFI* centered at *B*. Then as the velocity is greater in the second medium, the distance *BI* (whose precise position is to be determined) will be traversed in less time than the equal distance *AB* in the first. However, as the deflecting force is entirely perpendicular to the surface, there will be no change in velocity in the horizontal direction. And since *BI* is traversed in less time than *AB*, the horizontal distance *BE* will be *less* than *CB*. It follows that *GI* (=*BE*) and *AH* (=*CB*) will vary inversely with the velocities in the two media. Hence the ray is bent *toward* the normal *BG,* and obeys the empirical law of refraction, the "law of sines." For (recalling that in a right triangle the sine of an angle is the ratio of the side opposite to the hypotenuse):

$$\frac{\sin \angle ABH}{\sin \angle GBI} = \frac{AH/AB}{GI/BI} = \frac{AH}{GI} \quad (AH = BI)$$

but

$$\frac{AH}{GI} = \frac{\text{Velocity in glass}}{\text{Velocity in air}} = \text{Constant}$$

The constancy of the ratio of the sines is the observed "law."

Although this analogy is striking, it fails as a hypothesis; nature proves unkind to Descartes. There is one decisive difficulty: the particle must be

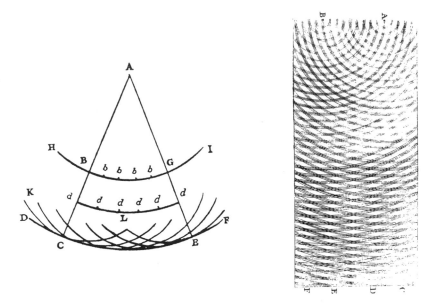

Fig. 2.19. (a) Huygens' diagram of a wave front; (b) Young's construction of an interference pattern.

*speeded up* in the glass. Unfortunately for the theory, measurements of the velocity of light made by Foucault and Fizeau in 1850, only a few years before Maxwell writes, had shown on the contrary that light is *slowed down* in a refracting medium such as glass. Should we say in such a case that the analogy may *help our thinking,* but has no force as an *explanation*? Or does it just mislead us?

## D2. On the Wave Theory of Light [line 54]

The "resemblance in form" in this case is between phenomena of light and the behavior of waves in an elastic medium—one which can be distorted but resists distortion with a proportional force. Huygens had offered an alternative theory of light as the motion of a wave through such an elastic medium, as for example a sound wave moves through steel. The "ray" of light now becomes a *wave front,* as Huygens suggests in Fig. 2.19a, taken from his *Treatise on Light.*

Figure 2.1b reproduces Huygens' diagram explaining refraction as a wave phenomenon. The light ray, now as a wave front, *slows down* in the water and wheels around to obey the same law of sines. *AC* is the wave at the point at which it first impinges on the glass. From points *K,* as each successively impinges on the surface, a new wave propagates in the glass at a lower velocity, with the result that a new wave front is formed, *NB,* whose direction *AN* has been swung toward the normal.

The wave theory, however, as Maxwell says "extends much farther," for it can much more readily account for phenomena such as the interference pattern which arises when a wave front passes through a pair of slits. Thomas Young's diagram of the reinforcement and cancellation which generates interference is reproduced in Fig. 2.19b. Finally, to account for the "polarization" of light, it is possible to take the waves as transverse, that is, vibrating at right angles to the line of motion, in which case waves on the surface of water turn out to make a very good "physical analogy."

Though no contradiction with experiment arises in this case—as it did between the predicted and observed velocities in the case of the analogy to particles in refraction—Maxwell nonetheless emphasizes that we have nothing to go on but a similarity *in form.* We have, after all, no evidence of anything "physical" that is "waving" in the case of light, which accomplishes its passage all the better when we evacuate the space it traverses.

Only in the second paper will it emerge that Maxwell's own study bears dramatically on this question of light "waves." And only in the twentieth century does the wave theory of light encounter phenomena, first in the form of the photoelectric effect, that it cannot explain.

## D3. On a Theory of "Transverse Alternations" [line 60]

This would presumably be a purely *formal* theory, which accounted for the observed phenomena and correctly predicted new ones—in terms of symbols and numbers only, with no reference to a "physical" medium. This is often taken as the only really legitimate goal of mathematical physics (think of Ampère's account of the method of "the learned men of France," cited in Chapter 1). Maxwell's objections to it—as lacking "vividness"—are therefore of special interest. Once again Maxwell seems to be thinking of a kind of mental "vision," which he says does admit "vividness," and now also "fertility." One wonders whether these criteria are pertinent to its *truth* or are concerned with something else.

It is striking that the two analogies Maxwell mentions in the case of light—on the one hand to particles and on the other to waves—remain inseparably joined and yet still mutually contradictory in the current quantum theory of light, in which light acts like a wave in one aspect and like a particle in another.

D4.  On Fourier's Theory of the Uniform Motion of Heat [line 78]

Maxwell is embarking here on a most striking example of physical analogy—between "the science of attraction" and that of the flow of heat by conduction through, for example, a rectangular iron plate heated along only one edge and cooled along the others. In speaking of the "science of attraction" I think we can understand him to have in mind *gravitational* attraction, as between the earth and the moon, that is, action of one body on another at a distance. The two physical systems are to all appearances completely different.

Jean-Baptiste Fourier had solved the problem of heat flow in conducting materials brilliantly more than fifty years before Maxwell writes, in his *Analytical Theory of Heat*. Figure 2.2 illustrates one paradigmatic problem which Fourier works through at length. A rectangular steel bar, finite in width but infinite in length, is constantly heated at one end (say, to temperature +1) and cooled along the two sides—should we add, and at infinity?—say, to temperature 0°. Fourier's problem is: what is the equilibrium temperature distribution throughout the plate?

His result is expressed in a converging series of terms, whose sum approaches the solution more and more closely as more terms are added to the summation. The approximation that arises when just the first four terms are taken is graphed, for the +1 end, in Fig. 2.20. The truly brilliant method that he developed for the solution of problems in the flow of heat, known today as "Fourier series," is very broadly applicable, and the foundation of a great deal of work in mathematics, modern physics, and technology. In its general form, it allows the expansion of any reasonably well-behaved single-valued function as a "trigonometric series," that is, a series of functions representing vibrations in an ordered relation of a fundamental to its successive overtones. If the "function" represents the vibrations generating a musical tone, the Fourier series becomes the fundamental and the overtones that give the distinctive character, the timbre, of the tone.

To digress altogether for just a moment from Maxwell's text—though not from its spirit—it is interesting to track down more closely this relation between the solution of a thermal problem and a musical tone. We have graphed Fourier's solution to the thermal problem in Fig. 2.2. We now, in Fig. 2.20, show the first four of the successive series terms which generate this solution. Speaking musically, we would call them the "fundamenatal" with its first three odd "overtones"—the third, the fifth, and the seventh, each with the amplitude Fourier calculated. At the

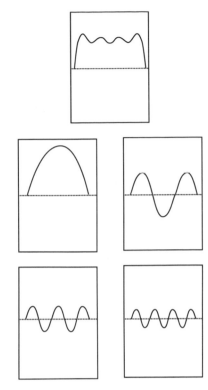

Fig. 2.20. The first four terms of Fourier's synthesis of a function as a series. The "fundamental" and the third, fifth and seventh "harmonics" are shown, each with its computed amplitude. Their summation is shown at the top.

top is shown their summation and we note that in this crude approximation, the oscillating character of the trigonometric function is showing through! If more terms were added, the solution would approximate more and more nearly to the "boundary condition"—a temperature of exactly +1 over the whole width—while the solution would become exact in the limit, as the number of terms added together exceeded all bounds. In our electronic era, a reader with a synthesizer could program these and produce a tone with a timbre which was the "sound" of Fourier's heated plate. That is how far the concept of "physical analogy" might take us!

To turn from music to the analogy of which Maxwell is speaking, however, we may take the case of electrostatics as the "attraction" in question. It helps if we think in terms of the electric *potential*, discussed in Chapter 1. The reader may recall that this associates with every point a quantity representing the amount of work it would take to bring a test charge from infinity to the point in question; thus is then thought of as

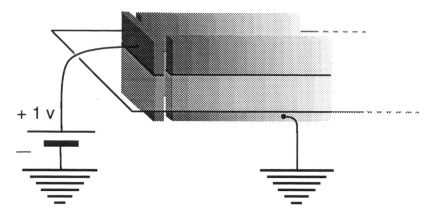

Fig. 2.21. Electrical interpretation of Fourier's heated bar.

stored or "potential." Energy is a scalar quantity, so the potential field is a scalar field. Every point around a charged body thus has such an associated potential (Fig. 1.24), which we found we could think of in terms of a gravitational model (Fig. 1.25). This in turn might be diagramed with equipotential contour lines in a kind of electrostatic "topographic map"

The analogy, Maxwell says, is now between *temperature* and *potential*. To put this insight to work: how can we read the analogy in the case of Fourier's problem, diagramed above? Evidently, the thermal source at the heated plate is transformed metaphorically into a source of attraction, the flow of heat becomes the electric force (the lines of heat flow become Faraday's "lines of force"), and the temperatures represent potentials. Thus, the *isothermal lines* become *equipotential lines.* As Thomson had discovered, in solving the thermal problem, Fourier had inadvertently solved an electric problem as well. Here, the "hot" wall of the heated bar becomes a charged plate of a simple two-dimensional configuration; the side walls become plates at zero (ground) potential. The equilibrium temperatures become the potential distribution between the plates, for which the thermal surface of Fig. 2.2 can be taken as a topographic representation. The equipotential lines would be contour lines on that surface.

What does it tell us about nature, that such formal likenesses should override substantive differences and link its phenomena in this way? Maxwell, reflecting on this puzzle, wrote a remarkable essay for the Cambridge "Apostles' Club," which he titled, curiously enough, "Are

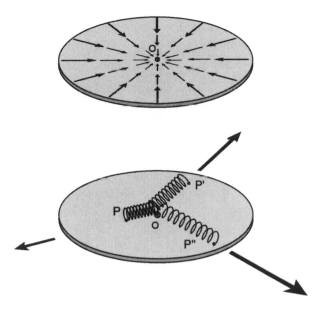

Fig. 2.22. A "washbasin model" of an elastic force: a model that does not work.

there Real Analogies in Nature?"[4] It seems to suggest clues to the cogitations which gave rise to the essay we are now reading.

D5.  On a Fluid-Flow Model That Does Not Work! [line 154]

Consider a spring, fastened at a center $O$ and thus defining a force at every point $P$ to which the other end is extended. In an ideal spring, the force obeys "Hooke's law," increasing in direct proportion to the length to which the spring is stretched. (It is a special idealization to suppose, as we will for convenience here, that the spring is capable of vanishing, so that zero force corresponds to zero length!) The force will be directed in every case toward the center $O$, and we might think we could model the spring with a fluid flowing toward a sink at the center—a sort of wash-basin model of the elastic force (Fig. 2.22).

But the model works backwards. As water flows inward toward the center, it must move faster and faster in order to get out of the way of the steadily converging flow inward from the periphery; yet if flow is to model *force*, this would imply that force must increase toward the center—the reverse of the case with the spring. Contrast the flow model of Fig. 2.10, in which the analogy is complete between fluid flow and an inverse-square force law, with the highest velocity in the *model* occurring at the center. It is this strict analogy which permits Maxwell's fluid images of Faraday's lines.

One might wonder whether a force could be imagined which would be accurately imaged by the "washbasin" flow of Fig. 2.22.

## D6.  Reflections on Method [line 172]

Maxwell seems more interested in introducing a new idea or shaping a point of view toward electrical and magnetic phenomena than in developing what is normally thought of as a scientific theory. At a time when we tend to emphasize the concept of mathematical physics and to embrace any theory which "works" in the sense of predicting phenomena, we may be struck by Maxwell's somewhat different sense of purpose. Although he immediately emphasizes his conviction that his fluid is "mathematical," his purpose in introducing it is rather thoughtful than predictive. In that sense it belongs at least in part to the art of rhetoric—the ability to use words, images, and metaphors to suggest ideas and to shape thought. This may be an aspect of the tradition of science too little stressed in our time.

With the aid of his analogies, Maxwell is taking aim at ideas, and at equations only as vehicles for ideas. It is interesting that he speaks of "mature" theory, and thus puts this in terms of the phases of a life. Since he is writing as a young man, perhaps it has some implications for his sense of his own prospects as a natural philosopher. In a "mature theory"—something unlike the present merely speculative enterprise—*physical facts will be physically explained.* Here "physical" may bear its root meaning of "natural," i.e., according to the order of truths about nature. Newton, one of Maxwell's principal masters in science, makes a similar distinction. His *Principia Mathematica Philsophiae Naturalis* sets forth only the mathematical prolegomena to the real philosophy, which will tackle physical questions about truths of the universe. Maxwell seems to see mature theory as an escape or graduation from the stage of mere mathematics, which can only frame the questions.

Is it not likely that Maxwell has seen in the elderly Faraday something of the image of the mature philosopher? Sending him this mathematical essay, as he will do, would be a little, then, like offering advice in a lesser mode—a suggestion from a younger man, who is not yet doing experiments and not yet quite ready to work altogether in Faraday's domain of physical truth.

## D7. On the Choice of Unit of Quantity of the Fluid [line 198]

Newton begins the *Principia* with the assertion that the fundamental quantity will be the *mass,* by which he says he means the *quantity ["mass"] of matter.* Maxwell's fluid is by contrast not made out of matter, and hence has no Newtonian "mass" or inertia. It must have quantity only in the geometrical sense of mere extension, as we use the geometrical term "volume" to refer to a space without regard to any content. Descartes made a similar move in founding his physics on a purely mathematical "extension." Though he calls it "matter," it is, like Maxwell's fluid, bereft of inertial mass. The success of Newton's physics and the relative failure of Descartes' would seem to certify the wisdom of founding a physics on good earth.

Why is Maxwell going the wrong way at this first turn, following Descartes and leaving matter out of his physics? It seems that he has something to show us about the value of thinking purely unreal thoughts—thoughts of something he can call with conviction "a collection of imaginary properties." Could it be that in the search for some latitude for thought, he is deliberately following in the footsteps of Descartes' rather willful retreat from the world as it appears to us? Do we gain room to maneuver in thought by eliminating the insistent intrusions of the surface of reality? And can such a move be a way to do revolutionary science?

We must acknowledge another consideration in this. If we *were* to be thinking "physically" of the problem of a space-filling substance which might mediate such effects as the electric force, it would have to operate without interruption in a vacuum. We would have to get the old matter out of the way to open a path for this mysterious new substance, *ether.*

It is interesting that when Maxwell set out to gain perspective on mechanics, for the sake of beginners, he named his work with the Cartesian pair, *Matter and Motion,* where by "matter," Descartes had meant just "extension."

## D8. On "Steady Motion" [line 211]

If there is *no time* in Euclid, we have to remark that by contrast our new geometry *is* curiously related to time. The term "line of fluid motion" suggests that this line is the path of destiny for any given element of the fluid lying along it; but this is in general not the case. Only if the pattern of the motion is itself constant does it determine the path of destiny of a fluid element within it. In this case, the motion is said to be *steady.*

However, if the pattern were changing, as it would be in general for a real fluid, the configuration at any moment would represent only a snapshot of the motion and the line of motion of an element of the fluid would be something else again. As beginners, we may be relieved to learn that it is only steady motion that Maxwell will be asking us to consider! We may want to return to this thought at the close of "Faraday's Lines," at which point a major question arises concerning the connectedness of our ideal fluid's motion in time.

We might remark that when "figure" is thus set in "motion," we have the very definition of what the classical tradition of the "seven liberal arts" called *astronomy*. The classical "quadrivium" of the mathematical arts consisted of arithmetic (number), geometry (figure), music (number in motion), and astronomy (figure in motion). In these terms, Maxwell, in his realm of pure thought, is doing something very like Archimedean mechanics or Ptolemaic astronomy. Indeed it is often remarked that Fourier's analysis of functions into a series of compounded functions, discussed in [D4] above, has a certain striking resemblance to the Ptolemaic analysis by way of compounded epicycles.

## D9. Further Reflections on the "Unit Tube" [line 225]

The following discussion may help in visualizing the "unit tube"; it goes on to raise a slightly metaphysical problem of the "unit" in the mental world!

Note that the cross-sectional area $s$ of the unit tube is not in general unity, but is adjusted to the velocity: high velocity, small $s$; low velocity, large $s$. The tube is so adjusted that in unit time, one unit volume will pass.

Imagine now that unit time has elapsed; then at velocity $v$ the fluid entering at the front will have advanced the distance equal to $v$ in unit time (that is what velocity is!). But the volume which has advanced in that time is $V = s \times v$, and by the definition of the "unit tube," $V = 1$. Therefore we have the idea of the unit tube expressed in a statement that will be helpful to us later:

$$vs = 1$$

This whole discussion may pose a certain conceptual problem. What benchmark can we place in this undifferentiated world of pure thought to be a *unit*? There can in principle be no counterpart to the platinum–iridium meter stick that was once formally deposited in Paris as the standard of

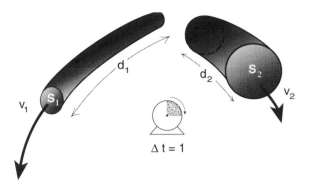

Fig. 2.23. Relationships in the unit tube.

length of the western world, nor have we introduced into our world of pure thought even one atom or wave of light whose dimensions might give us a "natural" unit. As for time: is there a clock in this Arden of our making? In what sort of "time" does our fluid move? If we can distinguish fast from slow or uniform from nonuniform motion, we must be referring to a governing concept of absolute uniformity. These problems are not new with Maxwell. They are there as well in different ways for Ptolemy, Archimedes, or Newton.

It seems that *every* tube is equally the "unit tube": if we can point to one in thought and hold it fixed long enough to take a ratio to it that will be our act of measurement in the domain of thought. The incompressibility of the fluid together with the "impenetrability" of a tube's walls guarantee that if the tube is *unit* at any point along its length, it will be *unit* throughout, that is, that this measure *belongs* to the tube as such. Only this fact permitted us to speak of a "unit tube" in the first place.

Choosing unit quantity of the ideal fluid is something like calibrating our heads. Maxwell makes good use of this curious freedom when the occasion arises to choose a definite scale for application of the fluid model to a specific physical problem.

D10.   The Law of Motion of the Fluid in Terms of Pressure
        [line 340]

We introduce here the powerful symbol $\Delta$ to denote "difference" (really an operator instructing us to "take the difference" of the quantity which follows). Thus $\Delta f$ may be written to denote the net force between two

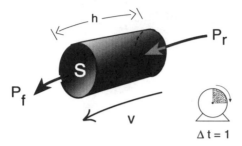

Fig. 2.24. The law of motion applied to a tube of flow.

surfaces in the fluid. In these terms for the net force on *unit volume* we have the law of motion

$$\Delta f = kv$$

with $v$ denoting velocity and $k$ a constant measuring the resistance of the medium through which the fluid is imagined to be moving. To get the net force $\Delta F$ on any other volume, we simply multiply this by the volume

$$\Delta F = (kv) \times (\text{vol})$$

Thinking of a fluid, however, in which the force in question is distributed over a surface forming a cross section of the tube, it will be more appropriate to express our law in terms of *pressure* rather than of force. The term *pressure* denotes *force per unit area* acting on a surface; we might say it measures the intensity of the force. (Think of the principle of the needle as a force multiplier, in which a modest force exerted at the point of the needle results in puncturing pressures.) Let us apply the law, rewritten in terms of pressure, to a segment of a tube of flow (Fig. 2.24).

If the cross-sectional area of the tube is $s$, and the front and rear pressures are $p_f$ and $p_r$, then the net force is (net pressure $\times$ area):

$$\Delta F = \Delta p \cdot s = (p_r - p_f)\, s$$

If the tube is $h$ units in length and is reasonably straight, then its volume will be $(h \times s)$, and the law of motion tells us that

$$\Delta F = (kv) \times \text{vol}$$

or here,

$$\Delta F = (kv) \times (h\, s)$$

If we divide by the area $s$, we get $F/s$, which is net pressure:

$$\Delta p = \frac{\Delta F}{s} = kvh$$

We could demonstrate our analytic versatility by solving for the velocity:

$$v = \frac{\Delta p}{kh}$$

which simply tells us what we would expect, i.e., that for a given length *h,* a greater pressure difference gives us a greater velocity, while the same pressure over a greater length or in a more resistant medium yields less velocity. Our analytic result makes sense!

## D11.  On Isobars and Lines of Flow [line 366]

The surfaces of equal pressure might be referred to as "isobaric surfaces," and would then rightly remind us of the isobars drawn on weather maps. The weather map, being drawn in two dimensions, shows plane cuts through the atmosphere's isobaric *surfaces.*

Since the motion is in the direction of the pressure gradient, the isobars (precisely the surfaces along which there *is* no gradient of pressure) will be perpendicular to the tubes of flow. With this recognition, the reader is invited to do some free-hand sketching of surfaces of equal pressure on any of the drawings of lines of force to which we have been referring.

It will be helpful to consider a particular case, such as that of Fig. 2.7, whose electrical interpretation will later turn out be especially significant for us. Electrically, it represents a vertical section through a parallel-plate capacitor, consisting of two horizontal rectangular plates, charged positive on top and negative below and separated by air as the insulator. Such a capacitor, but with vertical plates, is shown in Fig. 2.27.

In terms of a fluid model, the same figure depicts isobars viewed in two dimensions, the plane of the paper constituting a vertical plane section of flow between two parallel horizontal plates. Here *OA* is a linear section of a flat horizontal source (which we may think of as a rectangular "shower head"), infinite on the right and cut by the vertical plane of the paper. *PB* is a corresponding sink, that is, fluid flows out at a uniform rate along the length of *OA,* flows downward in the plane of the paper, and is absorbed along the length of *PB*. We may think of *OA* as a high-pressure line and *PB* as a vacuum.

We see that pressure drops steeply and uniformly in the region between *OA* and *PB*. If we were to visualize a gravitational model, the region *OAPB* would be a kind of shed roof or cellar door! (The reader is

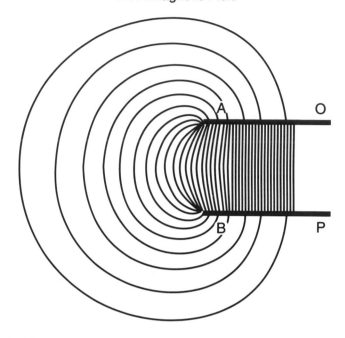

Fig. 2.25. Lines of flow for Fig. 2.7.

invited to invent a gravitational interpretation of the fringe lines on the left!) We may think of the horizontal midline or axis of the diagram as zero potential or ground level. We know, too, that we are licensed if we wish to interpret this as a thermal problem: then *OA* would be the section of a hot plate and *PB* the section of a plate kept cold, and the diagram would be of isotherms in a vertical section within a heat-conducting block joining them.

We know that the lines of fluid flow always move perpendicularly to the isobars (one speaks of the two sets of lines being "orthogonal"); Fig. 2.25 shows the flow lines that go with these isobars. Most of the flow is confined to the region of steep, uniform gradient between source and sink, since this is where the pressure gradient is intense. Some flow, however, takes the long way around, moving always orthogonally to the isobars, yet following a gentler gradient. We saw in Fig. 2.8, what this means in terms of an energy dissipation occurring throughout the fluid.

There is thus a relation between the spacing of the isobars and the spacing of the lines of flow. Maxwell helps us with this in the reasoning leading to the relation $s = kh$. As this argument shows, and our intuition confirms, the greater the sectional area $s$ of the unit tubes, that is, the lower the velocity of the fluid, the greater the spacing $h$ of the isobars or the less the pressure gradient.

## D12. On Unit Cells [line 385]

If we start with a unit tube—one conveying a flow of unit volume in unit time—and intersect with it a series of isobaric surfaces at unit increments of pressure, we divide the unit tube into cells called *unit cells*. Each such cell will have some length $h$ (not in general unity), and since the pressure difference $\Delta p$ is unity, the law of motion tells us that

$$kvh = 1$$

If fluid in a tube is moving with velocity $v$, in unit time it will advance a distance equal to $v$. If the cross-sectional area of the tube is $s$, the volume of the cylinder of fluid which passes in unit time is the height times the base, i.e., $v \times s$. But in a unit cell this volume is by definition unity, so for the unit cell,

$$vs = 1$$

Thus,

$$kvh = vs$$

Canceling $v$'s,

$$kh = s \qquad \text{for the unit cell}$$

Does this make sense? It is saying that a unit cell with large $s$ (a fat unit cell) will also have large $h$ (will be a long unit cell). The reader is invited to think that through!

## D13. On Work and Energy in the Unit Cell [line 388]

This remark of Maxwell's concerning "work spent in overcoming resistance" goes by quickly, but is really quite astonishing, as it reveals that *work* is being done—in our realm of pure thought!

Formally, "work" is defined as force times distance: a force does work in proportion to the distance through which it is exerted. A force lifting a weight does work measured in common units, in "foot-pounds" (the work of a pound-weight acting over the distance of a foot). The work you do in climbing one flight of stairs is your weight times the interval between floors.

Maxwell claims that one unit of work is done in each unit cell per unit of time. Let us think of a unit cell and watch the fluid advance through it for one unit of time. If the velocity is $v$, it will have advanced the distance $v$. We also know that the net force is the net pressure times the cross-sectional area, which we have been calling $s$. But the net pressure between isobars is unity. Therefore the net force equals $s$. What is the work?

$$\text{Work} = \text{Net force} \times \text{distance}$$

or

$$W = s \times v$$

But we concluded earlier (see Discussion [D7]) that for the unit tube,

$$s \times v = 1$$

Thus, not surprisingly Maxwell is right: unity of work is done in the unit cell per unit of time. Again formally, "work per unit time" is what we call *power,* so unit power is being dissipated per unit cell in our imaginary fluid.

What are we to make of this? In contemplating an ideal fluid in our mind's eye, are we generating ideal heat?

However we resolve this conundrum, we can see that our ideal fluid, which began as pure flow and then took on resistance, force, and configurations of pressure, now becomes a system of power and energy. If we envision the system fully divided into its unit cells, those cells are exhibiting a distribution of energy (Fig. 2.8). Where the cells are densely concentrated, the energy is as well. We may think of the entire system as containing a total flux of energy, the sum of the energies of its unit cells. We recall that Faraday called the system of lines about a bar magnet the *sphondyloid of power.* Further, when the lines get compressed and the cells become densely packed, this looked to Faraday like loading energy into a compressed spring.

### D14. Higher Ground: On the Scalar View of the Fluid Motion [line 403]

At this point, a major reversal is taking place. Thus far, we have begun in each case with the lines of fluid motion and built the systems of isobaric surfaces and cells upon them. Now Maxwell proposes that we might begin at the other end. If we are given the system as a whole delineated by the pressure surfaces, all else is implied, *including ulti-*

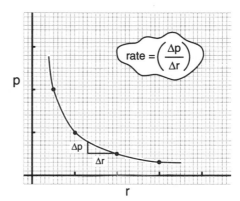

Fig. 2.26. Variation of pressure
with radius from a point source.

*mately the sources and sinks.* We might have thought of the source as primary: it is the origin of the fluid. But suppose we were to take the *system* as primary? All would work as well and the so-called "sources" would be seen, not as causes, but as incidental to the configuration of the whole. The *whole* may be taken as the first thought and the sources *derived.*

Pressure is a scalar quantity, which does not involve direction (as we have seen, we get direction from it by considering the direction and rate of change of its *gradient*). As scalars, pressures are conveniently additive. Then systems of pressures are equally, simply additive [compare Maxwell's Section (15)]. As we have now shown that the fluid motions are implicit in their pressure systems, this means that systems of motion can be analyzed and synthesized powerfully by way of their pressure representations (Fig. 2.9). This really represents "higher ground," in the sense of a theoretical position from which complexity and detail can be grasped in a single view.

### D15. Further Derivation of the Pressure Distribution in Flow from a Point Source [line 452]

In Fig. 2.26, we draw a picture of the pressure function, though we may not know how to characterize it algebraically. (We have already depicted it in fact in Fig. 1.25.) Maxwell tells us that the algebraic formula for this function is

$$p = \frac{k}{4\pi r}$$

That looks plausible in our figure. To confirm that it is correct we could take some sample values or, better, go through the formal process of relating the function to its rate of change, a process called *differentiation,* which lies at the foundation the differential calculus. We will have more to say about this when we meet a certain amount of analytic mathematics in reading "On Physical Lines of Force," in our next chapter. For the moment, let us be content to note that there is a general rule for algebraic functions, relating the function itself to its *rate of change.*[5] If the function is given by

$$p = ar^n \tag{1}$$

then the rate of change is given by

$$\frac{dp}{dr} = nar^{n-1} \tag{2}$$

To derive the rate of change of a function from the function itself is to do what is called *differentiating* the function or finding its derivative.

To apply this to the present case, Maxwell has proposed that the pressure be given by

$$p = \frac{k}{4\pi r} = \frac{k}{4\pi}\left(r^{-1}\right)$$

The rate of change of the pressure, where now we have pressure $p$ as a function of $r$, will then be the derivative, and by the rule above would be

$$\frac{dp}{dr} = -\frac{k}{4\pi}\left(r^{-2}\right) = -\frac{k}{4\pi r^2}$$

This rate of change is negative, telling us that pressure is *decreasing* with distance from the source. The *rate of decrease,* then, will be simply

$$\frac{k}{4\pi r^2}$$

in accord with Maxwell's result at line 458, the inverse-square law. We will deal with these questions of analytic math more thoroughly in Chapter 3.

If $k$ represents a source and a pressure, then as Maxwell adds at line 464, a sink will be represented by $-k$, which we may think of as a negative pressure or a suction.

## D16. On Electrical Interpretation of Our Fluid Models [line 514]

This is a good time to review some of the fluid models we have introduced, interpreting them now as electrical systems. On the whole, this may be left to the reader, but one or two deserve comment here.

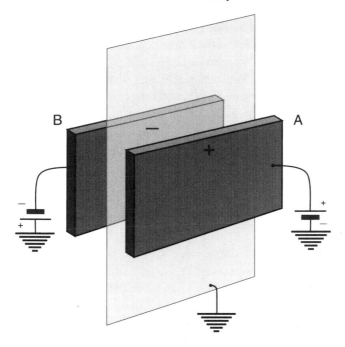

Fig. 2.27. The parallel-plate capacitor (a further interpretation of Fig. 2.7).

Figure 2.6, redrawn from Maxwell's *Treatise on Electricity and Magnetism,* now may be taken to represent two charges—let us say "positive charges," though we might wonder whether it would make any difference if they were negative. In any case, the two are of the same sign. They reside on two charged metal spheres, *A* and *B,* with the larger charge on *A.* Since we know there is no force *inside* a metal sphere (the Faraday cage principle!) the lines rightly go blank inside the two circles. As the charged spheres are metals, they must be equipotentials (electricity will flow to equalize any difference in potential) and they become the foci of the equipotential system of the diagram as a whole.

The lines of flow—now Faraday's lines of force—run everywhere orthogonally to the equipotential system and, in particular, meet the charged spheres at right angles. A small test charge carried about in the field will point everywhere along the lines of force. And what will it do at point *P?* We may be helped in answering that question by looking at Fig. 2.9, where we recognize the same situation diagramed as a potential problem (Note that there the finite spheres are not represented; if they were the mountains would be cut off as mesa tops!) Point $P$ now shows up vividly as a saddle point at which the test charge would rest in insecure equilibrium.

Anticipating a little, we had already seen an electrical interpretation of Fourier's heat-flow problem. This will now serve as a clue to the electrical interpretation we might offer of the series of drawings based on the sources and sinks of Fig. 2.7 (compare Fig. 2.8 and the Discussion [D11] above). The two-dimensional fluid flow shown in Figs. 2.7 and 2.8 can be simulated in the midplane between the parallel charged plates of Fig. 2.27. Here the configuration is very simple: just two parallel plates, this time placed vertically, equally and oppositely charged. $A$ (the "source") will be at a positive potential, $B$ (the "sink") will be at an equal negative potential, and the midplane (electrical "ground level") will be at zero potential, the potential ascribed to "infinity."

As anticipated in Discussion [D11], this is the device called the "parallel-plate capacitor," a fundamental element of electrical circuits which may be familiar to some readers. The way in which it functions to store electrical energy is vividly depicted in Fig. 2.8, where we see the concentration of unit cells between the plates. If a dielectric material such as glass is introduced, the effect is the equivalent of reducing the constant $k$ of the fluid medium. Then if the same potential difference is maintained between the plates, still more energy will be stored in the space between—the fluid interpretation of that remark being that the velocity of the fluid, for the same pressure difference, will be increased. Capacitors with dielectric insulators between the plates are widely used in electrical technology.

## D17.  On Calibrating the Model [line 530]

One electrical source, $dm$, gives rise to a potential (as Maxwell is modeling an attraction, he takes the potential difference as negative—think of a negative charge as a source):

$$dV = -\frac{dm}{r}$$

If there are many sources, their potentials add and we can simply indicate their arithmetic sum:

$$V = -\Sigma\left(\frac{dm}{r}\right)$$

Fig. 2.28. Real fluid. The Water of Urr at Glenlair, very familiar to Maxwell.

Fluid pressures similarly add as scalars (i.e., what matters is size not direction), so we can write an equation for the fluid analogous to this equation for *V*. The counterpart of *V* is fluid pressure *p*, while the increment to the source will be *dS* (in what follows we depart slightly from Maxwell's notation, since he seems to confuse the issue slightly by writing *S* rather than *dS*):

$$p = \Sigma \left( \frac{k}{4p} \right) dS$$

Drawing the bottom line of our analogy, we conclude

$$dS = \left(\frac{4\pi}{k}\right)dm$$

The calibration coefficient which links the fluid model with its electrical prototype must be

$$\frac{4\pi}{k}$$

Maxwell is playing games with equations, mixing two realms on the two sides of one equation, equating apples and oranges. What we get is a purely quantitative relation, really *calibrating our model,* which amounts (since our model is only a mental one) to adjusting our thoughts. In doing so, we have to remember to preserve order: the model is *temporal* and speaks of *flow;* electrostatics is, by definition, *timeless.* So motion is here a metaphor for stasis—the ideal lines are lines of flow, while those of electrostatics are going nowhere. Only our thoughts move.

# Chapter 3

## *On Physical Lines of Force*

Maxwell has embarked, it would seem, on an unrelenting course: it would be hard to say whether it is one of inquiry or of construction—is he *finding* something or *making a world* to suit some deep conviction? Even if the task is ultimately to find what is true and real in nature, the path is evidently by way of a very personal construction: conceiving alternative worlds, patiently and exactingly assembling their components, and weighing the result. The motivation for this cannot lie within what conventional discourse now calls "physics," for if it were simply a question of finding equations to account for the phenomena and predict accurately, Wilhelm Weber, William Thomson and others have offered a more efficient method.

Evidently, the conviction that is driving Maxwell through this series of works comes from somewhere else and lies close to his dedication to Faraday's vision. Maxwell in rejecting the stark but efficient formalism of action-at-a-distance equations is on the track of *being:* he wants to know what really *exists.* Even in this he is not simply a neutral inquirer. He is driven by a faith that the world must be of a certain sort, coherent in a way which admits understanding in terms of a complete chain of cause and effect. Here must lie the real significance of his search for a theory of the *field.* He is determined to tell a story of electromagnetism that will not only serve to predict all the phenomena, but will fill the world with intelligible connective links. To do this would be to restore the vision of the world as *cosmos,* in the old sense of a whole harmonious with its parts and commensurate with human understanding.

"Physical Lines" makes a fresh start in this direction. No longer will we deal with a mathematician's massless fluid; the account now is to be framed in terms that are strictly "physical." Maxwell conjures up a system of entities that have mass, obey Newton's laws of motion, and thus might in principle exist—there in the empty space of the vacuum through which electric and magnetic forces are seen to reach without hindrance.

Yet a playful mind is at work. Serious and disciplined though the argument must be in one sense, "Physical Lines" has nonetheless its comic aspect. As we shall see, Maxwell indulges in the invention of wheels and gears to fill the vacuum and do his bidding in a way that Rube Goldberg might envy. The reader must be the judge, but I submit that Maxwell has tongue in cheek here and does not seriously intend that the devices he conjures up would in fact be found to exist in the form he is describing. Maxwell was notorious among his friends for a ubiquitous and often near-inscrutable sense of humor. He is, I suspect, granting himself comic license on the scale of Aristophanes, who used the stage to fill out comic pictures of what the world would be like if peace prevailed or birds or women ruled. Like the comic poet, Maxwell is claimimg latitude to think things through to the end, to pursue the question whether any *imaginable* physical system could accomplish what electromagnets and pith balls are regularly observed to do. What is at stake may be the very concept of an intelligible relation of cause and effect in nature.

## On Analytic Math

It is ironic that in our excerpt from "Faraday's Lines," which was frankly housed in a *mathematical* mode of thought, Maxwell made every effort to avoid using analytic mathematics, the symbolic expressions of algebra and the calculus. He seemed resolved there to meet Faraday's request that the results of analysis be freed from their symbolic garb and expressed in a simpler mode, and indeed it was possible to construe that paper as a "letter to Faraday." Now, just when the mode turns from mathematical to *physical,* Maxwell, applying the laws of Newton, begins to make uncompromising use of formal mathematics! This means that while in the first paper it was sufficient to give at most rather short shrift to analytic math, we now must confront that art directly.

Some readers may be at home with the elements of the differential and integral calculus and more may be comfortable with algebra. But our aim is to make these papers of Maxwell's accessible to those for whom analytic math is anything but a comfortable medium. It is not possible here to begin altogether at the beginning, to build up algebra and the calculus from their foundations; but we can take certain measures for the sake of those who are in need of help. In the first place, we will make reference to one or two basic texts that are at the same time elementary and interesting: a combination that in principle should always go together. Further, with respect to what is called the differential and integral calculus, which Maxwell now uses freely, an introductory account is given in the first Discussion [D1] in this chapter. Though frankly only preliminary and intuitive, this should serve to get even the neophyte through a first reading of Maxwell's text. The reader is invited to turn to that now or whenever the moment seems appropriate. Finally, as Maxwell embarks on elaborations of his mathematical argument, I have taken the extreme liberty of intervening on behalf of the reader to reduce the argument to its simplest terms. Often this means that where Maxwell enjoys working with his equations in their full generality, I reduce them to the least case that will carry the point. An argument that Maxwell carries out in three dimensions may be reduced here to a more restricted case, in one dimension if that will do or two where necessary. This substituted account is printed distinctively, while the simplified version of Maxwell's equation bears his equation number followed by an asterisk. Even after return to Maxwell's text the abbreviated form of the equations has often been retained, and here again, the fact that the equation has been altered is signalled with an asterisk following Maxwell's equation number. Often, diagrams have been supplied as well, to invoke the help of visual intuition.

Maxwell, we might remark, is contributing as he goes along to the development of what is now known as the "vector calculus," a mathematical mode in which it is possible to address entities and operations in two or more dimensions with a single symbolic expression rather than a set of expressions. In this paper, however, Maxwell is still writing triples of equations, one for each direction in space. Even in his later *Treatise,* he uses a mixture of the two modes, "plowing" as he says "with an ox and an ass together." A discussion of the vector calculus will be included in Chapter 4.

## Newton's Laws of Motion

The fluid of "Faraday's Lines" was notable for its lack of inertial mass and was thus exempt from Newton's laws of motion. That fluid did not need mass because it was not called upon to exist; in this essay, by contrast, Maxwell is altogether concerned with the possible existence of a new fluid, the first characteristic of which is that it have mass. Newton's laws come into full play from the outset, and we must be aware of what they are. Fortunately, Newton's three laws are in principle extremely simple. To those for whom they are new or forgotten—or have always seemed enigmatic, as they might well, despite their simplicity—a discussion is given here [D2]. Again, the reader is invited to turn to that now or whenever the occasion seems to arise. And once more, excellent introductory works are available.

"On Physical Lines of Force," the essay to which this chapter will be devoted, was published in 1861 and 1862, at a time when Maxwell was establishing himself on the faculty at King's College, London. A great number of scientific topics, as disparate as color vision and statistical mechanics, are occupying his attention even as he works on "Physical Lines." By 1864, when he writes the last of the papers we will read, the "Dynamical Theory," both he and his world will have come a long way from the scene we described in Chapter 1. We will survey certain of these developments and reflect on their relations to the unfolding of his thought on electromagnetism, in conjunction with our reading of the "Dynamical Theory" in our final chapter. For the moment, we turn directly to "Physical Lines," and the question of the possible physical existence of the electromagnetic field.

[From the *Philosophical Magazine*, Vol. XXI]
XXIII. *On Physical Lines of Force*

## Part I. The Theory of Molecular Vortices applied to Magnetic Phenomena

In all phenomena involving attractions or repulsions, or any forces depending on the relative position of bodies, we have to determine the *magnitude* and *direction* of the force which would act on a given body, if placed in a given position.

5    In the case of a body acted on by the gravitation of a sphere, this force is inversely as the square of the distance, and in a straight line to the centre of the sphere. In the case of two attracting spheres, or of a body not spherical, the magnitude and direction of the force vary according to more complicated laws. In electric and magnetic phenom-
10    ena, the magnitude and direction of the resultant force at any point is the main subject of investigation. Suppose that the direction of the force at any point is known, then, if we draw a line so that in every part of its course it coincides in direction with the force at that point, this line may be called a *line of force* since it indicates the direction of
15    the force in every part of its course.

By drawing a sufficient number of lines of force, we may indicate the direction of the force in every part of the space in which it acts.

Thus if we strew iron filings on paper near a magnet, each filing will be magnetized by induction, and the consecutive filings will unite by
20    their opposite poles, so as to form fibres, and these fibres will *indicate* the direction of the lines of force. The beautiful illustration of the presence of magnetic force afforded by this experiment, naturally tends to make us think of the lines of force as something real, and as indicating something more than the mere resultant of two forces, whose seat
25    of action is at a distance, and which do not exist there at all until a magnet is placed in that part of the field. We are dissatisfied with the explanation founded on the hypothesis of attractive and repellent forces directed towards the magnetic poles, even though we may have satisfied ourselves that the phenomenon is in strict accordance
30    with that hypothesis, and we cannot help thinking that in every place where we find these lines of force, some physical state or action must exist in sufficient energy to produce the actual phenomena.

My object in this paper is to clear the way for speculation in this direction, by investigating the mechanical results of certain states of

35    tension and motion in a medium, and comparing these with the observed phenomena of magnetism and electricity. By pointing out the
mechanical consequences of such hypotheses, I hope to be of some
use to those who consider the phenomena as due to the action of a
medium, but are in doubt as to the relation of this hypothesis to the ex-
40    perimental laws already established, which have generally been expressed in the language of other hypotheses.

I have in a former paper* endeavoured to lay before the mind of the
geometer a clear conception of the relation of the lines of force to the
space in which they are traced. By making use of the conception of
45    currents in a fluid, I shewed how to draw lines of force, which should
indicate by their number the amount of force, so that each line may be
called a unit-line of force (see Faraday's *Researches,* 3122); and I
have investigated the path of the lines where they pass from one medium to another.

50    In the same paper I have found the geometrical significance of the
"Electrotonic State," and have shewn how to deduce the mathematical
relations between the electrotonic state, magnetism, electric currents,
and the electromotive force, using mechanical illustrations to assist
the imagination, but not to account for the phenomena.

55    I propose now to examine the magnetic phenomena from a mechanical point of view, and to determine what tensions in, or motions
of, a medium are capable of producing the mechanical phenomena
observed. If, by the same hypothesis, we can connect the phenomena
of magnetic attraction with electromagnetic phenomena and with
60    those of induced currents, we shall have found a theory which, if not
true, can only be proved to be erroneous by experiments which will
greatly enlarge our knowledge of this part of physics.

The mechanical conditions of medium under magnetic influence
have been variously conceived of as currents, undulations, or states
65    of displacement or strain, or of pressure or stress.

* * *

We come now to consider the magnetic influence as existing in the
form of some kind of pressure or tension, or, more generally, of *stress*
in the medium.

*See a paper "On Faraday's Lines of Force," *Cambridge Philosophical Trans-
actions,* Vol. X. Part I. [Chap. II, above.]

70    Stress is action and reaction between the consecutive parts of a
body, and consists in general of pressures or tensions different in different directions at the same point of the medium.

The necessary relations among these forces have been investigated by mathematicians; and it has been shewn that the most gen-
75    eral type of a stress consists of a combination of three principal
pressures or tensions, in directions at right angles to each other.

When two of the principal pressures are equal, the third becomes
an axis of symmetry, either of greatest or least pressure, the pressures at right angles to this axis being all equal.

80    When the three principal pressures are equal, the pressure is equal
in every direction, and there results a stress having no determinate
axis of direction, of which we have an example in simple hydrostatic
pressure.

The general type of a stress is not suitable as a representation of a
85    magnetic force, because a line of magnetic force has direction and intensity, but has no third quality indicating any difference between the
*sides* of the line, which would be analogous to that observed in the
case of polarized light.*

We must therefore represent the magnetic force at a point by a
90    stress having a single axis of greatest or least pressure, and all the
pressures at right angles to this axis equal. It may be objected that it
is inconsistent to represent a line of force, which is essentially dipolar,
by an axis of stress, which is necessarily isotropic; but we know that
*every* phenomenon of action and reaction is isotropic in its *results* be-
95    cause the effects of the force on the bodies between which it acts are
equal and opposite, while the nature and origin of the force may be
dipolar, as in the attraction between a north and a south pole.

Let us next consider the mechanical effect of a state of stress sym-
metrical about an axis. We may resolve it, in all cases, into a simple
100    hydrostatic pressure, combined with a simple pressure or tension
along the axis. When the axis is that of greatest pressure, the force
along the axis will be a pressure. When the axis is that of least pressure, the force along the axis will be a tension.

If we observe the lines of force between two magnets, as indicated
105    by iron filings, we shall see that whenever the lines of force pass from
one pole to another, there is *attraction* between those poles; and

*See Faraday's *Researches,* 3252.

where the lines of force from the poles avoid each other and are dispersed into space, the poles *repel* each other, so that in both cases they are drawn in the direction of the resultant of the lines of force.

110    It appears therefore that the stress in the axis of a line of magnetic force is a *tension,* like that of a rope.

If we calculate the lines of force in the neighbourhood of two gravitating bodies, we shall find them the same in direction as those near two magnetic poles of the same name; but we know that the mechani-
115    cal effect is that of attraction instead of repulsion. The lines of force in this case do not run between the bodies, but avoid each other, and are dispersed over space. In order to produce the effect of attraction, the stress along the lines of gravitating force must be a *pressure.*

Let us now suppose that the phenomena of magnetism depend on
120    the existence of a tension in the direction of the lines of force, combined with a hydrostatic pressure; or in other words, a pressure greater in the equatorial than in the axial direction: the next question is, what mechanical explanation can we give of this inequality of pressures in a fluid or mobile medium? The explanation which most read-
125    ily occurs to the mind is that the excess of pressure in the equatorial direction arises from the centrifugal force of vortices or eddies in the medium having their axes in directions parallel to the lines of force.

This explanation of the cause of the inequality of pressures at once suggests the means of representing the dipolar character of the line of
130    force. Every vortex is essentially dipolar, the two extremities of its axis being distinguished by the direction of its revolution as observed from those points.

We also know that when electricity circulates in a conductor, it produces lines of magnetic force passing through the circuit, the direction
135    of the lines depending on the direction of the circulation. Let us suppose that the direction of revolution of our vortices is that in which vitreous electricity must revolve in order to produce lines of force whose direction within the circuit is the same as that of the given lines of force.

140    We shall suppose at present that all the vortices in any one part of the field are revolving in the same direction about axes nearly parallel, but that in passing from one part of the field to another, the direction of the axes, the velocity of rotation, and the density of the substance of the vortices are subject to change. We shall investigate the resul-
145    tant mechanical effect upon an element of the medium, and from the

mathematical expression of this resultant we shall deduce the physical character of its different component parts.

PROP. I.-If in two fluid systems geometrically similar the velocities and densities at corresponding points are proportional, then the differences of
150  pressure at corresponding points due to the motion will vary in the duplicate ratio of the velocities and the simple ratio of the densities.

Let *l* be the ratio of the linear dimensions, *m* that of the velocities, *n* that of the densities, and *p* that of the pressures due to the motion. Then the ratio of the *masses* of corresponding portions will be $l^3n$,
155  and the ratio of the velocities acquired in traversing similar parts of the systems will be *m;* so that $l^3mn$ is the ratio of the momenta acquired by similar portions in traversing similar parts of their paths.

The ratio of the surfaces is $l^2$, that of the forces acting on them is $l^2p$, and that of the times during which they act is $\frac{l}{m}$; so that the ratio
160  of the impulse of the forces is $\frac{l^2p}{m}$, and we have now

$$l^3mn = \frac{l^3p}{m},$$

or

$$m^2n = p;$$

that is, the ratio of the pressures due to the motion (*p*) is compounded
165  of the ratio of the densities (*n*) and the duplicate ratio of the velocities ($m^2$), and does not depend on the linear dimensions of the moving systems [D3].

In a circular vortex, revolving with uniform angular velocity, if the pressure at the axis is $p_0$, that at the circumference will be $p_1 = p_0 +$
170  $(1/2)\rho v^2$, where $\rho$ is the density and *v* the velocity at the circumference [D4]. The *mean pressure* parallel to the axis will be

$$p_0 + \frac{1}{4}\rho v^2 = p_2.$$

If a number of such vortices were placed together side by side with their axes parallel, they would form a medium in which there would be
175  a pressure $p_2$ parallel to the axes, and a pressure $p_1$ in any perpendicular direction. If the vortices are circular, and have uniform angular velocity and density throughout, then

$$p_1 - p_2 = 1/4\,\rho v^2.$$

If the vortices are not circular, and if the angular velocity and the
180  density are not uniform, but vary according to the same law for all
the vortices,

$$p_1 - p_2 = C\rho v^2,$$

where $\rho$ is the mean density, and $C$ is a numerical quantity depending
on the distribution of angular velocity and density in the vortex. In fu-
185  ture we shall write $\mu/4\pi$ instead of $C\rho$, so that

$$p_1 - p_2 = \frac{1}{4}\mu v^2, \tag{1}$$

where $\mu$ is a quantity bearing a constant ratio to the density, and $v$ is
the linear velocity at the circumference of each vortex.

A medium of this kind, filled with molecular vortices having their
190  axes parallel, differs from an ordinary fluid in having different pressures in different directions. If not prevented by properly arranged
pressures, it would tend to expand laterally. In so doing, it would allow
the diameter of each vortex to expand and its velocity to diminish in
the same proportion. In order that a medium having these inequalities
195  of pressure in different directions should be in equilibrium, certain conditions must be fulfilled, which we must investigate.

PROP. II.-If the direction-cosines of the axes of the vortices with respect to the axes of $x$, $y$, and $z$ be $l$, $m$, and $n$, to find the normal and
tangential stresses on the co-ordinate planes.

200  * * *

Simplifying Maxwell's argument to the case of variation with respect to just one rather than all three axes, with a corresponding
simplification of the symbols employed, we may paraphrase Maxwell's text as follows.
205    The actual stress may be resolved into a simple hydrostatic pressure $p_1$ acting in all directions, and a simple tension $t = p_1 - p_2$ acting
along the axis of stress. Looking back at equation (1), we can
simplify still further by using a single letter for the constant term; let

$$k = \frac{\mu}{4\pi}$$

210  so that

$$t = p_1 - p_2 = kv^2$$

When the coordinate axes are aligned with the vortex, only this pressure and tension are involved (see Fig. 3.6 in our Notes to these lines). In the figure, a box has been drawn around point *P*, in order to
215 explore the stresses acting at that point. In this case, it finds only $p_1$ and *t*. If however the vortex were to lie at an angle to our coordinate system, as in Fig. 3.7a, the analysis would be somewhat more complex.

The analysis of this more general case is illustrated in Fig. 3.7b. There, $p_x$ and $p_y$ are the normal stresses parallel to the two axes, and
220 $s_x$ and $s_y$ are the tangential or shear stresses lying in the corresponding planes. The resolution of these stresses yields the following result, as explained in Discussion [D5]:

$$p_x = kv^2 \, l^2 - p_1$$
$$p_y = kv^2 \, m^2 - p_1$$
225
$$s_x = s_y = kv^2 \, lm$$

Since the shear stresses tend to rotate the fluid within the box, equilibrium requires that they be equal; $s_x = s_y$, and we may write them simply as *s:*

$$s = kv^2 \, lm$$

230 We may simplify further by following Maxwell in writing:

$$a = vl, \qquad b = v \, m$$

so that

$$p_x = k \, a^2 - p_1$$
$$p_y = k \, b^2 - p_1 \tag{2}*$$
235
$$s = k \, a \, b$$

PROP. III. To find the resultant force on an element of the medium, arising from the variation of internal stress.

\* \* \*

Again we paraphrase Maxwell by confining ourselves to the
240 much simpler case of phenomena in two dimensions. As before, we have added the figure and revised the notation.

We have in general, for the force per unit of volume in the direction of x, by the law of equilibrium of stresses [D6],

$$X = \frac{dp_x}{dx} + \frac{ds}{dy} \tag{3)*}$$

245 The task now is to carry out the indicated differentiations on the equations (2)*, which we derived earlier for the stresses in our fluid. This derivation is carried out in [D7], with this result:

$$X = \alpha\left[\frac{d}{dx}(k\alpha) + \frac{d}{dy}(k\beta)\right] + \frac{k}{2}\frac{d}{dx}(\alpha^2 + \beta^2)$$

$$+ k\beta\left(\frac{d\alpha}{dy} - \frac{d\beta}{dx}\right) - \frac{dp_1}{dx} \tag{4)*}$$

250 It will be useful in what follows to refer to the four terms of equation (4)*, respectively, as A, B, C, and D; thus:

$$\alpha\left[\frac{d}{dx}(k\alpha) + \frac{d}{dy}(k\beta)\right] \tag{A)*}$$

$$+ \frac{k}{2}\frac{d}{dx}(\alpha^2 + \beta^2) \tag{B)*}$$

$$+ k\beta\left(\frac{d\alpha}{dy} - \frac{d\beta}{dx}\right) \tag{C)*}$$

255
$$- \frac{dp_1}{dx} \tag{D)*}$$

Each of these will turn out in due course to have a crucial physical interpretation.

We have now to interpret the meaning of each term of this expression.

We suppose $\alpha$ and $\beta$ to be the components of the force that would act
260 upon a north magnetic pole, as usual simplifying Maxwell's discussion by taking $\gamma = 0$; $\mu$ represents the magnetic inductive capacity of the medium at any point referred to air (strictly, the vacuum) as unity; $\mu\alpha$ and $\mu\beta$ then represent the quantity of magnetic induction per unit area perpendicular to the $x$- and $y$-axes, respectively.
265    Earlier, we made the substitution

$$k = \frac{\mu}{4\pi}$$

to avoid carrying $p$ around in our calculations; we now translate back to terms of $\mu$ to connect with Maxwell's discussion.

The total amount of magnetic induction through a closed surface
270 surrounding the pole of a magnet depends entirely on the strength *m*
of that pole. We can show that if *dx dy dz* are the sides of an
infinitesimal test box, and we continue to take $\gamma = 0$,

$$\left[ \frac{d}{dx}(\mu\alpha) + \frac{d}{dy}(\mu\beta) \right] dx\,dy\,dz = 4\pi\,m\,dx\,dy\,dz \qquad (6)*$$

will represent the total amount of magnetic induction outward
275 through the surface of the box. This differential expression then in
turn will measure the amount of "imaginary magnetic matter" inside
the box [D8].

The *first term* of the value of *X*, which we earlier labeled (A)*,
looks like this when we substitute for *k* in terms of μ:

280
$$\alpha \left( \frac{1}{4\pi} \right) \left[ \frac{d}{dx}(\mu\alpha) + \frac{d}{dy}(\mu\beta) \right] \qquad (7)*$$

which may be written simply as α*m,* where α is the intensity of the
magnetic force, and *m* is the amount of north-pointing "imaginary
magnetic matter" in unit of volume, i.e.,

$$\alpha \left\{ \left( \frac{1}{4\pi} \right) \left[ \frac{d}{dx}(\mu\alpha) + \frac{d}{dy}(\mu\beta) \right] \right\} = \alpha\,m$$

285 As a result, we can now find the physical meaning of term (A)* of
our equation (4)*. Maxwell concludes:

> The physical interpretation of this term is that the force urging a
> north pole in the positive direction of *x* is the product of the intensity of
> the magnetic force resolved in that direction, and the strength of the
290 > north pole of the magnet.

> Let the parallel lines from left to right in fig. 1 represent a field of
> magnetic force such as that of the earth, *sn* being the direction from
> south to north. The vortices, according to our hypothesis, will be in the
> direction shewn by the arrows in fig. 3, that is, in a plane perpendicu-
295 > lar to the lines of force, and revolving in the direction of the hands of a
> watch when observed from *s* looking towards *n.* The parts of the vor-
> tices above the plane of the paper will be moving towards *e,* and the
> parts below that plane towards *w.*

> We shall always mark by an arrow-head the direction in which we
300 > must look in order to see the vortices rotating in the direction of the

Fig. 1.

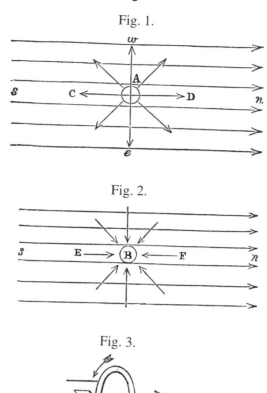

Fig. 2.

Fig. 3.

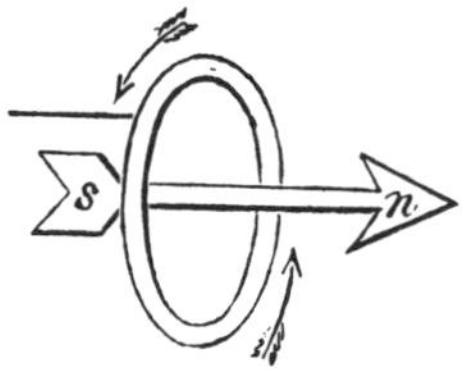

hands of a watch. The arrow-head will then indicate the (fig. 1) *north-ward* direction in the magnetic field, that is, the direction in which that end of a magnet which points to the north would set itself in the field.

305    Now let *A* be the end of a magnet which points north. Since it repels the north ends of other magnets, the lines of force will be directed *from A* outwards in all directions. On the north side the line *AD* will be in the *same* direction with the lines of the magnetic field, and the velocity of the vortices will be *increased.* On the south side the line *AC* will be in the opposite direction, and the velocity of the vor-

310    tices will be diminished, so that the lines of force are more powerful on the north side of *A* than on the south side [D9].

We have seen that the mechanical effect of the vortices is to produce a tension along their axes, so that the resultant effect on *A* will

be to pull it more powerfully towards *D* than towards *C;* that is, *A* will
315   tend to move to the north.

     Let *B* in fig. 2 represent a south pole. The lines of force belonging
to *B* will tend *towards B,* and we shall find that the lines of force are
rendered stronger towards *E* than towards *F,* so that the effect in this
case is to urge *B* towards the south.

320     It appears therefore that, on the hypothesis of molecular vortices,
our first term gives a mechanical explanation of the force acting on a
north or south pole in the magnetic field.

     We now proceed to examine the second term,

$$\left(\frac{\mu}{8\pi}\right)\frac{d}{dx}(\alpha^2 + \beta^2). \tag{B)*}$$

325     Here, $(\alpha^2 + \beta^2)$ . . . is the square of the intensity at any part of the
field, and $\mu$ is the magnetic inductive capacity at the same place. Any
body therefore placed in the field will be urged *towards places of
stronger magnetic intensity* with a force depending partly on its own
capacity for magnetic induction, and partly on the rate at which the
330   square of the intensity increases.

     If the body be placed in a fluid medium, then the medium, as well
as the body, will be urged towards places of greater intensity, so that
its hydrostatic pressure will be increased in that direction. The resultant effect on a body placed in the medium will be the *difference* of the
335   actions on the body and on the portion of the medium which it dis-
places, so that the body will tend to or from places of greatest magnetic intensity, according as it has a greater or less capacity for
magnetic induction than the surrounding medium.

     In fig. 4 the lines of force are represented as converging and be-
340   coming more powerful towards the right, so that the magnetic tension
at *B* is stronger than at *A,* and the body *AB* will be urged to the right. If
the capacity for magnetic induction is greater in the body than in the
surrounding medium, it will move to the right, but if less it will move to
the left.

345     We may suppose in this case that the lines of force are converging
to a magnetic pole, either north or south, on the right hand.

     In fig. 5 the lines of force are represented as vertical, and becoming
more numerous towards the right. It may be shewn that if the force increases towards the right, the lines of force will be curved towards the
350   right. The effect of the magnetic tensions will then be to draw any

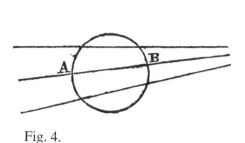

Fig. 4.

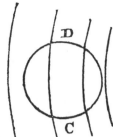

Fig. 5.

body towards the right with a force depending on the excess of its inductive capacity over that of the surrounding medium.

We may suppose that in this figure the lines of force are those surrounding an electric current perpendicular to the plane of the paper
355    and on the right hand of the figure.

These two illustrations will shew the mechanical effect on the
paramagnetic or diamagnetic body placed in a field of varying magnetic force, whether the increase of force takes place along the lines
or transverse to them. The form of the second term of our equation in-
360    dicates the general law, which is quite independent of the direction of
the lines of force, and depends solely on the manner in which the
force *varies* from one part of the field to another.

We come now to the third term of the value of $X$,

$$-\mu\beta\frac{1}{4\pi}\left(\frac{d\beta}{dx}-\frac{d\alpha}{dy}\right). \tag{C}*$$

365    Here $\mu\beta$ is, as before, the quantity of magnetic induction through unit
of area perpendicular to the axis of $y$ and $d\beta/dx - d\alpha/dy$ is a quantity
which would disappear if [ $\alpha dx + \beta dy$ ] were a complete differential,
that is, if the force acting on a unit north pole were subject to the condition that no work can be done upon the pole in passing round any
370    closed curve. The quantity represents the work done on a north pole
in travelling round unit of area in the direction from $+x$ to $+y$ parallel to
the plane of $xy$. Now if an electric current whose strength is $r$ is traversing the axis of $z$, which, we may suppose, points vertically up-
wards, then, if the axis of $x$ is east and that of $y$ north, a unit north pole
375    will be urged round the axis of $z$ in the direction from $x$ to $y$, so that in
one revolution the work will be $= 4\pi r$. Hence

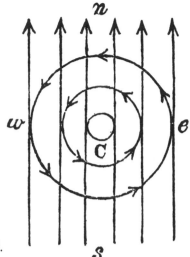

Fig. 6.

$$\frac{1}{4\pi}\left(\frac{d\beta}{dx} - \frac{d\alpha}{dy}\right) \qquad\qquad (9)^*$$

represents the *strength of an electric current parallel to z* through unit
of area.

380    * * *

The physical interpretation of the third term of *X*, −µβ*r*, is that if µβ
is the quantity of magnetic induction parallel to *y*, and *r* the quantity of
electricity flowing in the direction of *z*, the element will be urged in the
direction of −*x*, transversely to the direction of the current and of the
385    lines of force; that is, an *ascending* current in a field of force magnetized towards the *north* would tend to move *west*.
    To illustrate the action of the molecular vortices, let *sn* [fig.6] be
the direction of magnetic force in the field, and let *C* be the section
of an ascending magnetic current perpendicular to the paper. The
390    lines of force due to this current will be circles drawn in the opposite direction from that of the hands of a watch; that is, in the direction *nwse*. At *e* the lines of force will be the sum of those of the
field and of the current, and at *w* they will be the difference of the
two sets of lines; so that the vortices on the east side of the current
395    will be more powerful than those on the west side. Both sets of vortices have their equatorial parts turned towards *C*, so that they
tend to expand towards *C*, but those on the east side have the

greatest effect, so that the resultant effect on the current is to urge it towards the *west.*

400    * * *

The fifth term,

$$-\frac{dp_1}{dx} \tag{11)*}$$

merely implies that the element will be urged in the direction in which the hydrostatic pressure $p_1$ diminishes.

405    We may now write down the expressions for the components of the resultant force on an element of the medium per unit of volume, thus:

$$X = \alpha m + \left(\frac{\mu}{8\pi}\right)\frac{d}{dx}\left(v^2\right) - \mu\beta r - \frac{dp_1}{dx}. \tag{12)*}$$

* * *

The first term . . . . . refers to the force acting on the magnetic poles.
410  The second term to the action on bodies capable of magnetism by induction. The third . . . . to the force acting on electric currents. And the fifth to the effect of simple pressure.

* * *

## Part II. The Theory of Molecular Vortices applied to
415        ## Electric Currents

We have already shewn that all the forces acting between magnets, substances capable of magnetic induction, and electric currents may be mechanically accounted for on the supposition that the surrounding medium is put into such a state that at every point the pres-
420  sures are different in different directions, the direction of least pressure being that of the observed lines of force, and the difference of greatest and least pressures being proportional to the square of the intensity of the force at that point.

Such a state of stress, if assumed to exist in the medium, and to be
425  arranged according to the known laws regulating lines of force, will act upon the magnets, currents, &c. in the field with precisely the same resultant forces as those calculated on the ordinary hypothesis of direct action at a distance. This is true independently of any particular theory as to the *cause* of this state of stress, or the mode in which it can be sustained in the medium. We have therefore a satisfactory answer to

430 the question, "Is there any mechanical hypothesis as to the condition of the medium indicated by lines of force, by which the observed resultant forces may be accounted for?" The answer is, the lines of force indicate the direction of *minimum pressure* at every point of the medium.

The second question must be, "What is the mechanical cause of
435 this difference of pressure in different directions?" We have supposed, in the first part of this paper, that this difference of pressures is caused by molecular vortices, having their axes parallel to the lines of force.

We also assumed, perfectly arbitrarily, that the direction of these
440 vortices is such that, on looking along a line of force from south to north, we should see the vortices revolving in the direction of the hands of a watch.

We found that the velocity of the circumference of each vortex must be proportional to the intensity of the magnetic force, and that the den-
445 sity of the substance of the vortex must be proportional to the capacity of the medium for magnetic induction.

We have as yet given no answers to the questions, "How are these vortices set in rotation?" and "Why are they arranged according to the known laws of lines of force about magnets and currents?" These
450 questions are certainly of a higher order of difficulty than either of the former; and I wish to separate the suggestions I may offer by way of provisional answer to them, from the mechanical deductions which resolved the first question, and the hypothesis of vortices which gave a probable answer to the second.

455 We have, in fact, now come to inquire into the physical connexion of these vortices with electric currents, while we are still in doubt as to the nature of electricity, whether it is one substance, two substances, or not a substance at all, or in what way it differs from matter, and how it is connected with it.

460 We know that the lines of force are affected by electric currents, and we know the distribution of those lines about a current; so that from the force we can determine the amount of the current. Assuming that our explanation of the lines of force by molecular vortices is correct, why does a particular distribution of vortices indicate an electric current? A satisfac-
465 tory answer to this question would lead us a long way towards that of a very important one, "What is an electric current?"

I have found great difficulty in conceiving of the existence of vortices in a medium, side by side, revolving in the same direction about

parallel axes. The contiguous portions of consecutive vortices must
470 be moving in opposite directions; and it is difficult to understand how
the motion of one part of the medium can coexist with, and even produce, an opposite motion of a part in contact with it.

The only conception which has at all aided me in conceiving of this
kind of motion is that of the vortices being separated by a layer of par-
475 ticles, revolving each on its own axis in the opposite direction to that
of the vortices, so that the contiguous surfaces of the particles and of
the vortices have the same motion.

In mechanism, when two wheels are intended to revolve in the
same direction, a wheel is placed between them so as to be in gear
480 with both, and this wheel is called an "idle wheel." The hypothesis
about the vortices which I have to suggest is that a layer of particles,
acting as idle wheels, is interposed between each vortex and the next,
so that each vortex has a tendency to make the neighbouring vortices
revolve in the same direction with itself.

485 In mechanism, the idle wheel is generally made to rotate about a
*fixed* axle; but in epicyclic trains and other contrivances, as, for instance, in Siemens's governor for steam engines,* we find idle wheels
whose centres are capable of motion. In all these cases the motion of
the centre is the half sum of the motions of the circumferences of the
490 wheels between which it is placed. Let us examine the relations which
must subsist between the motions of our vortices and those of the
layer of particles interposed as idle wheels between them.

* * *

It appears therefore that, according to our hypothesis, an electric
495 current is represented by the transference of the moveable particles
interposed between the neighbouring vortices. We may conceive that
these particles are very small compared with the size of a vortex, and
that the mass of all the particles together is inappreciable compared
with that of the vortices, and that a great many vortices, with their sur-
500 rounding particles, are contained in a single complete molecule of the
medium. The particles must be conceived to roll without sliding between the vortices which they separate, and not to touch each other,
so that, as long as they remain within the same complete molecule,
there is no loss of energy by resistance. When, however, there is a

*See Goodeve's *Elements of Mechanism,* p. 118.

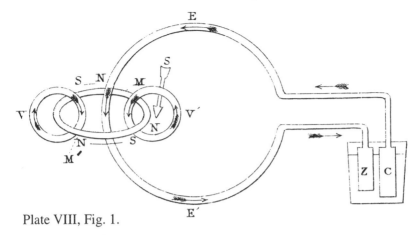

Plate VIII, Fig. 1.

505   general transference of particles in one direction, they must pass from
one molecule to another, and in doing so, may experience resistance,
so as to waste electrical energy and generate heat.

\* \* \*

510       In Plate VIII. . . . fig. 1, let the vertical circle *EE* represent an electric
current flowing from copper *C* to zinc *Z* through the conductor *EE'*, as
shewn by the arrows.
      Let the horizontal circle *MM'* represent a line of magnetic force em-
bracing the electric circuit, the north and south directions being indi-
515   cated by the lines *SN* and *NS*.
      Let the vertical circles *V* and *V'* represent the molecular vortices of
which the line of magnetic force is the axis. *V* revolves as the hands
of a watch, and *V'* the opposite way.
      It will appear from this diagram, that if *V* and *V'* were contiguous
520   vortices, particles placed between them would move downwards; and
that if the particles were forced downwards by any cause, they would
make the vortices revolve as in the figure. We have thus obtained a
point of view from which we may regard the relation of an electric current to its lines of force as analogous to the relation of a toothed wheel
525   or rack to wheels which it drives.
      In the first part of the paper we investigated the relations of the stati-
cal forces of the system. We have now considered the connexion of
the motions of the parts considered as a system of mechanism. It remains that we should investigate the dynamics of the system, and

530    determine the forces necessary to produce given changes in the motions of the different parts.

* * *

Maxwell here investigates carefully the energy relations within the system of the vortices and the particles, and concludes that the
535 kinetic energy of the vortical fluid per unit of volume will be:

$$\frac{1}{2}\left(\frac{\mu}{4\pi}\right)\left(\alpha^2 + \beta^2\right) \qquad (45)*$$

If we remember that in general the kinetic energy of a body of mass m moving with velocity $v$ is $1/2\ mv^2$, equation (45)* makes intuitive sense. Here $\alpha$ and $\beta$ measure the components of the velocity, while
540 $\mu$ incorporates a "shape constant" to adjust for the assumption that the vortex would not be circular and that its mass would not be uniformly distributed. At that point it was a matter of measuring the centrifugal force, but $\mu/4\pi$ plays the corresponding role here as the adjusted mass term.

545    In order to produce or destroy this energy, work must be expended on, or received from, the vortex, either by the tangential action of the layer of particles in contact with it, or by change of form in the vortex. We shall . . . investigate the tangential action between the vortices and the layer of particles in contact with them.

550 * * *

We find

$$\left(\frac{dP}{dy} - \frac{dQ}{dx}\right) = \mu \frac{d\gamma}{dt}. \qquad (54)*$$

From [this equation] we may determine the relation between the alterations of motion $d\gamma/dt$, &c. and the forces exerted on the layers of
555 particles between the vortices, or, in the language of our hypothesis, the relation between changes in the state of the magnetic field and the electromotive forces thereby brought into play.

In a memoir "On the Dynamical Theory of Diffraction" (Cambridge Philosophical Transactions, Vol. IX Part 1, section 6), Professor
560 Stokes has given a method by which we may solve equations (54), and find P, Q, and R in terms of the quantities on the right hand of those equations. I have pointed out . . . the application of this method to questions in electricity and magnetism.

\* \* \*

565    Here Maxwell, following as he says in the path of Stokes, introduces a rather formal device; we will see its physical significance
once we have set it out.
    We have the force–momentum equation we found earlier:

$$\left(\frac{dP}{dy} - \frac{dQ}{dx}\right) = \mu\frac{d\gamma}{dt} \qquad (54)*$$

570    We now *define a new quantity,* seemingly out of the blue; it will
be a vector, with (in our two dimensional account) two components,
*F* and *G*. We devise them so that:

$$\left(\frac{dF}{dy} - \frac{dG}{dx}\right) = \mu\gamma \qquad (55)*$$

    It is true that there are some important mathematical formalities
575 here, but the effect is that this can be done—under the interesting
condition that the "magnetic matter" $m = 0$. Now to keep things very
simple for the present discussion, let the electric force have only the
*P* component, acting in the *x*-direction, while the magnetic induction
has only the component *g*. Then *Q, R, a, b, G,* and *H* can all be set
580 equal to zero and even the remaining *P* and *g* will be subject to some
restrictions. We have left only the simple relations:

$$\frac{dP}{dy} = \mu\frac{d\gamma}{dt} \qquad (L)$$

$$\frac{dF}{dy} = \mu\gamma \qquad (M)$$

If we differentiate equation *M* with respect to *t*, we get

585
$$\frac{d}{dt}\left(\frac{dF}{dy}\right) = \mu\frac{d\gamma}{dt} \qquad (N)$$

*F* here gets differentiated twice—once with respect to distance
*y*, and then again with respect to time *t*. The equation thus
speaks of the *time*-rate of the *space*-rate of change of F, but the
order of differentiation does not matter. We can reverse that to
590 read:

$$\frac{d}{dy}\left(\frac{dF}{dt}\right) = \mu\frac{d\gamma}{dt} \qquad (O)$$

*Fig: 2.*

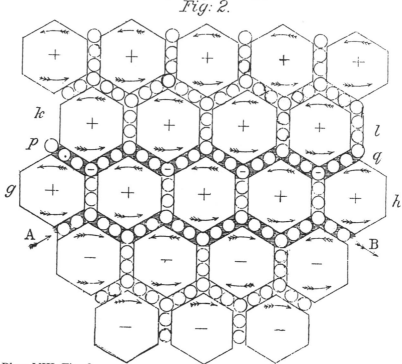

Plate VIII, Fig. 2.

and then we see that equation *O* will neatly correspond to equation
*L,* if we set

$$P = \frac{dF}{dt}$$
(58)*

595    This, in more general terms, is the result Maxwell wants: the
electromotive force is the time-rate of change of his newly constructed vector.

We have thus determined three quantities, *F, G, H,* from which we
can find *P, Q,* and *R* by considering these latter quantities as the rates
600    at which the former ones vary. In the paper already referred to, I have
given reasons for considering the quantities *F, G, H* as the resolved
parts of that which Faraday has conjectured to exist, and has called
the *electrotonic state.* In that paper I have stated the mathematical relations between this electrotonic state and the lines of magnetic force
605    as expressed in equations (55), and also between the electrotonic
state and electromotive force as expressed in equations (58). We

must now endeavour to interpret them from a mechanical point of
view in connexion with our hypothesis.

We shall in the first place examine the process by which the lines of
610 force are produced by an electric current.

Let *AB,* Plate VIII. . . . , fig. 2, represent a current of electricity in the
direction from *A* to *B.* Let the large spaces above and below *AB* represent the vortices, and let the small circles separating the vortices represent the layers of particles placed between them, which in our
615 hypothesis represent electricity.

Now let an electric current from left to right commence in *AB.* The
row of vortices *gh* above *AB* will be set in motion in the opposite direction to that of a watch. (We shall call this direction +, and that of a
watch −.) We shall suppose the row of vortices *kl* still at rest, then the
620 layer of particles between these rows will be acted on by the row *gh*
on their lower sides, and will be at rest above. If they are free to move,
they will rotate in the negative direction, and will at the same time
move from right to left, or in the opposite direction from the current,
and so form an *induced* electric current.

625 If this current is checked by the electrical resistance of the medium,
the rotating particles will act upon the row of vortices *kl,* and make
them revolve in the positive direction till they arrive at such a velocity
that the motion of the particles is reduced to that of rotation, and the induced current disappears. If, now, the primary current *AB* be stopped,
630 the vortices in the row *gh* will be checked, while those of the row *kl*
still continue in rapid motion. The momentum of the vortices beyond
the layer of particles *pq* will tend to move them from left to right, that
is, in the direction of the primary current; but if this motion is resisted
by the medium, the motion of the vortices beyond *pq* will be gradually
635 destroyed.

It appears therefore that the phenomena of induced currents are
part of the process of communicating the rotatory velocity of the vortices from one part of the field to another.

As an example of the action of the vortices in producing induced
640 currents, let us take the following case: Let *B,* Plate VIII. . . . , fig.3, be
a circular ring, of uniform section, lapped uniformly with covered wire.
It may be shewn that if an electric current is passed through this wire,
a magnet placed within the coil of wire will be strongly affected, but no
magnetic effect will be produced on any external point. The effect will
645 be that of a magnet bent round till its two poles are in contact.

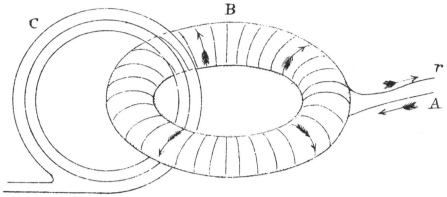

Plate VIII, Fig. 3.

If the coil is properly made, no effect on a magnet placed outside it can be discovered, whether the current is kept constant or made to vary in strength; but if a conducting wire *C* be made to *embrace* the ring any number of times, an electromotive force will act on that wire
650 whenever the current in the coil is made to vary; and if the circuit be *closed,* there will be an actual current in the wire *C*.

This experiment shews that, in order to produce the electromotive force, it is not necessary that the conducting wire should be placed in a field of magnetic force, or that lines of magnetic force
655 should pass through the substance of the wire or near it. All that is required is that lines of force should pass through the circuit of the conductor, and that these lines of force should vary in quantity during the experiment.

In this case the vortices, of which we suppose the lines of magnetic
660 force to consist, are all within the hollow of the ring, and outside the ring all is at rest. If there is no conducting circuit embracing the ring, then, when the primary current is made or broken, there is no action outside the ring, except an instantaneous pressure between the particles and the vortices which they separate. If there is a continuous con-
665 ducting circuit embracing the ring, then, when the primary current is made, there will be a current in the opposite direction through *C;* and when it is broken, there will be a current through *C* in the same direction as the primary current.

We may now perceive that induced currents are produced when
670 the electricity yields to the electromotive force—this force, however, still existing when the formation of a sensible current is prevented by the resistance of the circuit.

The electromotive force, of which the components are *P, Q, R,* arises from the action between the vortices and the interposed parti-
675 cles, when the velocity of rotation is altered in any part of the field. It corresponds to the pressure on the axle of a wheel in a machine when the velocity of the driving wheel is increased or diminished.

The electrotonic state, whose components are *F, G, H,* is what the electromotive force would be if the currents, &c. to which the lines of
680 force are due, instead of arriving at their actual state by degrees, had started instantaneously from rest with their actual values. It corresponds to the *impulse* which would act on the axle of a wheel in a machine if the actual velocity were suddenly given to the driving wheel, the machine being previously at rest.

685 If the machine were suddenly stopped by stopping the driving wheel, each wheel would receive an impulse equal and opposite to that which it received when the machine was set in motion.

This impulse may be calculated for any part of a system of mechanism, and may be called the *reduced momentum* of the machine for
690 that point. In the varied motion of the machine, the actual force on any part arising from the variation of motion may be found by differentiating the reduced momentum with respect to the time, just as we have found that the electromotive force may be deduced from the electrotonic state by the same process.

695 Having found the relation between the velocities of the vortices and the electromotive forces when the centres of the vortices are at rest, we must extend our theory to the case of a fluid medium containing vortices, and subject to all the varieties of fluid motion. If we fix our attention on any one elementary portion of a fluid, we shall find that it
700 not only travels from one place to another, but also changes its form and position, so as to be elongated in certain directions and compressed in others, and at the same time (in the most general case) turned round by a displacement of rotation.

* * *

705 We may now recapitulate the assumptions we have made, and the results we have obtained.

(1) Magneto-electric phenomena are due to the existence of matter under certain conditions of motion or of pressure in every part of the magnetic field, and not to direct action at a distance between the mag-
710 nets or currents. The substance producing these effects may be a

certain part of ordinary matter, or it may be an aether associated with matter. Its density is greatest in iron, and least in diamagnetic substances; but it must be in all cases, except that of iron, very rare, since no other substance has a large ratio of magnetic capacity to
715    what we call a vacuum.

(2) The condition of any part of the field, through which lines of magnetic force pass, is one of unequal pressure in different directions, the direction of the lines of force being that of least pressure, so that the lines of force may be considered lines of tension.

720    (3) This inequality of pressure is produced by the existence in the medium of vortices or eddies, having their axes in the direction of the lines of force, and having their direction of rotation determined by that of the lines of force.

We have supposed that the direction was that of a watch to a spec-
725    tator looking from south to north. We might with equal propriety have chosen the reverse direction, as far as known facts are concerned, by supposing resinous electricity instead of vitreous to be positive. The effect of these vortices depends on their density, and on their velocity at the circumference, and is independent of their diameter. The den-
730    sity must be proportional to the capacity of the substance for magnetic induction, that of the vortices in air being 1. The velocity must be very great, in order to produce so powerful effects in so rare a medium.

The size of the vortices is indeterminate, but is probably very small as compared with that of a complete molecule of ordinary matter.*
735    (4) The vortices are separated from each other by a single layer of round particles, so that a system of cells is formed, the partitions being these layers of particles, and the substance of each cell being capable of rotating as a vortex.

(5) The particles forming the layer are in *rolling contact* with both
740    the vortices which they separate, but do not rub against each other. They are perfectly free to roll between the vortices and so to change their place, provided they keep within one *complete molecule* of the substance; but in passing from one molecule to another they experi-

---

*The angular momentum of the system of vortices depends on their average diameter; so that if the diameter were sensible, we might expect that a magnet would behave as if it contained a revolving body within it, and that the existence of this rotation might be detected by experiments on the free rotation of a magnet. I have made experiments to investigate this question, but have not yet fully tried the apparatus.

ence resistance, and generate irregular motions, which constitute
745 heat. These particles, in our theory, play the part of electricity. Their
motion of translation constitutes an electric current, their rotation
serves to transmit the motion of the vortices from one part of the field
to another, and the tangential pressures thus called into play constitute electromotive force. The conception of a particle having its motion
750 connected with that of a vortex by perfect rolling contact may appear
somewhat awkward. I do not bring it forward as a mode of connexion
existing in nature, or even as that which I would willingly assent to as
an electrical hypothesis. It is, however, a mode of connexion which is
mechanically conceivable, and easily investigated, and it serves to
755 bring out the actual mechanical connexions between the known electro-magnetic phenomena; so that I venture to say that any one who
understands the provisional and temporary character of this hypothesis, will find himself rather helped than hindered by it in his search after the true interpretation of the phenomena.
760   The action between the vortices and the layers of particles is in part
tangential; so that if there were any slipping or differential motion between the parts in contact, there would be a loss of the energy belonging to the lines of force, and a gradual transformation of that energy
into heat. Now we know that the lines of force about a magnet are
765 maintained for an indefinite time without any expenditure of energy;
so that we must conclude that wherever there is tangential action between different parts of the medium, there is no motion of slipping between those parts. We must therefore conceive that the vortices and
particles roll together without slipping; and that the interior strata of
770 each vortex receive their proper velocities from the exterior stratum
without slipping, that is, the angular velocity must be the same
throughout each vortex.
  The only process in which electro-magnetic energy is lost and transformed into heat, is in the passage of electricity from one molecule to
775 another. In all other cases the energy of the vortices can only be diminished when an equivalent quantity of mechanical work is done by
magnetic action.
  (6) The effect of an electric current upon the surrounding medium is
to make the vortices in contact with the current revolve so that the
780 parts next to the current move in the same direction as the current.
The parts furthest from the current will move in the opposite direction;
and if the medium is a conductor of electricity, so that the particles are

free to move in any direction, the particles touching the outside of
these vortices will be moved in a direction contrary to that of the cur-
785  rent, so that there will be an induced current in the opposite direction
to the primary one.

If there were no resistance to the motion of the particles, the induced current would be equal and opposite to the primary one, and
would continue as long as the primary current lasted, so that it would
790  prevent all action of the primary current at a distance. If there is a resistance to the induced current, its particles act upon the vortices beyond them, and transmit the motion of rotation to them, till at last all
the vortices in the medium are set in motion with such velocities of rotation that the particles between them have no motion except that of
795  rotation, and do not produce currents.

In the transmission of the motion from one vortex to another, there
arises a force between the particles and the vortices, by which the particles are pressed in one direction and the vortices in the opposite direction. We call the force acting on the particles the electromotive
800  force. The reaction on the vortices is equal and opposite, so that the
electromotive force cannot move any part of the medium as a whole,
it can only produce currents. When the primary current is stopped, the
electromotive forces all act in the opposite direction.

(7) When an electric current or a magnet is moved in presence of a
805  conductor, the velocity of rotation of the vortices in any part of the field
is altered by that motion. The force by which the proper amount of rotation is transmitted to each vortex, constitutes in this case also an
electromotive force, and, if permitted, will produce currents.

(8) When a conductor is moved in a field of magnetic force, the vor-
810  tices in it and in its neighbourhood are moved out of their places, and
are changed in form. The force arising from these changes constitutes
the electromotive force on a moving conductor, and is found by calculation to correspond with that determined by experiment.

We have now shewn in what way electro-magnetic phenomena
815  may be imitated by an imaginary system of molecular vortices. Those
who have been already inclined to adopt an hypothesis of this kind,
will find here the conditions which must be fulfilled in order to give it
mathematical coherence, and a comparison, so far satisfactory, between its necessary results and known facts. Those who look in a dif-
820  ferent direction for the explanation of the facts, may be able to
compare this theory with that of the existence of currents flowing

freely through bodies, and with that which supposes electricity to act at a distance with a force depending on its velocity, and therefore not subject to the law of conservation of energy.

825   The facts of electro-magnetism are so complicated and various, that the explanation of any number of them by several different hypotheses must be interesting, not only to physicists, but to all who desire to understand how much evidence the explanation of phenomena lends to the credibility of a theory, or how far we ought to regard a coinci-

830   dence in the mathematical expression of two sets of phenomena as an indication that these phenomena are of the same kind. We know that partial coincidences of this kind have been discovered; and the fact that they are only partial is proved by the divergence of the laws of the two sets of phenomena in other respects. We may chance to find, in the

835   higher parts of physics, instances of more complete coincidence, which may require much investigation to detect their ultimate divergence.

## Note

Since the first part of this paper was written, I have seen in *Crelle's Journal* for 1859, a paper by Prof. Helmholtz on Fluid Motion, in which

840   he has pointed out that the lines of fluid motion are arranged according to the same laws as the lines of magnetic force, the path of an electric current corresponding to a line of axes of those particles of the fluid which are in a state of rotation. This is an additional instance of a *physical analogy,* the investigation of which may illustrate both

845   electro-magnetism and hydrodynamics.

## Part III. The Theory of Molecular Vortices applied to Statical Electricity

In the first part of this paper I have shewn how the forces acting between magnets, electric currents, and matter capable of magnetic induc-

850   tion may be accounted for on the hypothesis of the magnetic field being occupied with innumerable vortices of revolving matter, their axes coinciding with the direction of the magnetic force at every point of the field.

The centrifugal force of these vortices produces pressures distributed in such a way that the final effect is a force identical in direction

855   and magnitude with that which we observe.

In the second part I described the mechanism by which these rotations may be made to coexist, and to be distributed according to the known laws of magnetic lines of force.

I conceived the rotating matter to be the substance of certain
860   cells, divided from each other by cell-walls composed of particles
which are very small compared with the cells, and that it is by the
motions of these particles, and their tangential action on the substance in the cells, that the rotation is communicated from one cell
to another.
865     I have not attempted to explain this tangential action, but it is necessary to suppose, in order to account for the transmission of rotation from the exterior to the interior parts of each cell, that the
substance in the cells possesses elasticity of figure, similar in kind,
though different in degree, to that observed in solid bodies. The undu-
870   latory theory of light requires us to admit this kind of elasticity in the
luminiferous medium, in order to account for transverse vibrations.
We need not then be surprised if the magneto-electric medium possesses the same property.
According to our theory, the particles which form the partitions be-
875   tween the cells constitute the matter of electricity. The motion of these
particles constitutes an electric current; the tangential force with which
the particles are pressed by the matter of the cells is electromotive
force, and the pressure of the particles on each other corresponds to
the tension or potential of the electricity.
880     If we can now explain the condition of a body with respect to the
surrounding medium when it is said to be "charged" with electricity,
and account for the forces acting between electrified bodies, we
shall have established a connexion between all the principal phenomena of electrical science.
885     We know by experiment that electric tension is the same thing,
whether observed in statical or in current electricity; so that an electromotive force produced by magnetism may be made to charge a
Leyden jar, as is done by the coil machine.
When a difference of tension exists in different parts of any
890   body, the electricity passes, or tends to pass, from places of
greater to places of smaller tension. If the body is a conductor, an
actual passage of electricity takes place; and if the difference of
tensions is kept up, the current continues to flow with a velocity proportional inversely to the resistance, or directly to the conductivity
895   of the body.
The electric resistance has a very wide range of values, that of the
metals being the smallest, and that of glass being so great that a

charge of electricity has been preserved* in a glass vessel for years without penetrating the thickness of the glass.

900     Bodies which do not permit a current of electricity to flow through them are called insulators. But though electricity does not flow through them, the electrical effects are propagated through them, and the amount of these effects differs according to the nature of the body; so that equally good insulators may act differently as dielectrics.**

905     Here then we have two independent qualities of bodies, one by which they allow of the passage of electricity through them, and the other by which they allow of electrical action being transmitted through them without any electricity being allowed to pass. A conducting body may be compared to a porous membrane which opposes

910 more or less resistance to the passage of a fluid, while a dielectric is like an elastic membrane which may be impervious to the fluid, but transmits the pressure of the fluid on one side to that on the other.

    As long as electromotive force acts on a conductor, it produces a current which, as it meets with resistance, occasions a continual trans-

915 formation of electrical energy into heat, which is incapable of being restored again as electrical energy by any reversion of the process.

    Electromotive force acting on a dielectric produces a state of polarization of its parts similar in distribution to the polarity of the particles of iron under the influence of a magnet,[†] and, like the magnetic polariza-

920 tion, capable of being described as a state in which every particle has its poles in opposite conditions.

    In a dielectric under induction, we may conceive that the electricity in each molecule is so displaced that one side is rendered positively, and the other negatively electrical, but that the electricity remains en-

925 tirely connected with the molecule, and does not pass from one molecule to another.

    The effect of this action on the whole dielectric mass is to produce a general displacement of the electricity in a certain direction. This displacement does not amount to a current, because when it has at-

930 tained a certain value it remains constant, but it is the commencement of a current, and its variations constitute currents in the positive or negative direction, according as the displacement is increasing or

*By Professor W. Thomson.
**Faraday, *Experimental Researches,* Series XI.
[†]See Prof. Mossotti, "Discussione Analitica," *Memorie della Soc. Italiana* (Modena), Vol. XXIV.

diminishing. The amount of the displacement depends on the nature
of the body, and on the electromotive force; so that if $h$ is the displace-
935 ment, $R$ the electromotive force, and $E$ a coefficient depending on the
nature of the dielectric,

$$R = -4\pi E^2 h; \qquad\qquad (\alpha)^*$$

and if $r$ is the value of the electric current due to displacement,

$$r = dh/dt. \qquad\qquad (\beta)^*$$

940 These relations are independent of any theory about the internal
mechanism of dielectrics; but when we find electromotive force producing electric displacement in a dielectric, and when we find the dielectric recovering from its state of electric displacement with an equal
electromotive force, we cannot help regarding the phenomena as
945 those of an elastic body, yielding to a pressure, and recovering its
form when the pressure is removed.

According to our hypothesis, the magnetic medium is divided into
cells, separated by partitions formed of a stratum of particles which
play the part of electricity. When the electric particles are urged in any
950 direction, they will, by their tangential action on the elastic substance
of the cells, distort each cell, and call into play an equal and opposite
force arising from the elasticity of the cells. When the force is removed, the cells will recover their form, and the electricity will return to
its former position.

955 In the following investigation I have considered the relation between the displacement and the force producing it, on the supposition
that the cells are spherical. The actual form of the cells probably does
not differ from that of a sphere sufficiently to make much difference in
the numerical result.

960 I have deduced from this result the relation between the statical
and dynamical measures of electricity, and have shewn, by a comparison of the electro-magnetic experiments of MM. Kohlrausch and Weber with the velocity of light as found by M. Fizeau, that the elasticity
of the magnetic medium in air is the same as that of the luminiferous
965 medium, if these two coexistent, coextensive, and equally elastic media are not rather one medium.

\* \* \*

# Interpretive Notes

1   "the Philosophical Magazine, vol. XXI"
    Date of publication: March 1861.

3   "the magnitude and direction of the force"
    A quantity incorporating *"magnitude* and *direction"* has become known as a *vector* quantity—a term that we have used freely in earlier notes, in distinction from a *scalar.* Thus in terms of *energy,* magnetism is described by a space-filling scalar quantity; while in terms of *force,* the same space is filled with a vector. We have used "contour maps" to delineate equal values of the former; the lines of force show the directionality of the latter. Are these then simply two ways of telling exactly the same story?

    The term *vector* as a formal term denoting such a multiple mathematical quantity was just coming into use, though it had been introduced by William Rowan Hamilton some twenty years earlier.

7   "the case of two attracting spheres"
    Can one draw lines of force for *gravitational* problems? Maxwell had explored this question in a letter to Faraday, following publication of "Faraday's Lines," and following as well Maxwell's receipt of a paper Faraday had sent him on the "Conservation of Force." Maxwell includes a sketch of the lines of force spreading out from the sun but "diverted" by a planet, and comments:

> You have also seen that the great mystery is, not how like bodies repel and unlike attract, but how like bodies attract (by gravitation). But if you can get over that difficulty, either by making gravity the residual of the two electricities or by simply admitting it, then your lines of force can 'weave a web across the sky,' and lead the stars in their courses without any necessarily immediate connection with the objects of their attraction.

It seems that a gravitational diagram cannot be read like a magnetic one as here all "charges" are "like," and "like charges attract"! Maxwell suggests: "Now conceive every one of these lines ... to have a *pushing* force instead of a *pulling* one, and then sun and planet will be pushed together with a force which comes out as it ought." The reader is invited to attempt such a diagram and to have a look at Maxwell's sketch, which is reproduced together with the letter itself in Campbell's biography.

19    "magnetized by induction"

This is a more circumstantial account of the physics of the iron filings than we had been given before. It seems to accord with the spirit of this paper, turning from the lines as geometrical to the lines as physical—whatever that distinction will ultimately turn out to mean!

23    "as indicating something more"

The iron filings are considered now as an index, or pointer, calling our attention to something: but to what? Not something remote, but something immediate; something, Maxwell says, present and *real*. The filings indicate not remote sources but the lines of force; and now the lines of force are proposed as real entities, present whether the filings have been introduced to indicate them or not. In the title of Maxwell's paper, then, shall we understand the word *physical* to mean simply *real*? And is this the same as the question of substance or *being*? It would seem that Maxwell is embarking on a quest that, taken seriously, is not only "physical," but metaphysical. He really wants to know what exists. He is clearly on the philosophical side of the watershed between "natural philosophy" and "modern science."

The question in these terms may be inherited directly from Faraday, who in 1852 had written a paper with the title, "On the Physical Lines of Magnetic Force."[1]

33    "to clear the way for speculation"

"Speculation" is the term that introduced "Faraday's Lines"—but now it seems that mirror is to catch an image of a different order. In the earlier paper, we were involved with a play of *thought*. Now we are on a search for *being*. One name for such being is "a medium," an entity that would mediate between bodies exerting forces on one another. Speculation will now move in the direction of an intelligible linkage of cause and effect or in the broadest sense, a *mechanism*.

37    "such hypotheses"

In the first paper, it was clear that we would not create *hypotheses*. Now, it appears that we will. In the "Dynamical Theory" they will be gone again—they seem to be way stations along the course of Maxwell's investigations. All of this, of course, serves to raise the question of the role of hypothesis in the development of our understanding of nature.

42    "the mind of the geometer"
There, we were being "geometers." It seems that now we have put on different thinking caps and are to think more earnestly about reality. What, if not geometers, are we now? Have we become mechanics, with a concern for mere matter? Or graduated, to become natural philosophers?

50    "the 'Electrotonic State'"
Our selection from "Faraday's Lines" closed with an allusion to the "electrotonic state," an elusive concept that seems, however, to be close to the center of Faraday's thinking about the field. Maxwell acknowledged at line 828: "The idea of the electrotonic state . . . has not yet presented itself to my mind in such a form that its nature and properties may be clearly explained without reference to mere symbols." That seemed to have something to do with the lack of momentum or energy in the merely "geometrical" medium. Now that these have been introduced, the prospect in the present paper is much better! Maxwell turns to the question of the "eletrotonic state" at line 603.

60    "we shall have found a theory"
In the previous paper, we were specifically *not* seeking a "theory." Now, by contrast, it seems that we *are* and that an experimental test may be possible "which will greatly enlarge our knowledge." We might wonder whether that will prove true: whether *any* experimental test could decide between Maxwell's hypothesis of a medium and a thoroughly consistent theory of action at a distance.

66    * * *
Maxwell discusses several alternative physical hypotheses, reluctantly omitted here for sake of the economy of our enterprise.

68    "stress in the medium"
The concept of *stress* is a generalization of that of *pressure*. We think first of all of a *force* as applied to a single object as a rope

may be tied at a point of attachment to, let us say, a block of wood. Force, however, may be applied over an area, in which case we speak of *pressure* or *force per unit area*. We have followed Maxwell in "Faraday's Lines" in thinking of a medium that was dense with pressures with, however, a single pressure corresponding to each point. At every point in the fluid, the pressure was acting equally in all directions; such nondirectional pressure may be called *isotropic*. Since it may be denoted by a single number associated with each point, isotropic pressure is a scalar quantity.

Now, in "Physical Lines," we are to begin thinking of a much richer medium, with not just a single pressure at a point, but a *set* of pressures and tensions, which differ according to the direction we choose to consider within the medium. Familiar examples would be interior points in a twisted elastic solid or in a turbulent fluid. At each such point we must specify a whole set of pressures, each with the direction in which it is acting. It is such a complex of directional pressures that is denoted by the term *stress*. The problem becomes quite complicated when we think of a point under stress in a space of three dimensions, but fortunately much of what Maxwell is about to develop can be grasped in the simpler case if *variations* are confined to just two dimensions. We will reduce it to these simpler terms whenever we can.

Since each pressure making up a pattern of stress is associated with a direction, each is a vector, and stress at each point will involve a set of vectors. If we think of a vector as a kind of quantity involving a set of numbers and hence more complex than a scalar, we may in turn think of stress at a point as a yet more complex quantity, a set of vectors; such a quantity is termed a *tensor,* though we need not trouble ourselves here with the term or its formal properties.

Think, then, of a point within a medium incorporating stresses (Fig. 3.1). To catch the distinctions we will need, we must surround the point with a tiny box—we give it depth in the $z$-direction—but if we suppose there is no variation in the $z$-direction, we need only consider relations in the $xy$-plane. Then, in general, there will be *normal* forces on the four sides (i.e., forces perpendicular to the sides) and also *shear forces* that act tangentially, within the faces of the box. There are balancing relations among these, which assure that the box in equilibrium neither moves in the $x$- or $y$-direction nor rotates on its axis. These are the "neces-

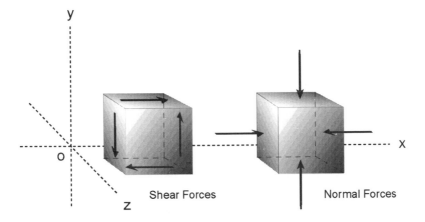

Fig. 3.1. The "stress box."

sary relations," Maxwell says, that the mathematicians have worked out.

75   "three principal pressures or tensions"
In what we have just said, we have taken the general case in which shear forces must be considered. But if we choose our coordinate axes adroitly, we can orient the box so as to avoid shear forces: the normal forces then fully represent the stress at the point and are called the "principal" tensions or pressures. If "the three principal pressures are equal" (line 80), we return to the simplifying assumption we made throughout "Faraday's Lines" and do not need the concept of "stress."

89   "represent the magnetic force"
The "axis" becomes the line of force, and the pressures at right angles will represent the mutual actions among the lines, as we shall see below.

94   "isotropic in its results"
Is a *line of force* appropriately represented as a *stress*? Maxwell anticipates the objection that a line of force is "essentially dipolar," i.e., it has a directionality from one pole to the other, while stress is "isotropic." We have used the term "isotropic" in a more

general sense, but here it is used simply with respect to the line of force: stress has no directionality one way or the other along the line. Note that although the specific direction of the line of force, from north pole to south, is conventional, the fact of its *directionality* is not. The force that the line traces would be substantively otherwise if the two poles were reversed: it *runs* directionally whether we choose to draw the arrow one way or the other.

The line of force has two aspects. One, which is directional, has to do with the selection of the target—selective action on a north or south pole. Once the target is established, however, the line acts like any rope, and the stress in a rope is isotropic. Think of a horse pulling a loaded cart up a steep hill by means of a rope. By Newton's third law of motion, the action of the horse on the cart is matched by an equal and opposite force of the cart on the horse—and the rope impartially conveys them both!

At this point, we might meet an objection concerning horses and carts. At first thought, it may seem that the force *does* run in one direction, from the horse to the cart, and would be very different, and run in the other direction, if the cart were pulling the horse. But this is not so as by the very nature of the concept every force as such is strictly symmetric: the cart pulls the horse exactly as much as the horse pulls the cart. See the reflections on Newton's third law in Discussion [D2].

While Maxwell has now established the isotropy of the line of force in its guise as a rope, he has up his sleeve as well a mechanical answer to the separate problem of its directionality.

99  "We may resolve it"

As we approach the analysis of stresses within a medium, we are free to orient our coordinate axes in any way we please and each orientation will yield a corresponding stress box as in Fig. 3.1. The theory of stresses demonstrates that there will always be some orientation of the axes such that the actual stresses can be expressed as a set of components acting perpendicularly upon the faces of the box, termed *normal* stresses. In this orientation of the coordinate axes, the analysis is simplified, for there will thus be no *shear* components, which are components that act within the surface of a face. Furthermore, if the stresses are radially symmetric about an axis, they can be expressed as the sum of a simply

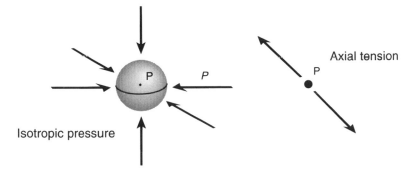

Fig. 3.2. Isotropic (hydrostatic) pressure and axial tension.

isotropic, hydrostatic pressure and a tension or pressure along the axis (Fig. 3.2).

Maxwell is about to apply this concept to the lines of force, which will lie along just that axis. But the lines of force, we have seen, function in their mechanical effect like ropes, that is, they act under tension. Such tension in the axis will correspond to a "sucking" effect—pressure along the lines must be less than pressures perpendicular to them. Thus Maxwell says that in this case the axis will be "that of least pressure."

112   "gravitating bodies"

Maxwell is still wondering at the distinction between lines of magnetic attraction and those of gravity. The meaning of the term "line" seems to depend crucially on its application!

122   "the next question is"

In "Faraday's Lines," we were dealing with the idea of a perfect fluid, which Maxwell emphasized was without mass; it was more like a thought than an entity. Such a fluid was isotropic in three-dimensional space—it could not sustain a stress that was not the same in all directions.

The new domain is much richer: still a fluid but shaped with more complex properties. It has mass (or it could not generate the centrifugal force of which Maxwell is speaking), and having mass, it can both be perfectly fluid and sustain anisotropy: there can be suctions and pressures that are not equalized despite its

fluidity and shear forces as in a solid. It is to be densely permeated
with threads or fibers of vortices: the spinning fluid will generate
a *pressure* outward from its axis, while a suction—familiar in the
spinning of a vortex as water is drawn out of a bathtub—will
define the lines of the magnetic field. Its very structure, then, is
expressed and maintained in its state of motion. The richness of its
inner shape is the consequence of its inertial character together
with its state of intense activity. We have to remember as we say
this that we are still speaking of a *static* magnetic field, such as
that of the earth in the room in which you sit: the presumed
activity is utterly latent.

All this may be a work of the merest scientific fantasy. But as
we shall see, it is fraught with consequences!

129    "the dipolar character of the line of force"
We saw earlier that the tension along the fluid's axis is strictly
isotropic: we may add now that the tension would be unaffected if
the vortex revolved in the opposite sense. Now we see however
that the sense of rotation of the vortex, clockwise or counterclockwise, does serve Maxwell perfectly to reflect the anisotropy of the
line in its aspect as conveyor of a polar force. We will have to see,
as the theory unfolds, how the sense of rotation alters the medium's mechanical effects.

136    "the direction of revolution"
There is no connection between the (utterly conventional) "direction" of motion of electricity and that of these vortices. Yet
by keying the one to other, we have a way of uniquely defining
the sense of rotation we have equally arbitrarily chosen for the
vortices.

136    "vitreous electricity"
We recall from Chapter 1 that "vitreous" electricity, i.e., that
produced upon a glass rod when rubbed with silk, has been
dubbed by convention "positive."

146    "we shall deduce"
Maxwell has defined his project: so far as magnetism alone is
concerned, we have before us the entire idea of his hypothesis.

What does it entail? All we have to do is think it through: the properties of magnetism must emerge, like the rabbit from the magician's hat.

149    "are proportional"
We are asked to think of two systems of vortices, which are scale models of one another. They are similar in shape—one is simply scaled in size with respect to the other, the ratio being *l*. They are in motion at different speeds, and the ratio of their velocities is independent of the ratio of their sizes—a second parameter of the modeling. They preserve that ratio throughout, however, so that everywhere their speeds bear a constant ratio. They also have, independently, different densities—one fluid is light and the other heavy—but, again, the ratio of the densities is constant throughout.

Maxwell speaks now in the antique language of ratios, language established in Euclid's book of magnitudes, the Fifth Book of his *Elements,* and used throughout the mathematical literature of the West well into the modern era. Maxwell had learned it, if not otherwise, in his study of Newton's *Principia.* A ratio is not the same thing as a fraction: the ratio is the relationship; the fraction is a number. To *duplicate* a ratio is to take it with itself twice, as to double the double. Translated into the language of number, the *duplicate ratio* is then the *square*. This is a lovely mode of reasoning, which belongs to an era in which life was not yet suffused with number. It does not culminate in a formula by which we can directly compute anything but in ratios that tell us how the different factors—density, velocity, and size—weigh with one another in affecting a result. In the end, we find here that *size* ceases to be a consideration.

152    "let *l* be the ratio of the linear dimensions"
It may help to picture the two systems, though in doing so we specify their shapes, while Maxwell deliberately leaves this question open (Fig. 3.3). For convenience, we may show them as circular, though Maxwell will ultimately envision them in a different form, as we shall see. This does not matter to our reasoning in ratios. For further notes on Maxwell's argument here, see the Discussion [D3].

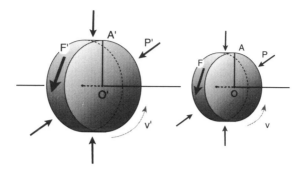

Fig. 3.3. Two vortices.

167    [D3] On the Ratio of Pressures in Two Vortices

171    [D4] On Centrifugal Force

171    "the mean pressure"
There might seem to be a problem in averaging the pressure in this way. But for a body rotating as a whole—that is, with a common angular velocity about an axis—the centrifugal pressure works out to be proportional to the radius, and thus as we go out from the axis to the circumference, the average pressure $p_2$ is just the arithmetic mean, half the maximum.

183    "C is a numerical quantity"
It is interesting to watch Maxwell in his efforts to keep his argument utterly general: he did not want to specify shape, yet had to revert to circular vortices to keep his mechanical reasoning simple. Now, however, he retreats and lets a single constant $C$—which immediately reverts to $\mu/4\pi$ for future purposes—absorb the effects of varying the shape. We should remember that $\mu$, though it bears other implications, has come in by way of the "shape constant" of this vortex. (This shape constant will show up as $k$ in our reasoning below.) The actual shape Maxwell adopts in diagraming his vortices is the hexagon, evidently in order to achieve close-packing. Anticipating his own illustration, we may draw his space-filling "beehive" diagram of the vortices of a uniform magnetic field (Fig. 3.4).

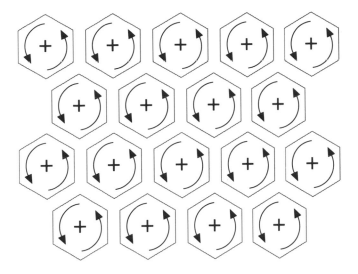

Fig. 3.4. The hexagonal vortices of a uniform magnetic field. ( + notation denotes counterclockwise rotation.)

190   "differs from an ordinary fluid"

This is indeed "no ordinary fluid," for, infused with these vortices and thus with pressures that are different in different directions, it will be full of a new order of complexity. The intention is, of course, that it will now have exactly that level of complexity commensurate with the electromagnetic phenomena. A fluid thus replete with pressures and tensions will tend to fly apart. Once the preliminary formalities that follow are taken care of, Maxwell will proceed by writing the equations that impose equilibrium on his unruly fluid. If his hypothesis works, those equations of equilibrium should turn out to be the equations of electromagnetism itself!

197   "direction-cosines"

The "direction-cosine" is a useful device in analytic geometry. If a line is tilted with respect to an axis through an angle $\theta$, its direction-cosine with respect to that axis will be simply cos $\theta$. This can be thought of as the projection of a unit vector lying along the oblique line onto the axis (Fig. 3.5). These projections are traditionally named as follows:

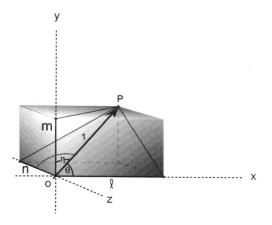

Fig. 3.5. Direction-cosines.

$$\cos \theta = l \ (x\text{-axis})$$
$$\cos \eta = m \ (y\text{-axis})$$
$$\cos \zeta = n \ (z\text{-axis})$$

218    "This more general case"

Here we give the results in two dimensions (compare Figs. 3.6 and 3.7.), while Maxwell gives them for three. It might seem that we could avoid complexity by always choosing the coordinate axes so that they lie along the principal stresses—no shear would arise. But even if we could, we might miss the value of the exploratory "box." By rotating the axes, we are in effect mentally investigating the fluid, finding out what is going on amid the pressures and tensions.

222    [D5] On the Concept of Stress

236    "Prop. III"

Again, to keep things as simple as possible in a first reading, we substitute for Maxwell's general treatment the cases with the least complexity. We are now interested in *dis*equilibrium of the fluid: that is, we are concerned with those net forces which will act upon parts of its body, or, in other words, those forces that must be applied to restore equilibrium caused by unbalanced forces within the fluid. These arise, Maxwell says, "from the variation of inter-

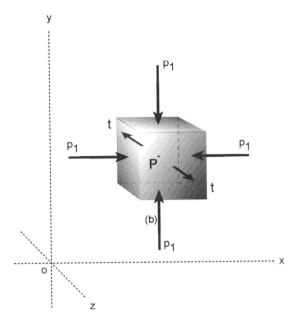

Fig. 3.6. Hydrostatic pressure $p_1$ and tension $t$ in the vortical fluid.

nal stress"—that is, from change of stress from point to point in the fluid as through curvature of the lines of force. We are speaking, then, of variation in *space*. We are not yet involved with variation in *time*, which will come in due course.

242 "We have in general, for the force"
Equation (3)* is derived in Discussion [D6]. In a general way, however, we can see that net forces will arise in the fluid when the stresses on two opposite faces of our exploratory box are unequal (Fig. 3.8). This will arise in turn when the pressure in the very near neighborhood of the point $P$ differs from the pressure at point $P$ itself (i.e., when the derivative of $p_x$ with respect to $x$, $dp_x/dx$, is nonzero.) It will equally arise when there is an unbalanced shear force in the $x$-direction, which will occur if $ds/dy$ is not zero. The specific terms of equation (4)* then arise rather magically from the algebra, as Maxwell carries out the differentiations artfully to fetch out the special interpretations he has in view (see [D6])!

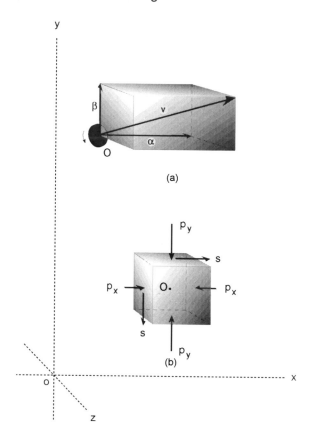

Fig. 3.7. Tension and shear at O due to a vortex aligned obliquely to the coordinate axes: (a) orientation of a vortex with velocity $v$; (b) stresses exerted on a test box around point O (only those acting in the $x$- and $y$-directions are shown).

243    [D6] On the Law of Equilibrium of Stresses

247    [D7] Derivation of Equation (4)*

259    "We suppose $\alpha$ and $\beta$ to be"
        Here is the first step in giving magnetic interpretation to the
        vortex: let $\alpha$, which was defined as $vl$, correspond to the $x$-compo-
        nent of the *magnetic intensity* $H_x$, and $\beta$ to $H_y$. The intensity **H** is
        thus a vector, the force per unit pole on a test pole placed at any
        point in a magnetic region. Let us represent the velocity of the

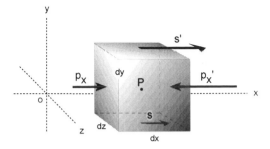

Fig. 3.8. Resultant forces on the exploratory box. Stress varies from point to point within the fluid; here, it is assumed that only the *x*-component $p_x$ varies.

vortex with a vector **v** along its axis. Then the interpretation recognizes **v** as **H,** and α and β as $H_x$ and $H_y$. Note that it is not the tension *t* that is the magnetic intensity, as we might have expected.

261   "μ represents the magnetic inductive capacity"
      The symbol μ represents that property that iron has and most other materials do not: the ability to become strongly magnetized under "induction." If *H* is the number of lines of force through a surface in air, that number will increase by the factor μ when iron is introduced. The quantity **B** = μ**H** is termed the *magnetic induction.*

269   "The total amount of magnetic induction"
      If a box is placed around a magnetic pole, the net number of lines of induction inward or outward through the surface measures the strength of the pole, *m.* More generally, if a test box is placed in a region in which there are magnetic lines of force, the net induction through the surface measures the total magnetic strength of all the sources within the box. The sources may be concentrated or distributed continuously. For the derivation of *m* in equation (6)*, see Discussion [D8].

277   [D8] On Total Magnetic Induction from a Source

291   "in fig. 1"
      The equation in Proposition II has reduced to just two basic factors, which Maxwell's "fig. 1" is meant to illuminate. We think

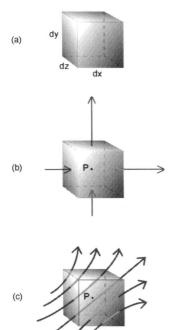

(a)

(b)

(c)

Fig. 3.9. The origination of lines from a source, termed divergence: (a) the divergence box; (b) divergence in two dimensions; (c) divergence in three dimensions.

of a field whose magnetic intensity is $\alpha$, and place within it a magnetic pole of magnitude proportional to $m$. Of course, thinking in terms of the hypothesis, there is only one fluid, which can take on only one complex motion, but we can perceive it analytically as compounded of two elements, each of which involves its own quite distinct motion.

311    [D9] Note on the Logic of Physical Speculation

325    "the square of the intensity"
Maxwell writes $\alpha^2 + \beta^2 + \gamma^2$, but we are sticking to our simplification, and omit $\gamma$. If the intensity **H** is a vector with components $H_x$ and $H_y$, the Pythagorean theorem tells us that the square of the total intensity is the sum of the squares of the components.

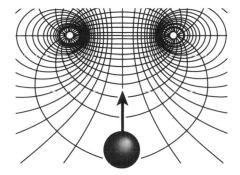

Fig. 3.10. Force drawing the fluid into regions of higher intensity.

327 "will be urged"

Expression (B), which we are now considering, is, we should remember, the second term in the long equation for the force in the *x*-direction on a piece of the fluid medium. This *derivative* is telling us that if the quantity $(\alpha^2 + \beta^2)$ is increasing, there will be a positive force on the fluid (Fig. 3.10). Again, however, as in (A), there are two factors involved: the expression is multiplied by the constant $\mu$. That fact becomes crucial in the interpretation of the result: the force we are speaking of will be greater or less, in proportion to $\mu$, i.e., great in a substance such as iron.

331 "the medium, as well as the body"

Again, the analysis has yielded two factors, and we look for a corresponding pair of entities by way of interpretation. Here—thinking first of the fluid—we have the derivative of the square of the velocity multiplied by a quantity proportional to $\mu$. What does that mean? The velocity will be that of the medium, but now we take a portion within it and consider the force on *it* due to this velocity. That portion is thought of as belonging to a "body" *within* the medium. What distinguishes it as a "body" and not simply as a piece of the medium? The answer lies in $\mu$, for this portion may have a $\mu$ different from that of its surroundings—*only* that difference could at this point distinguish a *body* as such.

The medium—taking it as air or vacuum—has a reference value of $\mu = 1$. A body within this medium may have $\mu > 1$ or $\mu < 1$. The action on a body, that is, really, that portion of the medium with

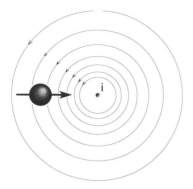

Fig. 3.11. Iron ball attracted to a current (current *i* is flowing vertically out of the paper).

its own μ that we are identifying as a "body," will experience a differential force as it is drawn more or less than the surrounding medium. It is like a diver, whose density dictates sinking or rising, as it is greater or less than that of the water.

341  "If the capacity for magnetic induction"
The capacity for induction is greater than that of air (really, vacuum) in paramagnetic materials and less in diamagnetic. This then conforms fully to Faraday's insight concerning the differential "conduction" of the lines of force—as illustrated in the figure we reproduced earlier (Fig. 1.31). It is interesting to compare the present theory with the account of para- and diamagnetism Maxwell gave in "Faraday's Lines."

346  "either north or south"
The mathematics tells us that the body of μ > 1 is drawn in the direction of greater intensity, without regard to polarity. And of course that is true: a piece of soft iron magnetized by induction is drawn indifferently to either magnetic pole.

347  "In fig. 5"
It may be useful to redraw Maxwell's figure (Fig. 3.11). The field might be that of a vertical conductor, which is confusing because we know that a magnetic *needle* will point *tangentially*. The mathematics teaches us, though, that a *neutral* piece of soft iron will move toward the conductor and experiment confirms that conclusion.

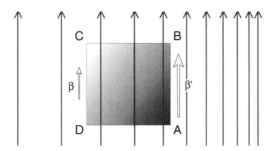

Fig. 3.12. Work around a closed loop.

360    "quite independent of the direction of the lines of force"
       Maxwell is impressed, as we might be too, to see that the theory
       teaches us to free ourselves from an over-literal notion of the
       "lines of force." Here, motion can equally well be *along* them
       (fig. 4) or *across* them (fig. 5). Things are being guided by some
       higher principle lying at the foundations in which the fluid hy-
       pothesis finds its unity.

366    "a quantity which would disappear"
       Figure 3.12 illustrates the situation Maxwell describes. For sim-
       plicity, let the magnetic field lie first in just the *y*-direction,
       represented as usual by the symbol $\beta$. It is however not uniform,
       but a function of *x,* shown here as weaker on the left and increas-
       ing toward the right. A unit pole is now imagined carried around
       the loop ABCD and, as it goes, we compute the work done by the
       field *on* the pole. Remember that "work" is, by definition, force
       times distance.
            As the field works for us on the right but against us on the left,
       the net work around the loop would be zero if the field were
       uniform. Net work arises because of the difference in the nonuni-
       form field between the intensity $\beta'$ on the right and $\beta$ on the left.
       In this case, work is done only when the pole traverses *AB* and
       *CD,* segments of the path which lie along the direction of the
       intensity. (Passages perpendicular to a force, like contouring
       around a hillside, come at no cost!)
            Positive work is done on the pole in the segment *AB* and
       negative work (against it) in *CD.* Net work arises only as a

consequence of the difference in these two, measured by the differential expression $(d\beta/dx)\Delta x$. The total work is thus

$$\Sigma W = \left(\frac{d\beta}{dx}\right)\Delta x \Delta y \qquad (a)$$

If $\alpha$ also varied, its effect would be reckoned by analogous reasoning, and the total work per unit area would be

$$\Sigma\frac{W}{\Delta x \Delta y} = \left(\frac{d\beta}{dx} - \frac{d\alpha}{dy}\right) \qquad (b)$$

We will deal more formally with this expression (as the operation "curl") in Chapter 4.

We saw in Chapter 1 that this *work around a closed loop* surrounding a current is proportional to the current and inversely proportional to the radius (the Oersted effect, Fig. 1.38). If the current is *r,* then at radius *R* the intensity is proportional to $(r/R)$. We define the unit of current very simply so as to make this proportionality an equation: unit current will produce unit field at unit distance. Then the work to carry a unit pole around this path will be just this intensity multiplied by the circumference of the path, as work is done uniformly all the way:

$$(r/R) \times (4\pi R) = 4\pi r \qquad (c)$$

It is striking that the radius cancels out, so the work around any closed path will be a direct measure of the current enclosed.

In fig. 6 of his text, Maxwell inserts a wire carrying current *r* and pointed out of the page into the earth's field, seen as lying in the page (one should place the book on a table top and orient the figure properly!) Since the uniform field of the earth contributes nothing to the work around the loop, the work remains $4\pi r$ as before. This becomes the situation to which Maxwell will now apply the third term, (C)*, of his force equation.

Note that for any path which does not enclose currents, the total work will be zero, no matter how complex the magnetic field in the region may be. The field in a region without currents is said to be *irrotational,* and from an analytical point of view the quantity in parentheses becomes the *complete differential* to which Maxwell refers.

377    "represents the strength of an electric current"
       Here is a striking instance of the strategy Maxwell is employing in
       interpreting his equations.
          We can put equation (b) and equation (c) of the previous note
       together. Equation (b) gave us *work per unit area* around a loop in
       a region of the magnetic field, while (c) gives us that same *work*
       in terms of a current that is generating such a field. To equate
       them, we must divide (c) by $\Delta x \Delta y$ to obtain work per unit area.
       Then

$$\frac{4\pi r}{\Delta x \Delta y} = \left( \frac{d\beta}{dx} - \frac{d\alpha}{dy} \right) \tag{d}$$

$$\frac{r}{\Delta x \Delta y} = \frac{1}{4\pi} \left( \frac{d\beta}{dx} - \frac{d\alpha}{dy} \right) \tag{e}$$

What equation (e) is telling us is that if there is a current density
(current per unit area) anywhere, there will be a magnetic field in
the same region with the certain special rotational structure that
equation (b) prescribed.
   Then, assuming the reasoning reverses, the presence of that
special structure will be a sure sign of the presence of a current.
The magnetic field expression

$$\frac{1}{4\pi} \left( \frac{d\beta}{dx} - \frac{d\alpha}{dy} \right)$$

which has emerged in his equations as his term (9)*, *represents* a
current—as surely as a bear track represents the presence of a bear
or smoke the presence of a fire!

381    "The physical interpretation"
       Maxwell now proceeds to use his fig. 6 to support an interpreta-
       tion of the third term of equation (4)*, the force equation. We
       wrote the (C)* term at line 254 as

$$k\beta \left( \frac{d\alpha}{dy} - \frac{d\beta}{dx} \right)$$

If we now insert the value of $k$ (line 266) and reverse the sign of
the binomial to conform to the present discussion, we may write

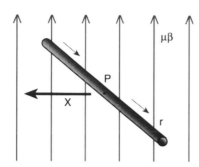

Fig. 3.13. Interpretation of equation (C)*:
force on a current *r* in a magnetic field *mb*.

$$X_C = -\left(\frac{\mu}{4\pi}\right)\beta\left(\frac{d\beta}{dx} - \frac{d\alpha}{dy}\right) \tag{a}$$

The work to carry a unit pole around the current *r* of fig. 6 is $4\pi r$ (see the previous note), and this we know is just the value of the binomial,

$$\left(\frac{d\beta}{dx} - \frac{d\alpha}{dy}\right) = 4\pi r \tag{b}$$

Substituting, we get Maxwell's result for the force, which we see, by its sign, acts to the left in the figure:

$$X_c = -\left(\frac{\mu}{4\pi}\right)\beta(4\pi r) = -\mu\beta r \tag{c}$$

Once again, Maxwell's vortices play all roles: initially vortices with uniform, parallel axes provide the earth's magnetic field, while here they are disturbed by an overlaid motion generating the circular field of the current. And again we must remind ourselves of the completeness of this world of vortices: in it at this point there is—or need be—nothing else, so that the "vertical current" can be it would seem nothing but that configuration of vortex motions that yields a nonzero value of the expression in (b). It is not necessary to have a flowing electric substance running in the wire!

387  "To illustrate the action"
     There seems to be a genuine distinction between "physical inter-
     pretation" and "illustration of an *action*." In the former, it seems
     that we can identify the magnetic and electric effects but do not

concern ourselves with the processes in the fluid medium; while in describing the "action" of the vortices, it seems that we give a complete account of the connection of events in the fluid world. Here the action is described in terms of the tendency to *expand,* while earlier it was the *tension* along the axis we invoked. These two, corresponding to $p_1$ and $t$ in our first analysis, have been from the beginning two sides of the same coin, that is, two ways of looking at the effect of the centrifugal force of the vortices.

389   "magnetic current"
Maxwell intends "electric current"?

400   * * *
A short section is omitted concerning a term which arises only when our two-dimensional version of the equations is extended to include the third dimension. It is identical in signficance to the third term we have just examined.

401   "The fifth term"
We note that, for the moment at least, this last term *has no interpretation.* That is, it has an important mechanical effect in the domain of the fluid (a pressure gradient, giving rise to a real and operative stress), unrecognized as having any counterpart in the electromagnetic domain. Is this an important problem?

405   "We may now write"
Equation (12)* neatly summarizes our results. The symbol $m$ now represents the magnetic source which gave rise to the first differential expression. The second term is now written in terms of the gradient of the square of the velocity $v$, which we know in the interpretation is the magnetic intensity. In the third, the symbol $r$ stands for the current in the $z$-direction, which was indicated by the circular character of the vortical field. And as we have just indicated, the last term stands at present without interpretation.

411   "the force acting on electric currents"
It is noteworthy that electric current has entered thus far quite mysteriously—*only* by way of an "interpretation." Its presence has been assured by the existence of a magnetic field for which the

total work around a loop is nonzero: we know that there must be a current, but in terms of the vortex model it remains simply a missing link. It is interesting, too, as we try to understand the role of analytic mathematics in Maxwell's thought process, that it is a term [namely, (D)*] generated through the operations of a mathematical formalism that is serving now as the index to an omitted element of the physical model. Oracle-like, it points the way to the next stage of Maxwell's construction. There must be more to the forms of analytic mathematics than *mere* "formalism"!

414  "PART II"
*Philosophical Magazine,* April and May 1861

422  "proportional to the square of the intensity"
It would be interesting to consider what would have resulted if we had set the magnetic intensity itself, rather than the *square* of the intensity, proportional to the difference of the greatest and least pressures. This seems to have been an arbitrary decision of the modeling process, asserting an analogy between otherwise unrelated domains. Is there some inherent reason why the correspondence should go one way rather than the other?

In "Faraday's Lines," the lines of force were thought of as directions of *pressure,* in the flow of a mathematical "fluid." Now they have become lines of *tension*! Is it acceptable for them to be in one story lines of pressure and in the next account, lines of tension?

428  "independently of any theory as to the cause"
Maxwell here very carefully disentangles the distinct stages, or layers, of his reasoning. He settles on calling the first stage, which we have just followed, a *mechanical deduction,* as if it were a mathematical consequence free of hypothesis. It has shown that a distribution of pressures in a continuous medium *could* model the complete set of most basic electromagnetic phenomena. This answers the overriding question, left unaddressed in "Faraday's Lines," whether a physical system filling the space between two poles *could* in principle account for their interactions, and thus replace a mathematical theory based on sheer action at a distance.

Intermixed with this reasoning, yet separable in thought from it, is the suggestion that this system of pressures could be accounted for as the consequence of an infusion of vortex motion throughout a highly energetic physical fluid. This suggestion gets called a "hypothesis," though at this point it surely remains highly tentative, especially as it is confronted with a set of unresolved difficulties: (1) Adjacent vortices will conflict with each other at their tangencies, a difficulty Maxwell confronts at the outset of Part II. (2) No causal theory of the vortices has been introduced: what sets them in motion or serves to alter their motions? (3) Electricity has entered this magnetic system only as a ghost, with no physical presence in the model. It turns out, in Part II, that all three difficulties will be resolved with one dramatic stroke of the imagination.

447    " 'How are these vortices set in rotation?' "
If the vortices have been introduced as causes of the tensions, we now go further to ask about causes of the vortices; and this seems a more questionable move, qualifying as a "suggestion" rather than a hypothesis. By contrast, the hypothesis of vortices Maxwell here goes so far as to rate as "probable".

456    "in doubt as to the nature of electricity"
It is very easy to assume that electricity is a substance of some sort—however subtle—resting on charged conductors or flowing through wires. For most theorists, the question was limited to that of the sign of the substance: is it positive charge that flows (from + to –) or negative charge (from – to +) when a current "flows through a wire?" Or both, flowing symmetrically in both directions? Maxwell, however, showed in "Faraday's Lines" that nothing would be lost to thought if there were *no substance at all* on a "charged" conductor—but merely a state of affairs in the space around it.

The time now arrives to fill the gap left in Part I of this paper. There, the pattern of stress seemed to indicate the presence of an electric current, though there was nothing as yet in the fluid to correspond to it. Maxwell himself points out the insecurity of his reasoning. He is about to insert a new element into his account, an element of the machine to correspond in some way to "electricity"

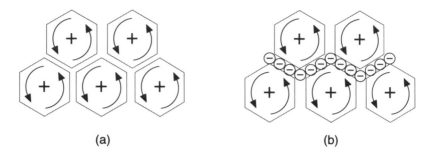

(a)                                        (b)

Fig. 3.14. Maxwell's problem and its solution: (a) Problem—To fill space with nonconflictive vortices. (b) Solution—Insert counterrotating idler wheels! (– signs indicate clockwise rotation).

itself—though it is in truth unsafe to assume that the term actually refers to a substance of any kind at all. He concludes nonetheless that if this "conception" were to succeed in its own terms, it might "lead us a long way" toward resolving the question, "What is an electric current?" Maxwell thus feels free to play with yet another device of the imagination.

467    "I have found great difficulty"
The problem Maxwell confronts is illustrated in Fig. 3.14a in which it is evident that the surfaces of adjacent vortices will move in conflicting directions.

473    "The only conception"
Figure 3.14b illustrates the proposed solution to Maxwell's problem. Idler particles fill the spaces between vortices, their motion mediating in such a way that the conflict is resolved.

493    * * *
We omit a lengthy mathematical study of the kinematics of the idler wheels. The outcome is that the motion of the particles can be made proportional to the current and thus fills the gap we noticed earlier in the hypotheses. The symbol "$r$," that previously denoted the electric current per unit area of a surface perpendicular to the $z$-axis (but had no counterpart in the domain of the fluid)

now becomes simply the number of particles passing per unit time through that surface.

497   "these particles are very small"
Maxwell supposes that the particles are very small and of negligible mass and that they spin constantly without dissipating energy, even in an unvarying magnetic field such as that of the earth surrounding you even as you read! The reader to whom that seems unreasonable might consider that Maxwell is not making any assumptions that have not subsequently been made concerning the "electron." Electrons are part of the assumed composition of all atoms and are characterized as in some sense having "spin" and "orbital" motions without dissipation of energy. The flow of electrons, as that of Maxwell's particles, constitutes the electric current in familiar circumstances. I point this out, not to suggest that Maxwell "anticipates" the electron theory, but only to note that his assumptions concerning perpetual motion without loss of energy are reasonable—if electron theory is!

509   * * *
(One paragraph has been omitted at this point.)

511   "flowing from copper C to zinc Z"
In a battery (an electric cell) with copper and zinc electrodes, it is the copper that is positive. Current is arbitrarily taken as flowing from the positive terminal through the external circuit to return to the negative.

552   "$(dP/dy - dQ/dx) = \mu d\gamma/dt$"
This equation is not as mysterious as it may appear. As we shall show, the left-hand side measures the net force that the particles exert on the periphery of a vortex. A simple example of these peripheral forces accelerating a vortex is shown in Fig. 3.15. Here $P$, $Q$, $P'$, and $Q'$ are the forces of four particles, all acting on the vortex with the same rotational sense, producing a combined effect tending to turn the vortex clockwise.

The right-hand side is the change of momentum of the vortex. Behind the equation lies the fundamental principle of Newton's mechanics, which we met earlier: the force on any body equals the

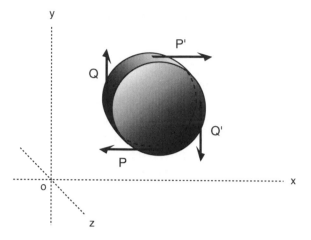

Fig. 3.15. Electromotive force around a loop.

time-rate of change of momentum. Here, we recognize $\mu$ as the effective mass per unit volume of the vortex in rotation, and $\gamma$ as its velocity in rotation about a line parallel to the $z$-axis. (For Maxwell's purposes, we do not need to spell out details of this rotational motion. By wrapping his reasoning in the broader terms of proportionalities and constants of proportionality, he spares us the necessity of defining specific measures of the theory of rotational motion, in which the counterpart of force becomes "torque" and the rotationally effective mass becomes the "moment of inertia.") As $\mu$ is constant, the expression $\mu \, d\gamma/dt$ measures the rate of increase of the vortex's momentum, which is proportional to the net force acting on it.

The problem remains to see the expression on the left as the total (i.e., *net*) force of the particles on the periphery of the vortex. $P$, $Q$, and $R$ are the three components of force per particle, that is, the force exerted per particle in the $x$-, $y$-, and $z$-directions. Now, if $P$ and $P'$ are both positive in sign, they act in the same direction at the top and bottom of the vortex. If they were equal in magnitude, we see that their rotational effects would—in contrast to the case of Fig. 3.15—simply cancel (they would be attempting to turn the vortex in opposite senses). Any rotational effect must arise, then, only from the *difference* in their magnitudes, which will be due to some nonzero value of $dP/dy$.

Similarly, any rotational effect of $Q$ will arise from a nonzero value of $dQ/dx$. Finally, a little consideration will show that a positive value of $dQ/dx$ ($Q$ increasing to the right) will be subtractive from the rotational effect of a positive $dP/dy$ ($P$ increasing upward). A positive $dP/dy$ will have a clockwise turning effect, while a positive $dQ/dx$ will act counterclockwise. We conclude that the net turning effect of the particles is given by the left-hand side of our equation, namely ($dP/dy - dQ/dx$).

As this expression, measuring net turning effect, gained significance for Maxwell, he cast about to find an appropriate name for it. His choice, influenced by his familiarity with a popular Scottish game, was *curl*. It has subsequently come to denote the corresponding formal operation in the vector calculus. We shall meet it often from this point on.

A useful notion, suggested by Skilling in his *Electric Waves*, is that of the *curl meter*. Imagine a paddle wheel mounted on a vertical axis and inserted from above into the middle of a stream. If the current is uniform, the stream's overall force on the paddles will balance out and the wheel will not turn (zero curl). However, if the current is swifter on one side than on the other, the paddle will turn, indicating a net value of curl around its perimeter.

Finally, as the force $P$ is "per particle," and the particles are lined up as we have seen in rows with a certain number per unit distance, we can take $P$ as proportional to the force per unit length along the particle row. As this is the force of the particles tending to set the vortex in motion, it is called the *electromotive force*.

562    "I have pointed out"
Maxwell refers here to "Faraday's Lines."

594    "$P = dF/dt$"
What is this $F$ that we have made? Remembering that $P$ is electromotive force and seeing that it equals the time-rate of change of something, we conclude that that *something* must be of the nature of a *momentum*. Maxwell goes on to identify it with the "electrotonic state" Faraday had been searching for. In Fig. 3.16, there is no apparent connection between two coils, $A$ and $B$, mounted in

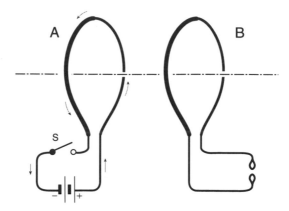

Fig. 3.16. Electromagnetic induction.

separate, parallel planes. Yet if the current in coil *A* is suddenly interrupted by opening the switch *S,* an electrical impulse will be produced in coil *B;* if the two coils are wound with many turns, a powerful spark can be produced at the gap. It seemed to Faraday that effects such as these attested to the presence of something like a state of tension in the space surrounding coil *B,* whose sudden relaxation when the current in *A* ceased caused the impulse in *B;* hence the name *electrotonic.*

Now, however, Maxwell shows that this "state" may also be thought of as having the properties of *momentum.* It is as if a powerful flywheel lurked in the empty space and had been stopped in midcourse when switch *S* was opened. It reacts with a strong counterforce tending to resist the change and continue its motion, evidenced by the spark in the gap. The spinning flywheel is a storehouse of momentum and the field is strikingly like that—infused with rotational momentum. Rotational momentum is a vector whose direction is that of the axis about which it spins. The new vector quantities *F* and *G* are the components of this momentum vector, which has a definite magnitude and direction at every point throughout the field.

616    "The row of vortices *gh*"

This initial state of affairs is depicted in Fig. 3.17. Here, shaded figures are nonrotating (though in the case of the electric particles, they may be moving in translation).

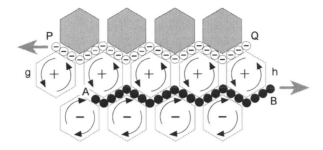

Fig. 3.17. Induced current: an interpretation of Maxwell's Fig. 2.

625   "If this current is checked"
      If *PQ* exists simply in space, with no circuit through which it can
      flow, vortices *kl* will be set in motion in their turn, as in Fig. 3.18.

663   "except an instantaneous pressure"
      Maxwell has contrived here a dramatic crisis for the theory of
      vortices—one which evidently interests him especially, as it
      brings into vivid silhouette Faraday's concept of the "electrotonic
      state." Here the vortices are entirely confined: by bending the
      hitherto open-ended coil into a closed ring, he has kept the field
      lines bottled-up inside the iron of the torus (a torus is a shape like
      a doughnut). There may then be intense vortex energy inside the
      coil, but outside . . . what? No magnetic field can be detected. In
      this sense, there is no *activity*—yet it seemed to Faraday that there
      must be some sort of stress or tension.
      This tension may be exhibited dramatically whenever the cur-
      rent in the iron-cored coil is interrupted—for if a loop in the outer
      region encircles the coil, it will function like coil *B* in Fig. 3.16
      and exhibit a spark. This seems to demonstrate that the "electro-
      tonic state" is real, yet is not reducible to a simple magnetic field,
      since none is detectable in the space around the torus when the
      current in the coil, however strong, is constant!
      This may be the appropriate point at which to complete our
      interpretation of the phenomenon of induction. We have shown in
      Fig. 3.17 how a current, *PQ,* is induced when the current in the
      "primary" circuit is initiated. In Fig. 3.18, we see how the field
      would spread through a nonconducting medium, such as air. Now

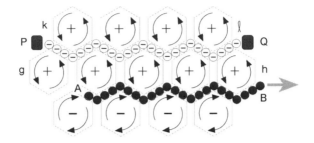

Fig. 3.18. Spreading magnetic field. Current PQ is blocked by the absence of any conductor, so the magnetic field spreads to vortex row *kl*.

in Fig. 3.19 we look at events when the current in the primary *AB* is stopped, as by suddenly opening a switch. The vortices around current *AB* are brought to an abrupt stop but those beyond the conductor *PQ*, which are actively spinning, now deliver that momentum to *PQ*, which persists in the motion that has been interrupted by *AB*—a true flywheel effect. The sudden pulse of motion to the right in the secondary *PQ* is a symptom of this transfer.

681    "It corresponds to the *impulse*"

"Impulse" is a strict term in Newtonian mechanics, but it means pretty much what one would expect intuitively. If a force acts for a time, the *impulse* is the product of the force and the time. Thus, for a given force, the longer it acts, the greater its impulse. Hence, the impulse is in some sense the "total effect." But this is true for a brief impact as well and a great force for a short time will have the same effect. A hammer blow can deliver an impulse equivalent in changing momentum to a long, slow force. Here, then, Maxwell is interested in the impulse of a sudden force that would bring a machine abruptly into rotation. "Impulse" becomes surrogate for a quantity of momentum.

By Newton's laws, a steady force yields a constant acceleration, that is, a constant rate of change of velocity:

$$f = m\,a = m\,(\Delta v / \Delta t)$$

If we want to think of this from the point of view of "impulse," we may multiply both sides by the time interval $\Delta t$:

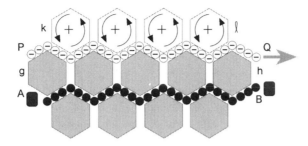

Fig. 3.19. Current induced in the secondary PQ when the primary current *AB* is interrupted.

$$f \Delta t = m \Delta v$$

*f*$\Delta t$ *is* impulse and $m \Delta v = \Delta(mv)$ or the change in momentum. Hence the rule

Impulse = Change in momentum

If a wheel has a certain momentum, we may think of that as having been imparted suddenly by an impact. Then as Maxwell has identified the "electrotonic state" as a momentum, he is right to say that it corresponds to the impulse that would serve to start it suddenly from scratch, and that impulse would be in turn a shock delivered to its axle.

692    "the reduced momentum"

This is a concept that will have important consequences in Maxwell's next paper—the "Dynamical Theory." For the moment, it is enough to see that impulses to different "parts of the machine" may contribute differently to its overall momentum. Looking at the same thing the other way around, we may think of a sudden stopping of the whole machine: the corresponding portion of the shock of stopping will be delivered to each part. Thus the contribution of each to the momentum of the whole would be measured by the impulse delivered there when the machine is brought to a sudden stop. Here the "machine" is the system of vortices with their idlers and a "part" might be a conductor or a current anywhere in the system.

704    \* \* \*

Maxwell at this point enters upon an extensive development of the theory of vortices in which, as he says, "all varieties of fluid motion" must be taken into account. Unfortunately, we cannot follow him here in that fascinating investigation, from which a general theory of forces upon moving conductors arises. We must skip, then, to the summary with which he concludes Part II of "Physical Lines." A major new turn of the theory, however, is still to come—for as yet no account has been given of electric *charge,* which will be the subject of Maxwell's Part III.

707    "Magneto-electric phenomena are due"

The thought may be hypothetical, but in this restatement it takes bold, declarative form! Maxwell allows the alternative of "motion or pressure"—and we remember that he distinguished earlier between a system of distribution of pressures and the assumption of vortices as its cause. It would thus seem that "direct action at a distance" is altogether eliminated. The term "magnetic field" seems to be coming into its own as well; earlier, it might have been thought to apply to the earth's "field," but now, it is coextensive with magnetism.

This account envisions the magnetic field really consisting of inertial matter, though it must not be "ordinary matter." It must instead be an *ether,* coupled in some way to ordinary matter, with which the field exchanges very evident mechanical energy. How would such an ether differ from ordinary matter? It certainly has "mass" in the sense of bearing large quantities of momentum and energy. Yet it has no detectable weight. It is a fundamental outcome of Newton's work that all matter gravitates in proportion to its inertial mass, but it seems the ether may have to be an exception.

Maxwell discusses the ether extensively in the opening sections of our third paper, the "Dynamical Theory," and raises the possibility of the existence of matter that does not gravitate in proportion to its inertial mass both at the end of his little book, *Matter and Motion,* and in his article "Ether" in the *Encyclopaedia Britannica.* It may occur to readers familiar with the findings of modern physics that the "photon" is the answer to this conundrum: in action, photons bear the field's momentum and energy,

Fig. 3.20. Maxwell's electromagnetic top.

but when they are no longer in motion they have no "rest mass" and, hence, as Maxwell required, *no weight.*

### 733 "The size of the vortices"

For the same magnetic field intensity, the rotational inertia or *angular momentum* of the vortices will be greater if their diameter is greater: this means that they will exhibit correspondingly greater effects such as those of a flywheel or a spinning top. Maxwell refers in the footnote to "experiments to investigate this question." Although as it turns out the parameters of nature are such that he would not have achieved measurable effects and he never published his results, this remains a fundamental, crucial experiment. We are familiar with the apparatus that he used, which is illustrated in the *Treatise on Electricity and Magnetism.* Figure 3.20 is a photograph of a counterpart in the Smithsonian Institution. Maxwell's sense of the reality of the vortices is attested to by the care with which he designed this apparatus to find them experimentally.

In Maxwell's apparatus, at the same time that the coil is being rotated at high speed a current is being passed through it. The iron core is thus filled with vortices and if they possessed measurable angular momentum, this would add to or subtract from the mechanical momentum of the rotating coil. The reader is invited to speculate as to what might have happened as the current through the spinning coil was reversed in direction—if nature had included large vortices. What sort of measurable effects would have been observed? And would Maxwell's apparatus, now virtually forgotten, have taken instead a respected place as one of the heroic experiments of modern physics? Is an experiment designed to decide between possible universes less fundamental for having yielded a null result?

752    "would willingly assent as an electrical hypothesis"
Once again, Maxwell pushes our understanding of the nature of "hypothesis" and the possibilities of an imaginative conception—however "provisional and temporary"—as an instrument in the interpretation of phenomena. One senses that he is trying to gain latitude for speculative, uncommitted thought.

776    "an equivalent quantity of mechanical work"
The possibility of converting electrical energy into thermal form suggests a strategy for calibrating electrical instruments in units compatible with standardized mechanical measures (Fig. 3.21), and thereby building a solidly quantitative bridge between these two realms. A meter designed to indicate electrical potential difference (a "voltmeter") measures work per unit charge, while an instrument to indicate current (an "ammeter") measures charge passing per unit time. If the two are connected as shown, the product of their readings will be:

(Work per unit charge) × (Charge per unit time)

or work per unit time, which is *power*. If the circuit is allowed to run for a measured period of time, with readings of the voltmeter and the ammeter held constant, the product of all three will be the work done:

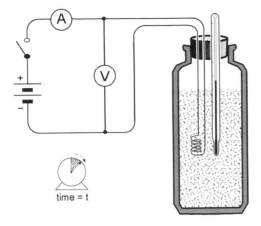

Fig. 3.21. A circuit for electrical calorimetry: A = ammeter (current in amperes); V = voltmeter (emf in volts); and V × A = power (watts)—V × A × *t* = electrical energy = heat energy measured by the calorimeter.

(Work per unit charge) × (Charge per unit time) × (Time)

= (Voltage) × (Current) × (Time)

= (Work per unit time) × (Time) = Work

That is, the total electrical work will have been measured. If all of this electrical work is transformed into heat in a heater coil *H* in a calorimeter and the heat produced is measured, then the combined readings of the electrical instruments will have been verified against the thermal measurement. A few principles of calorimetry are assumed to convert the observed increase in temperature to thermal energy in standard units. The electrical instruments will then have been calibrated with a thermometer and a clock!

787  "no resistance"

If there were as Maxwell imagines *no* resistance to the passage of electricity, we would have the phenomenon of *superconductivity*—no more than a thought in Maxwell's time but of vast practical interest a century later, as it holds the lure of amazing efficiencies in the manipulation of energy. If the particles, which, as massless, are *utterly light,* also met no resistance and so were utterly free, no purchase could be achieved to transmit a motion beyond them, and they would cancel all transmission of effects.

801    "move any part of the medium as a whole"

In this difficult sentence, Maxwell seems to be pointing to the fact that the particles *have no mass* and hence can possess *no momentum*. All the momentum belongs to the vortices, while the currents, whatever their motions, can neither add to nor subtract from this total. Thus if a motion is imparted by the emf to a series of particles *AB,* they gain no momentum and can impart none. Their only effect must be that of a set of massless gears—a strictly *kinematic* constraint.

One immediate conclusion from Newton's third law of motion is that momentum is conserved in any mechanical transaction: since reactions are equal and opposite, total initial momentum is equal strictly to total final momentum. Thus if the vortices were initially at rest and the particles make only a kinematic contribution to momentum transactions, the final momentum of the vortices must similarly be zero. But that is an *algebraic* sum: it would be quite acceptable for half to go one way and half the other. Hence Maxwell's proposition is rightly put—that the emf has no leverage on the vortical medium *as a whole.*

This is reminiscent of the situation at the end of "Faraday's Lines": endowed with no mass, the electric particles are sealed in a world of their own, radically decoupled from the mechanical world "as a whole," and have no part in the reckoning of energies.

Maxwell's particles seem to be precursors of the electron, which we now know carries a certain mass. But this mass is very small, so that in all the immense energy transactions in modern power grids, the energies involved in the currents *per se* are negligible, and Maxwell's point still holds. The energies are never *in* the wires but fill the surrounding spaces. Then, is the interaction of electrical currents and mechanical energies any less of a mystery for us than it seems to be for Maxwell?

802    "When the primary current is stopped"

When the flow of particles in the primary is blocked, the vortices geared to it are immediately locked. All the momen-

tum, not a jot of which can be lost, must be instantly trans-
ferred to the secondary.

814 "We have now shewn"
In the terms Maxwell is using, we have covered one large part of
the subject he is calling "electromagnetism." We have not looked
at electrostatic phenomena; their incorporation will be the subject
of the next section and will stretch the imagination much further.
It appears that he originally intended to terminate the paper at this
point and only after an interval of some eight months ventured to
publish the part that follows.

Maxwell's strain of rather ironic skepticism is well represented
in this formulation: "we have *imitated* the phenomena by an
imaginary system. . . . " We are reminded how far he is from
claiming that he is offering a "true" theory. Would he regard any
theory as more than an "imitation"? This cautious page, one
suspects, might have been occasioned by the discovery mentioned
in Maxwell's "Note," to the effect that Helmholtz had used the
concept of the vortex to account for magnetism in a quite different
way.

836 "to detect their ultimate divergence"
"Faraday's Lines" was made up of fragmentary "convergences,"
as between heat flow and electrostatics or between electrostatics
(applying the mathematical image of the "lines" one way) and
magnetism (applying the same image another way). Here, a much
broader coherence is sought: but it will have its limits. The vortices may be made to agree with electromagnetism and (in the
next section) with electrostatics, but they offer no coherence with
gravity.

As we have seen in "Faraday's Lines," behind these remarks
lies the dream of an "ultimate" complete theory, which would
give rise to no divergence when applied to all the known phenomena of nature. In the earlier passage, Maxwell had spoken of "a
mature theory, in which physical facts will be physically explained." He does not seem to have a firm and consistent goal
toward which each step is an advance; rather the nature of the goal
itself remains a real question. The whole structure of the inquiry
will change, and the question of the goal will be posed anew when

we shift from the "physical" criterion of the present paper to something more starkly formal in the "Dynamical Theory" of the next chapter.

840    "lines of fluid motion"
Helmholtz's proposal is pretty nearly the inverse of Maxwell's, if electric currents rather than magnetic lines correspond to axes of rotation of a fluid medium!

846    "PART III"
*Philosophical Magazine,* January and February 1862.

868    "possesses elasticity of figure"
At this point, Maxwell takes a major new step in the construction of an imagined world. To this point, it has included mass and dissipative resistance but there has been no allusion to any ingredient bespeaking *elasticity*. Even in "Faraday's Lines," for which thermal diffusion was a paradigm borrowed from Fourier's work, it was necessary to acknowledge that while a steel plate might *gradually* rise in temperature and in some formal sense "absorb heat," Maxwell's fluid by contrast everywhere always totally filled its space. Effects at a distance in the fluid might be dispersed spatially and hence decrease with distance but had to be felt at all distances instantaneously. The fluid there was strictly "hard" in this sense. Might we not have supposed up to this point that the fluid of the world of "Physical Lines" was similarly unyielding?

Maxwell asserts rather that "it is necessary to suppose" that the substance of the fluid is elastic. The source of the necessity appears to be the fact of transmission of a tangential force, that is, a force in *shear*. It is thus not simple *compressibility* that is in question but an "elasticity of figure" that permits the transmission of *tangential action*—a force causing a deformation in shear through the mediation of an elastic constant in such a way that Hooke's law for a spring (a tangential stress will give rise to a *proportional* tangential strain) governs (Fig. 3.22). It may, incidentally, be difficult to judge, and perhaps quite unnecessary to decide, whether at this point we are still dealing with a liquid or are thinking of something more like a solid! Does it matter that we have certainly never seen a material like this? Maxwell seems to

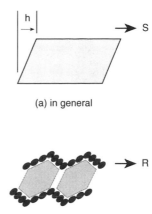

(a) in general

(b) applied to a vortex

Fig. 3.22. Elasticity of figure.

be pushing the limit of thinkability, of the mind's capacity to conceive a material that turns freely in the vortices while rigidly sustaining the tensions of static electricity. Has he gone too far?

880   "If we can now explain"
      Once before, in a note to line 555 of "Faraday's Lines," we discussed an account of electric charge but that was in the world of "Faraday's Lines." There it appeared that the "charge" was not on the surface of the electrified body but was rather a state of the lines of force entirely external to the body. Now, however, we have to start over and our task is more difficult. In "Faraday's Lines" we addressed phenomena one by one without demanding the unity of a single theory. Now we must fit our concept of charge into one embracing account in which the electrical and magnetic concepts cohere. As we come to the theory by way of magnetism, *charge* remains a concept to be added in some new way at the end of the construction.

885   "electric tension is the same thing"
      Faraday had arranged a careful weave of experiments to investigate this point using electrostatic charges, artfully guided, to accomplish things normally done with galvanic currents—and, conversely, using the galvanic cell (or an electromagnetic generator, the "coil machine") to produce "electrostatic" effects.[2] The

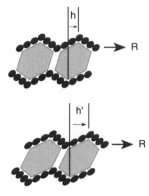

Fig. 3.23. Two perfect insulators with different dielectric properties. They are subject to the same emf *R*, but they yield in proportion to their dielectric constants.

notions of "electromotive force" and "electric intensity" are, however, conceptualy so distinct that Maxwell has his work cut out to bring them together in a single theory. The units of what Maxwell has termed "electromotive force" are not, as we have seen, literally those of "force" at all but rather work per unit charge.

904    "may act differently as dielectrics"
A way in which two different media, each a complete insulator, might "act differently as dielectrics" is suggested in Fig. 3.23. Each, as Maxwell says below, "transmits the pressure" without allowing any continuing current flow, but one transmits more readily and the other less so. The two differ in the "elasticity of figure" of their vortices: the more yielding figure "transmits the pressure" more fully, while the other, resisting, permits less displacement of the electric matter. It seems that the greater or lesser "transmission" amounts to a greater or lesser displacement of the electric matter.

917    "a state of polarization"
It appears that "polarization" consists first of a displacement of the idler wheels from their initial position. We should beware, however, of imputing too much significance to the idlers themselves. They have no powers and bear no energy. In themselves, they cannot serve as the "matter of electricity." Their function is always strictly kinematic, as keys linking the cells, while it is in the cells themselves that all the power associated with electricity or magnetism inheres. In the absence of a magnetic field, the

vortices are at rest; yet a displacement of this resting matter will load it with all the energy of a lightening blast!

`With the release of such electrostatic energy, a discharge current will flow as the idlers surge back to their original position, and by way of these keys the vortices will be set spinning. One of the most powerful images in Melville's *Moby Dick* arises when the discharge of a lightening bolt flows down a mast of the *Pequod* and Captain Ahab wonders to see his compass needle reversed. Thus does an unrecognized fundamental experiment reveal the coupling, through the idlers, of these two powers of the ambivalent medium, at once fluid and rigid, of the field!

931    "its variations constitute currents"

This is an astounding claim and one on which the rest of the theory will depend. The implication of Maxwell's account is that even in absolutely empty space, the variation of the electric strain is equivalent to a current and will give rise to the same magnetic effects as if actual currents were flowing. These in turn, of course, will induce new electromotive forces, so all the interactive processes of electromagnetism must actually be going on—without a need for circuits or conductors—in a total vacuum! Unfortunately, the effect must be so slight that Maxwell saw no prospect of a direct experimental test of the predicted effect.[3] Less directly, however, as he shows next, it has an enormous implication—that of a propagating electromagnetic wave. We will explore these consequences in our discussion of the third paper, on the "Dynamical Theory," in Chapter 4.

In equation (α)* Maxwell writes his Hooke's law relation for the elastic medium with a minus sign, evidently indicating that the displacement $h$ gives rise to a restoring force in the medium in the reverse direction: if the displacement is to the right, the restoring force will be to the left. As has been extensively discussed by commentators, he should probably not have called this restoring force $R$. The $x$-component of the electric field that *caused* the displacement, $R$ acts to the right, not the left. In the subsequent argument in his text, Maxwell seems to have made a second sign error, the two canceling to give a correct result. Fortunately, we will follow the corresponding reasoning in the "Dynamical Theory" paper instead, where this problem does not arise.

960    "I have deduced"

Here Maxwell reveals one of the most fundamental and transforming discoveries in the history of science: his clue to the
identity of light and electromagnetism, two realms which would
seem to have nothing to do with each other. The imagined world
of the vortices has led almost inexorably to this revelation, for to
make it cohere, Maxwell has had to introduce "elasticity of figure"—and the combination of the momentum of the vortices with
the elasticity of their forms has given rise to a vibrating medium.
What could not have been anticipated was that the computed
velocity of propagation of their vibrations would be strikingly
close to newly measured values of the velocity of light.

967    * * *

It behooves us to leave our study of "Physical Lines" at this crucial
point and to pick up the argument in the context of the "Dynamical
Theory" in Maxwell's third paper. What follows in the remainder of
the paper we have just read is a complex sequence of reasoning that
is unfortunately very heavily committed to the detail of his particular
physical model. He must demonstrate the existence of transverse
vibrations in the model, and then use arguments derived from the
physics of vibrating solids, together with certain assumptions about
the nature of his specific medium, in order to arrive at a predicted
velocity of propagation. His attention focuses on two recently obtained sets of data. One concerns the values of electrical units, which
set critical parameters for his physical medium. The other constitutes
the latest figures for the measured velocity of light. The resulting
agreement between his predicted value for the speed of propagation
of his elastic waves and the measured velocity of light is dramatic.
He concludes:

> The velocity of transverse undulations in our hypothetical medium, calcu-
> lated from the electro-magnetic experiments of MM. Kohlrausch and Weber,
> agrees so exactly with the velocity of light calculated from the optical
> experiments of M. Fizeau, that we can scarcely avoid the inference that *light
> consists in the transverse undulations of the same medium which is the cause
> of electric and magnetic phenomena.*[4]

This conclusion, wonderful as it is, is flawed by the fact that it
is so intricately involved with the physics of Maxwell's rather

peculiar imaginary mechanism. The argument is both intricate and insecure. His envisioned mechanism, implausible as it may be, has nonetheless led him to the crucial consequence of the identity of light and electromagnetism. Such a sweeping insight concerning the natural world cries out to be established on less contingent grounds.

In Chapter 4 we turn with Maxwell to a new mode of thought, that which he calls "dynamical." In this next paper there will be no physical hypotheses but reference instead to a mechanical system in only the broadest and surest of terms. The result will be a new, logically much firmer derivation of the electromagnetic theory of light. In these new terms, we will follow Maxwell's argument all the way to the end.

# Discussions

## D1.  On Analytic Mathematics

In Chapter 2 Maxwell was intent on achieving a vision of the lines of force that depended primarily on intellectual intuition and only secondarily on methods of analytic mathematics. It was clear that he was trying to get away from analytic methods as much as possible. Here, in "Physical Lines," it is different. Maxwell's new aim is a construction in *physical* terms, but the road to the physics is inevitably through equations and analytic relations: they will often point the way. So we must come to terms with them to an extent that will permit us to read Maxwell's text with some degree of satisfaction. Your guide's obligation is now very especially focused on that reader for whom analytic methods are new or long forgotten. This is an obstacle that would ordinarily prevent a prudent reader from even thinking of approaching an essay such as "Physical Lines." How can we get such a stumbling block out of our way?

Let us suppose that *algebra* as such is not our central problem. The rules of algebra are not complicated and the elementary techniques are not hard to acquire once one feels any interest in the project. Most people who fail to learn algebra avoid the effort precisely because it seems such a tedious or meaningless affair. In application to reasoning such as Maxwell's, however, it becomes the key to surprising insights. We will assume the basic rules here. However, for anyone stuck at this point, the necessary groundwork can be covered in a few sections of a sensible introductory text of the sort referenced in our Bibliography.

Our real problem is what is called the *calculus,* which has two parts, the *differential* and the *integral* calculus. Real study of the calculus is a major task and lies close to the heart of the serious study of classical mathematics: we will have to approach it quite otherwise, merely to get a sense of what is going on. Fortunately, there are once again excellent texts well adapted to our purposes to which the interested reader may turn.

## I. The Derivative

We begin with the differential calculus by way of a simple example. Suppose you are driving your car and on an open highway have occasion to start up from zero speed. From that starting point, you cover a total distance that increases with time. In the first minute or two, the increase when graphed might look something like Fig. 3.24a. (If your car has an odometer that registers tenths of a mile, the data for the graph could be thought of as its readings plotted against those of a stopwatch. But in any experimenting, keep your eye on the road!)

Now we ask a crucial question: at what *rate* has the distance been increasing? Graphically, that question can be posed in terms of a triangle whose height is $\Delta s$ and whose base is $\Delta t$ (see Fig. 3.24b). The *rate* in this case is *distance increment divided by time increment,* or $\Delta s/\Delta t$—which is the height of the triangle divided by its base. (Here we are using the very convenient delta notation, in which $\Delta s$ denotes simply an *increment in s* or the difference between two *s*-values, and $\Delta t$ a corresponding increment in time.) We can see graphically that $\Delta s/\Delta t$ approximates the *slope* of our $(s,t)$-graph, and that this slope is not constant but increases with time. After all, this *rate* is just our *velocity,* and we are well aware that the velocity increases with time as we accelerate from rest; the graph accords intuitively with our experience.

We note further that the "rate" we are speaking of is only a rough average: smaller triangles would give more accurate measures. At this point, we come to the crucial insight of the differential calculus: as our intuition tells us very clearly, there must be an exact velocity at every instant. *In the limit,* as we made our measuring triangles smaller and smaller, we would arrive at that specific number for the velocity at any moment that was not an average or approximation, but the exact value. This limiting value of $\Delta s/\Delta t$ as $\Delta t$ becomes as small as we please is defined as the *derivative of s with respect to t.*

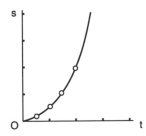

(a) Distance as a function of time.

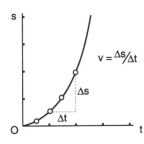

(b) The rate triangle.

Fig. 3.24. Distance *s* (for space) as a function of time.

Let us make our example more definite. Suppose that the distance function in our graph is given by the equation

$$s = kt^2 \qquad (1)$$

with *k* as a constant. Now we can actually calculate the velocity—first as an average and then precisely in the limit—to see how the limiting process works. Be assured, reflection on the reasoning in this example will pay dividends in understanding as we go on! Substitution in equation (1) spells out the following expression for Δs:

$$\Delta s = (s + \Delta s) - s = k[(t + \Delta t)^2 - t^2] \qquad (2)$$

If we expand the expression $(t + \Delta t)^2$ on the algebraic principle that for any *a* and *b*,

$$(a + b)^2 = a^2 + 2ab + b^2$$

and substitute properly, we get

$$\Delta s = k[(t^2 + 2t\Delta t + \Delta t^2) - t^2] \qquad (3)$$

which reduces neatly, since the $t^2$'s cancel out,

$$\Delta s = k(2t\Delta t + \Delta t^2) \tag{4}$$

$$\Delta s/\Delta t = k(2t + \Delta t) \tag{5}$$

This is only the *average* rate, taken over the time interval $\Delta t$. We will zero in on a more precise value if we take successively smaller time intervals. Thus to make our value of $\Delta s/\Delta t$ more accurate, we want to make $\Delta t$ smaller; and to make it *exact,* we want to find its value as $\Delta t$ goes to zero.

*Will* there be any number to represent a value of the quotient $\Delta s/\Delta t$ as numerator and denominator approach as nearly as we please to zero? This is where the mathematician pulls the rabbit out of the hat: "yes," the ratio of these two quantities, each of them vanishing separately, *may* have a finite value—and in this case we feel that it must because we know the car really is going at *some* speed at *every* instant!

It is remarkably easy, algebraically, to accomplish this mathematical miracle in our case. In (5), we merely *let $\Delta t$ go to zero,* and substitute zero for it in the expression. If $v$ represents the velocity at any moment, then we may evidently write

$$v = \text{limit}\left(\frac{\Delta s}{\Delta t}\right) = 2kt, \qquad \text{as } \Delta t \to 0 \tag{6}$$

(Read: "The limit, as $\Delta t$ becomes as small as we please of the quotient $\Delta s/\Delta t$ is $2kt$.")

We may write this more elegantly by introducing a notation for the derivative. Denote the derivative by the single symbol

$$\frac{ds}{dt}$$

and abbreviate "limit" as "lim." Equation (6) then becomes, more elegantly,

$$v = \frac{ds}{dt} = \lim_{\Delta t \to 0}\left(\frac{\Delta s}{\Delta y}\right) = 2kt \tag{7}$$

or simply,

$$\frac{ds}{dt} = 2kt \tag{8}$$

As simple as that! Note that here the notation $\frac{ds}{dt}$ does not denote a quotient of two quantities, but the single quantity that is the limit of a quotient.

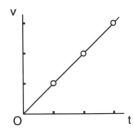

Fig. 3.25. The velocity function $v = 2kt$.

We can now draw an accurate graph of the velocity as a function of time (Fig. 3.25).

## II. Leibniz's Notation

Now we come to an extremely interesting piece of insightful trickery in the realm of notation—a device we owe to Leibniz. Above, we have written $ds/dt$ to denote the derivative, emphasizing that *this is a single symbol:* $ds$ and $dt$ here do not have any meaning separately—the combined symbol means one number.

If we agree never to forget the above ultimate stipulation, we can work very effectively with $ds$ and $dt$ *as if* they were separate numbers. Our agreement takes this form: when we write $ds$ by itself, we will mean a very small but finite $\Delta s$ (not an "actual infinitesimal"!), and we solemnly agree that any expression that results will be carried through to the limit, where it will be strictly true. This $ds$ is a small quantity (called a *differential*)—but when we speak of it, we affirm on oath that the resulting expression will refer to an expression in the limit. *Not* the limiting value of $ds$ itself—for that is always zero—but of an expression such as the derivative $\Delta s/\Delta t$, which as we have seen may have a definite value even though both the top and the bottom of the fraction go relentlessly to zero. With this preparation, we will find that differentials can be extremely convenient to work with.

Just to pull all of this together, we might note that our velocity function has a rate, too. Since it is a straight line, its rate is constant:

$$\frac{dv}{dt} = k \tag{8}$$

The rate of change of velocity is *acceleration,* so we conclude that we must have kept our foot remarkably steady on the accelerator. Our set of three graphs may be conveniently collected, as in Fig. 3.26, to show the three stages of our analysis: the distance function, its *first derivative* (the

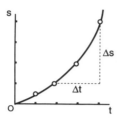

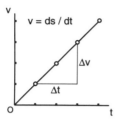

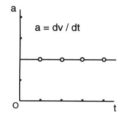

Fig. 3.26. Three stages of differentiation: (1) equation of motion (distance): $s = kt^2$; (2) first derivative (velocity): $v = 2kt$; and (3) second derivative (acceleration): $a = 2k$.

velocity function), and its *second derivative* (the derivative of its derivative), the acceleration function.

## III. General Rules of Differentiation

THE PRODUCT RULE: We have looked at one particular function as an example, but it would be nice to have this result as a more general rule—and in fact we can readily demonstrate one that will take us a long way with our use of the calculus. All we have to do is to substitute $n$ for 2 in the reasoning we used above. In general, it will not always be *time* that is the independent variable for our rates (in fact, most of Maxwell's equations involve space-rates), so let us now generalize our relation. In place of $s$ as a function of $t$, let us write $y$ as a function of $x$:

$$y = b\,x^n$$

in which $b$ represents any constant multiplier.

Using the same methods (and following the same steps) as in our earlier example, we find that

$$\frac{dy}{dx} = nbx^{(n-1)}$$

as you might want to verify for yourself. With the aid of one algebraic device, we can put this to immediate use and repay a debt to ourselves from Chapter 2. The trick concerns negative exponents. We know what it means to raise a number to a power, denoted by an exponent, but what will it mean to raise a number to a *negative* exponent? The answer is a matter of definition. By *definition*,

$$x^{-n} = \frac{1}{x^n}$$

Now, if we go back to Discussion [D15] of Chapter 2, concerning the variation of pressure in flow from a point source, we see that what we really wanted to do there was to differentiate the function

$$p = \left(\frac{k}{4\pi}\right)\left(\frac{1}{r}\right)$$

in which the quantity $k/4\pi$ is constant. By our definition of the negative exponent, this can just as well be written

$$p = \frac{k}{4\pi r} = \frac{k}{4\pi}\left(r^{-1}\right)$$

Differentiating with respect to $r$ and applying our rule with $n = -1$, we get

$$\frac{dp}{dr} = -\frac{k}{4\pi}\left(r^{-2}\right) = -\frac{k}{4\pi r^2}$$

which was the result we had to accept with less explanation in Chapter 2.

THE CHAIN RULE: We will need to know what is called the *chain rule* for differentiation: If $w$ is a function of $v$, while $v$ in turn is a function of $x$, and we wish to find the derivative of $w$ with respect to $x$, we have to go through two stages: If $w = f(v)$ while $v = g(x)$, then

$$\left(\frac{dw}{dx}\right) = \left(\frac{dw}{dv}\right)\left(\frac{dv}{dx}\right)$$

In other words, we first differentiate $w$ with respect to $v$, and then multiply by the derivative of $v$ with respect to $x$. This is provable by an argument based on the concept of the limit, but the demonstration involves no surprises and we may accept it provisionally here as an intuitively reasonable principle.

RULES FOR SUMS AND PRODUCTS:   Straightforward and rather uninspiring application of the limiting process yields rules for algebraic forms:

DERIVATIVE OF A SUM:

$$\frac{d}{dx}[f(x) + g(x)] = \frac{df}{dx} + \frac{dg}{dx}$$

DERIVATIVE OF A PRODUCT:

$$\frac{d}{dx}(u \cdot v) = v\frac{du}{dx} + u\frac{dv}{dx}$$

It follows from this that a constant multiplier factors out in differentiation. If $b$ is a constant and

$$\frac{db}{dx} = 0$$

then

$$\frac{d}{dx}[b \cdot f(x)] = b\frac{d}{dx}f(x)$$

We will make use of these rules freely as needed in all that follows.

IV. Partial Differentiation

Things get a little more complicated when one thing is a function of *two or more* other things (more formally: one dependent variable is a function of two or more independent variables), as tends to be the case in the real world. In fact, we have been dealing with many cases of this sort; For example, in Fig. 2.2, we diagramed the temperature over the surface of a rectangular plate. If we call the temperature $z$, and the dimensions of the plate $x$ and $y$, then

$$z = f(x,y)$$

where $f(x,y)$ is the function Fourier was looking for, while the quantity $z$ is here a variable dependent on both $x$ and $y$. Thus for any fixed value of $x$, there will be various values of $z$ as a function of $y$, and for a fixed value of $y$, there will be a range of values of $z$ as a function of $x$. To analyze the behavior of such a function, we may first hold $y$ constant and vary $x$ and then hold $x$ constant and vary $y$. The rate

$$dz/dx \text{ [}y\text{ const]}$$

is called the *partial derivative* of $z$ with respect to $x$, while

$$dz/dy \; [x \; const]$$

is the partial derivative of $z$ with respect to $y$. Algebraically, the variable being held constant is treated as a fixed number, so no new problem is presented. Geometrically, holding one variable constant at a chosen value can be thought of as slicing the three-dimensional figure with a two-dimensional knife. In modern notation, partial derivatives are usually written with a curly notation

$$\frac{\partial z}{\partial x}, \qquad \frac{\partial z}{\partial y}$$

but Maxwell does not use a distinct notation for them—so perhaps it is as well if in this commentary we do not either.

V. Integration: the Inverse Process

In our sample problem of the accelerating automobile, we began in a sense with the outcome and analyzed the function to find its rates of variation. Suppose we had started the other way—begun with the rate of acceleration, found from it the velocity, and finally from the velocity function found the distance covered? That would be more like the order of events in reality—we press the accelerator, achieve as a consequence a succession of velocities, and thus cover gradually increasing distances. The constant acceleration is in a way the secret of it all. Each velocity step contributes something to the distance covered, and the actual distance at any moment is the summation of all these increments.

In this sense, we would be seeking the *limit of a sum,* as the intervals are again taken smaller and smaller. This inverse process of building a function by summation is called *integration*—it amounts simply enough to reading the graphs of Fig. 3.26 in the reverse order, from bottom to top. In addition of course, the algebraic rule for differentiation of a given function can be inverted to become a rule of integration. We will need this rule, and a notation as well. We can begin with a simple function for which we already know the derivative: If

$$y = x^2$$

then

$$dy/dx = 2x$$

Inversely, we say that the *integral* of the function $2x$ is $x^2$ and write it in this way:

$$y = \int 2x\,dx = x^2$$

which we may read: "the integral of the function $2x$ with respect to $x$ is $x^2$, as we can verify by differentiating again. To be a little more general, we note that it would not make any difference to the derivative if we were to add a constant to our integral, as the derivative of a constant is zero:

$$y = \int 2x\,dx = x^2 + c$$

for

$$\frac{d}{dx}\left(x^2 + c\right) = \frac{d}{dx}\left(x^2\right) + \frac{d}{dx}(c) = 2x + 0 = 2x$$

The general rule for integration will thus be

$$\int x^n\,dx = \frac{x^{(n+1)}}{(n+1)} + c$$

This represents the limit of a sum:

$$\lim(\Delta x \to 0) = \sum x^n \Delta x = \int x^n\,dx$$

As the summation is taken between the least and greatest values of $x$, a "definite" integral must be evaluated between these as upper and lower limits. The value of the definite integral is the difference between the values with the upper and the lower limits substituted for $x$.

It would be interesting and valuable to say more now about integration, but our little discussion of the calculus is already becoming rather long, and Maxwell writes his expressions primarily in terms of derivatives. They are thus *differential equations.* If the constant acceleration was as we said the "secret" of the highway experience, ferretted out by differentiation, perhaps Maxwell is similarly writing equations in the form of derivatives in order to speak most directly of the secrets, or principles, underlying the phenomena of electromagnetism. Differentiation feels like a form of unveiling of the manifest phenomena. We can continue this discussion in terms of the actual examples as we meet them in Maxwell's essays.

VI. Operator Notation
It is sometimes helpful to think of the derivative as an "operator"—that is, invoking a mathematical operation to be carried out on a function, just

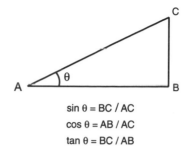

$$\sin \theta = BC / AC$$
$$\cos \theta = AB / AC$$
$$\tan \theta = BC / AB$$

Fig. 3.27. The trigonometric functions.

as squaring or taking the square root is an operation. In this sense, the operator $d/dx(\ )$ invokes *differentiation* of the function on which it operates. Thus,

$$\frac{d}{dx}\left(x^2\right) = 2x$$

## VII. A Note on Trigonometry

It may be useful to add a note on the definitions of the trigonometric functions. The simplest definitions are in terms of the right triangle $ABC$, with its two sides $AB$ and $BC$, and the hypotenuse $AC$. Then three functions of the angle $\theta$ are defined, the sine, cosine, and tangent of $\theta$ (written sin $\theta$, cos $\theta$, and tan $\theta$).

The values of these functions of $\theta$ are computed by methods that are in truth quite tricky, yet the results can be extracted readily from most pocket calculators. A pioneer trickster was Ptolemy, who, from scratch, using what must have been a roomful of human computers, generated a table of the sine function, essential for his computations of arcs observed in the heavens.

The trigonometric functions, once tabulated, permit finding one part of the triangle when two others are known. Thus we can write

$$BC = AC \sin \theta$$

$$AB = AC \cos \theta, \qquad \text{etc.}$$

It is also important to keep firmly in mind the *Pythagorean theorem* (as a geometrical theorem to be found in Euclid's *Elements*, Book I, Prop. 47):

$$AC^2 = AB^2 + BC^2$$

The above account gives a rather static view of the trigonometric relations. They become much more interesting when viewed in their aspect as generating *functional relations*. The angle θ is then taken as an independent variable, while sin θ becomes a continuous function of that variable; thus: $y = k \sin θ$. The values of $y$ constitute a periodic function, oscillating between extremes of +1 and –1.

## D2.  On Newton's Laws of Motion

In "Faraday's Lines," Maxwell deliberately defined a medium that had no mass—and that was thereby exempted from Newton's laws of motion. Now, by contrast, he will be conjuring a medium that includes mass and as a consequence becomes subject to Newton's laws. We should, therefore, look at those laws before we begin reading "Physical Lines." Again, as in the case of analytic math, excellent books are available for those who want to give more attention to the foundations of mechanics (see the Bibliography). Newton's first law reads as follows[5]:

> LAW I: Every body continues in its state of rest, or of uniform motion in a right line, unless it is compelled to change that state by forces impressed upon it.

Simple as it seems, this law defines the universe we experience directly: everything that we familiarly call "matter," larger than the quantum world and smaller than the astronomical orbits, rigidly obeys it. It is enough for us to recall by contrast the imagined world of "Faraday's Lines," in which matter was not thus massive and came to rest immediately when a force was removed. Newton speaks for that real world hidden behind the world of commonsense and revealed by strategic experimentation. In truth, no force is required to *keep* a body in motion. Once started, a body continues in motion with no help; a force would be required to bring it to rest. When we see a moving body seemingly bring itself to rest, we know that some force, an unseen friction, has acted upon it.

> LAW II: The change of motion is proportional to the motive force impressed; and is made in the direction of the straight line in which that force is impressed.

It seems we must understand Newton to mean: *in a given time.* Then the *change of motion* becomes the motion or momentum acquired *per unit time.* Momentum is the product of a body's *mass* times its *velocity;* Newton's "motive force" is today simply termed "force." Then the second law becomes, in words:

LAW II': The time-rate of change of momentum is proportional to the force impressed, or in symbols:

$$f = \frac{\Delta(mv)}{\Delta t}$$

where $\Delta(mv)$ is a change in momentum and $\Delta t$ is the time interval in which this change occurs.

This gives an average value of the force over the time $\Delta t$; we could make the expression precise by going to the limit as $\Delta t$ shrinks to zero. The second law would then be formulated as a derivative giving the value of the force at any instant:

$$f = \frac{d(mv)}{dt}$$

Since mass is normally constant, this can be written

$$f = m\frac{dv}{dt}$$

However, we have saw in Discussion [D1], $dv/dt$ is simply the *acceleration*. A third wording of the second law will thus be simply:

LAW II'': Force is equal to mass times acceleration,

or

$$f = ma$$

In the case of our automobile in [D1], the constant acceleration was due to the application of a constant force. We must always mean *net* force, which here will be the force applied by the engine minus such forces as those of wind or road friction. Law II, similar to Law I, is universal: no body within the broad size range of its application can escape it. Having made that claim, we must acknowledge that in "Physical Lines," Maxwell will introduce, as we shall see, one class of particles whose mass he supposes to be negligible. Their physics must have a dubious relation to Newton's world.

If we multiply both sides of the equation above by a length of time $\Delta t$ during which the force $f$ acts, we will get

$$f \Delta t = m(a\Delta t)$$

Note that since

$$a = \Delta v / \Delta t$$

by the definition of acceleration, if we solve for $\Delta v$:

$$\Delta v = a\Delta t$$

and

$$f\Delta t = m\Delta v = \Delta(mv)$$

The right-hand side here is the change in momentum and the left-hand side—a force acting through a time—is given the name "impulse." If we combine the effects of force and time in a single concept, an impulse is like a shove, a nudge, or a kick, indeed, anything from the flick of a finger to a hammer blow, Hence the second law can be stated:

Impulse = Change in momentum

Finally:

LAW III: To every action there is always opposed an equal reaction: or, the mutual actions of two bodies upon each other are always equal, and directed to contrary parts.

Our first thought must be that if every action is opposed by an equal and opposite reaction, nothing can ever happen in Newton's world! Some misunderstanding must be involved here. Newton means rather that if a horse pulls a cart by means of a rope, the rope pulls backward on the horse as much as the horse pulls forward on the rope. Similarly, the cart pulls back on the rope as much as the rope pulls forward on the cart. But the cart accelerates nonetheless. Let us suppose for clarity that the cart moves without friction: it will resist the pull of the rope, according to Law II, by the amount $ma$. That is, far from being kept at a standstill, the frictionless cart resists the rope precisely *by virtue of its acceleration.*

Newton's three laws together, with all sorts of corollaries and theorems that follow from them (the burden, that is, of Newton's *Principia*), express the connectedness of the material world. They apply, we know now, only to the limited domain of bodies that are neither too small (in which case we need quantum mechanics) nor too large or too fast (in which case relativity theory takes over). In "Physical Lines"—in contrast with "Faraday's Lines," whose world is massless—Maxwell constructs a world strictly governed by Newton's laws.

We should take note of an important outcome of the argument of the *Principia,* the principle of universal gravitation: at a given distance, any two bodies attract one another with a force proportional to the product of their masses. In turn two bodies are attracted to any third body, such as the earth, each with a force proportional to its mass. This proportionality of gravity to mass accounts for the outcome of Galileo's Pisa experiment—if in the second law force is rigorously proportional to mass, then acceleration is the same for all bodies. The argument of "Physical Lines," however, seems to be bringing Maxwell into confrontation with a body, the ether, to which the law of universal gravitation cannot apply, for it functions by virtue of being massive but we do not observe it to have any weight. This will be discussed in our note to line 707.

## D3. On the Ratio of Pressures in Two Vortices [line 167]

Maxwell first demonstrates that the *ratio of pressures* in two vortices is independent of their linear dimension, *l.* This means that the size of the vortices in his fluid model will not matter for present purposes—it is a factor he will not need to worry about in constructing a mechanical model for electromagnetism. (Later, in another context, the question of the size of the vortices becomes very interesting; Maxwell investigates it with the instrument illustrated in Fig. 3.20.) The argument here seems straightforward as he gives it: the point is that *l,* the ratio of the linear dimensions, has canceled out and thus does not appear in the final expression.

The next expression, for the pressure at the circumference, requires more cautious reasoning. The question is this: if a mass is swung on the end of a string, how much tension is there in the cord—centripetal force if we are thinking of the string holding the mass to the center or centrifugal if we are thinking of the mass straining outward? For Maxwell, this is the first question that has to be answered as he sets his fluid medium spinning. Interestingly, it was also the first question Newton had to answer at the threshold of celestial mechanics: for him, the *mass* was that of the moon and the *tension* was exerted by that invisible cord, the pull of gravity. Newton's argument is given in Proposition 4 of the *Principia,* whose enunciation reads:

> The centripetal forces of bodies, which by equable motions describe different circles, tend to the centers of the circles; and are to each other as the squares of the arcs described in equal times divided respectively by the radii of the circles.

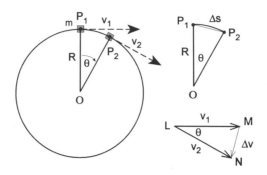

Fig. 3.28. Centrifugal force.

Algebraically,

$$f = \frac{mv^2}{R}$$

Despite the far-reaching significance of this result, it is not too hard to derive. Let the mass $m$ move with velocity $v$ on the circle whose radius is $R$ and whose center is $O$ (Fig. 3.28). As the weight follows the curve of the circle from $P_1$ to $P_2$, its velocity changes in direction so that $\Delta v$ will be the vector as drawn—one side of triangle $LMN$. At the same time, the weight traverses the distance given approximately by the line $\Delta s$ in the triangle $OP_1P_2$. However, the two triangles are similar, making their sides proportional. Letting $v$ represent the common magnitude of $v_1$ and $v_2$, we have

$$\frac{\Delta v}{v} = \frac{\Delta s}{R}$$

But by the definition of acceleration, a change in velocity is the product of the acceleration and the time elapsed, i.e.,

$$\Delta v = a\Delta t$$

and, similarly, the distance traveled is the velocity multiplied by the time interval:

$$\Delta s = v\Delta t$$

If we substitute these two relations in our proportion above, we find that the time interval $\Delta t$ cancels out:

$$\frac{a\Delta t}{v} = \frac{v\Delta t}{R}$$

$$\frac{a}{v} = \frac{v}{R}$$

$$a = \frac{v^2}{R}$$

Now we know the *acceleration* toward the center, and by the second law,

$$f = ma$$

or

$$f = \frac{mv^2}{R} \qquad \text{"QED"} !$$

This becomes the basis for all our calculations on the dynamics of Maxwell's complex fluid medium. The reader may be concerned— probably ought to be—by the fact that we reasoned about the *chord* of the arc above when we ought to have used the *arc* itself. Actually, our argument applies strictly to instantaneous velocities, accelerations, and forces, i.e., applies strictly *in the limit,* and one of the demonstrations at the outset of the *Principia* is that in the limit, the arc and its chord are equal.

## D4.  On Centrifugal Force [line 171]

We can translate the result of the previous Discussion [D3] to the question of pressure in a *vortex* by means of a diagram that looks at a portion of the vortex at any distance $r$ from the center as in Fig. 3.29a. We take a pie slice cut out by an angle at the center that we label $d\theta$; we can give it some arbitrary thickness $dz$ and call the interval it marks out along the radius $dr$.

Before we go further, we may take advantage of an insight beloved of mathematicians. Everything is simplified if we measure angles in *radians*. The measure of an angle in radians is defined as the arc length cut off, divided by the radius. Thus here, the measure of angle $d\theta$ is $ds/r$, as shown in Fig. 3.29b. In the present case, this device makes it possible for us to write

$$ds = r\, d\theta$$

very much simplifying our work.

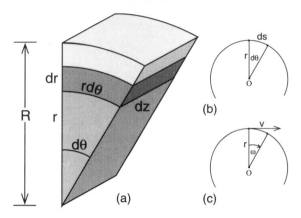

Fig. 3.29. Relations in the vortex: (a) the "wedge" to determine pressures; (b) radian measure of an angle; and (c) angular velocity.

The small mass to which we now can apply our equation for centripetal force thus has the sides $dr$, $dz$, and $r\,d\theta$. That being the case, if the density of the fluid is $\rho$, then the mass will be the *density* times the *volume* of the small box:

$$dm = \rho(r\,d\theta\,dz\,dr)$$

We can apply our equation to find the centripetal force on this mass; call $df$ the net force on the small mass $dm$:

$$df = (dm)\frac{v^2}{r}$$

$$df = (\rho r\,d\theta\,dz\,dr)\frac{v^2}{r}$$

and canceling the $r$'s:

$$df = (\rho\,d\theta\,dz\,dr)\,v^2$$

This being a fluid continuum, what we really want is pressure not force. We remember that pressure is *force per unit area*:

$$dp = \frac{df}{Area} = \frac{(\rho r\,d\theta\,dz\,dr)v^2}{r\,d\theta\,dz}$$

$$dp = \frac{\rho v^2 dr}{r}$$

The mass to which we apply our equation for centripetal force is now the small segment whose sides are *dr, dz,* and *r dθ.*

We now extend our "radian" trick in the way shown in Fig. 3.29c. We saw that in radians

$$d\theta = \frac{ds}{r}$$

Now we define the *angular velocity* ω of a revolving body in *radians per second,* thus dividing *dθ* by *dt:*

$$\omega = \frac{d\theta}{dt} = \frac{1}{r}\left(\frac{ds}{dt}\right)$$

$$\omega = \frac{v}{r} \qquad \text{[radians/second]}$$

The value of this device is that although the velocity varies everywhere in the vortex, the whole vortex has a common angular velocity ω. We do well, then, to replace *v* by ω in our pressure expression above, substituting the appropriate relation:

$$dp = \frac{\rho dr v^2}{r} = \frac{\rho dr (r\omega)^2}{r}$$

$$dp = \rho \omega^2 r dr$$

This gives us the small increment of pressure due to each mass *dm,* but the total pressure at the circumference will be the sum of all these elementary pressures. Our use of the differential notation *dp,* etc., reminds us that we are making an argument that is understood to be valid *in the limit.* Our summation must then be an *integration:*

$$p = \int \rho \omega^2 \, r dr$$

where the limits of the summation will be between the center *O* and the full length of the radius *R.*

Since ρ is constant and ω is the same for the whole vortex, we can take $(\rho \omega^2)$ as a multiplying constant, outside the integral, and then apply the power rule for integration:

$$p = \rho\omega^2 \int r\,dr = \rho\omega^2 \left(\frac{r^2}{2}\right)$$

To evaluate the integral between limits as a definite integral, we substitute the lower and upper limits in the expression for the integral. Here the integration is from the center, at which $r = 0$, to the circumference, where $r = R$. Our lower and upper limits are then 0 and $R$:

$$\int_0^R r\,dr = \frac{r^2}{2}\Bigg|_0^R = \left[\frac{R^2}{2} - 0\right] = \frac{R^2}{2}$$

Finally, remembering that $\omega = \frac{v}{r}$, we have our answer:

$$p = \rho\omega^2 \left(\frac{R^2}{2}\right) = \tfrac{1}{2}\rho(\omega^2)R^2 = \tfrac{1}{2}\rho\left(\frac{v}{R}\right)^2 R^2$$

$$p = \tfrac{1}{2}\,\rho v^2$$

which is Maxwell's result. This is the total centrifugal pressure at the periphery, radially outward from the axis, which he calls $p_1$.

He contrasts to this the pressure parallel to the axis, denoted $p_2$, which he represents by an average value. A rather fine distinction must be made in order to evaluate this average. We have just shown that the outward pressure varies with $r^2$, and this is the pressure that exists in each "box" bounded by $(r\,d\theta)$, $dz$, and $dr$ in the "wedge" of Fig. 3.29. In the axial direction, this pressure is exerted on the face $(r\,d\theta)$, $dr$; we need to think of the total effect of all such faces. Fortunately, we have already developed a simple relation:

$$dp = \rho\omega^2 r\,dr$$

For uniform steps $dr$, $dp$ is directly proportional to $r$. For such a linear progression, the value at the midpoint, at $R/2$, will be the average or mean. But the axial pressure $dp$ at that point will be just one-half the axial pressure at the circumference. This argument pertains to the way in which the axial pressure contributions $dp$ vary with $r$. The actual pressure at the rim of the vortex will be the value we obtained above

$$p_1 = \tfrac{1}{2}\,\rho\,v^2$$

one-half of this, the mean axial pressure, will be:

$$p_2 = \tfrac{1}{4}\,\rho\,v^2$$

Throughout the medium, the vortices will generate radial pressures $p_1 - p_2$ above this ambient value at their boundaries, so that the distinctive centrifugal pressure will be

$$p_1 - p_2 = \tfrac{1}{4}\,\rho\,v^2$$

This completes the construction of the dynamical fluid medium as far as circular vortices are concerned.

## D5. On the Concept of Stress [line 222]

If we think of a perfect fluid as flowing freely in all directions, then when the fluid is at rest, pressure similarly equalizes in all directions—this is the kind of pressure Maxwell is calling "hydrostatic." A single, scalar quantity measures such pressure at any point. Such was the fluid of "Faraday's Lines," which lacked any physical complexity and accordingly sustained only scalar, hydrostatic pressures.

Stressed solids and physical fluids that are undergoing accelerative motions can on the other hand sustain internal forces that do not reduce to such a simple situation. At a given point in the body of the fluid, the value of the pressure we measure may then depend on the direction in which we make the measurement. This is just the situation Maxwell has very deliberately constructed by setting the vortices spinning—he now has a situation which he analyzes (lines 167 ff) as a tension $t$ along the axis of the vortex superimposed on a hydrostatic pressure $p_1$ acting uniformly in all directions (Fig. 3.2).

The problem of resolution of the resulting directional stress is more complex than the resolution of a simple vector. Think of a test plane, the wall of a small sampling volume aligned with the coordinate axes, to which the axis of the vortex may have any orientation (Fig. 3.7a). In the figure, the axis of the vortex is represented by the vector **v** Two stresses now act on the test plane: a uniform pressure $p_1$ acts inward, while a tension $t$ that varies with the orientation of the vortex acts outward. The net stress on the plane, taking tension as positive and inward pressure as negative, will be the difference $t - p_1$. Since here $p_1$ is constant, it is the resolution of the term $t$ that we must consider.

Maxwell shows (line 211) that when the axis of the vortex is normal to the test plane, $t$ has its maximum value

$$t = kv^2 \qquad\qquad (a)$$

and the net tension on the test face will thus be $kv^2 - p_1$. Now suppose that the plane is turned so that the axis of the vortex makes the angles $\theta$ $\eta$, and $\gamma$ with the coordinate axes. The pressure $p_1$ will remain the same, but we must determine how $t$ will vary, that is, we must resolve the tension $t$. Its resolution is different from that of a vector. The resulting net stress on the surface will be

$$p = t - p_1$$

where $t$ is a quantity whose variation we must now determine.

A vector acts at a point, but a stress is distributed over a surface as a force *per unit area*. We may think of it as a stream in which our test plane is inserted. We must then ask how much of that stream the test plane intercepts in any given orientation—edge-on, it would intercept none at all. Looking once more at Fig. 3.7a, we must project the area of the the vortex, on which the tension acts directly, onto our measuring surface. To do so, we must multiply by the cosine of the angle that $\mathbf{v}$ makes with the normal to that surface (if they align, $\cos 0 = 1$; edge-on, $\cos 90° = 0$).

It is useful to name the three measuring planes, which are the faces of the test box of Fig. 3.7, by the axes normal to them. Thus the plane whose edges lie along the $y$- and $z$-axes is the $x$-plane, that with edges along the $x$- and $z$-axes is the $y$-plane, and the remaining face is the $z$-plane. The total tension acting on each of these planes may be given as:

$$kv^2 \cos \theta \qquad \text{on the } x\text{-plane}$$

$$kv^2 \cos \eta \qquad \text{on the } y\text{-plane} \qquad \text{(b)}$$

$$kv^2 \cos \zeta \qquad \text{on the } z\text{-plane}$$

In addition to this reduction of the stress because of reducing the intercepted area, we have also the usual resolution of the remaining stress into components, since it is acting obliquely to the axes. For this, we resolve the stress into components just as we would a vector. Thus, to get the normal stress on each plane, we must multiply again by the cosine of the angle between the stress and the normal to the surface—the same factor by which we projected the area. Thus the normal component of the tensile stress on the three planes is given by:

$$kv^2 \cos^2 \theta \qquad \text{on the } x\text{-plane}$$

$$kv^2 \cos^2 \eta \qquad \text{on the } y\text{-plane} \qquad \text{(c)}$$

$$kv^2 \cos^2 \zeta \qquad \text{on the } z\text{-plane}$$

These are the normal stresses. The remaining components generate what are termed *shear* stresses—stresses that act not perpendicularly to the surface as pressures do but within the surface, parallel to its edges. These are computed by taking the components in the directions of the edges, just as we would take the corresponding components of a force—to get the y-component we will multiply by the factor (cos η), and to get the z-component. by the factor (cos ζ). All told, the three components of the stress *t* acting on the x-plane will be:

$$kv^2 \cos^2 \theta \qquad \text{in the } x\text{-direction}$$

$$kv^2 \cos \theta \cos \eta \qquad \text{in the } y\text{-direction} \qquad \text{(d)}$$

$$kv^2 \cos \theta \cos \zeta \qquad \text{in the } z\text{-direction}$$

We may now return to the original problem of writing the total stress on a surface, namely

$$p = t - p_1$$

in which *t* varies in the way we have just seen. A double-subscript notation helps to keep track of the resulting situation. We write a stress such as *p* with two subscripts, the first of which identifies the sampling plane (which plane the stress is acting on) and the second, along which axis this action is being resolved:

$$p_{xy}$$

The normal stresses are thus denoted $p_{xx}$, $p_{yy}$, and $p_{zz}$, and using Maxwell's notation $\cos \theta = l$, $\cos \eta = m$, and $\cos \zeta = n$ we have for the normal stresses:

$$p_{xx} = kv^2 l^2 - p_1$$
$$p_{yy} = kv^2 m^2 - p_1$$
$$p_{xx} = kv^2 n^2 - p_1 \qquad \text{(e)}$$

The shear stress acting in the x-plane in the y-direction will be

$$p_{xy} = kv^2 lm \qquad \text{(f)}$$

where *l* projects the area onto the x-plane, while *m* takes the component of *v* in the y-direction. Note that the hydrostatic pressure $p_1$ does not act in shear, Similarly, if we project onto the y-plane but take the x-component, we will have

$$p_{yx} = kv^2 ml \qquad (g)$$

Here the equality

$$p_{xy} = p_{yx} \qquad (h)$$

results from the algebra. Maxwell introduces the single symbol $s$ for this shear force, so that

$$s = p_{xy} = p_{yx} = kv^2 lm$$

## D6. On the Law of Equilibrium of Stresses [line 243]

Maxwell's equation (3)* develops fairly readily from Fig. 3.7. The problem is to express the conditions for equilibrium of the test box drawn around point $O$ in the figure. In general, any portion of a fluid *per se* is not necessarily in equilibrium, but if it is not, the fluid will tend to move and deform. Maxwell needs a medium that—however much it is bursting with activity within—will be stable as a whole, for the ether has nowhere to go! Thus any portion larger than the vortices themselves must be balanced by exactly those external forces that will hold it in equilibrium.

Note that although we draw a finite box, we denote its sides as differential elements $dx$, $dy$, $dz$, indicating that we will carry the argument to the limit. We are interested, then, in the state of affairs at a point $O$: this may be hard to believe as it is hard to envision such complexity, including the rates we are about to speak of, concentrated at a point. But that is just the richness of this dynamical fluid!

In Fig. 3.8, in which we assume for simplicity that all forces act only in the x-direction, there is a difference between the pressures on the right and left faces of the box. That means that the pressure $p_x$ is not constant, but changing throughout the space. Its space-rate of change with $x$ is

$$\frac{dp_x}{dx}$$

and the actual difference in pressure from the left to the right sides of the box will be that rate multiplied by the interval $dx$:

$$dp = \left(\frac{dp_x}{dx}\right) dx$$

This will give rise to a *force* that is this pressure times the area *dydz*, over which it acts:

$$df_p = (dp)(dydz) = \left(\frac{dp_x}{dx}dx\right)(dydz) = \left(\frac{dp_x}{dx}\right)(dxdydz)$$

Note how rearrangement of the bracketing expresses the force as acting on the volume element *dx dy dz*.

Shear forces in the two horizontal faces also act in the *x*-direction. Since shear acts on the whole face, in this case the difference *ds* will be that between the top and bottom faces, that is, the variation will be with *y*, rather than with *x*:

$$ds = \left(\frac{ds}{dy}\right)dy$$

Again, this shear will give rise to a net force that is *ds* times the area of the face over which it acts, namely *dxdz*:

$$df_s = \left(\frac{ds}{dy}\right)(dxdydz)$$

Note that *s* may also vary with *x*, while $p_x$ may also vary with *y*, but these so-called second order variations vanish in the limiting process and thus do not contribute to the present calculation.)

Thus the total unbalanced force in the *x*-direction, $f_x$, is the sum of these two terms, the volume element neatly factoring out:

$$f_x = df_p + df_s = \left(\frac{dp_x}{dx}\right)(dxdydz) + \left(\frac{ds}{dy}\right)(dxdydz) = \left(\frac{dp_x}{dx} + \frac{ds}{dy}\right)(dxdydz)$$

The total force *X* of which Maxwell speaks is the *force per unit volume* of the fluid at point *P*. We arrive at this by dividing our force $f_x$ by the volume of the test box, *dxdydz*:

$$X = \frac{f_x}{dxdydz} = \frac{dp_x}{dx} + \frac{ds}{dy} \tag{3)*}$$

To appreciate the true complexity of these stress relations, we need to remind ourselves that this relation, written for the *x*-direction, holds at the same point for the *y*- and *z*-directions as well.

As we already have expressions for $p_x$ and *s*, equation (3)* tells us to perform the indicated differentiations on them—a derivation that as it turns out involves some interesting mathematical trickery. It is carried out in the following Discussion [D7].

## D7. Derivation of Equation (4)* [line 247]

The derivation of (A)*, (B)*, (C)*, and (D)* is instructive not only as an interesting application of the rules of differentiation, but as an example of creative use of the rules of algebraic transformations. Manipulation of algebraic terms often seems dull and routine work, but employed with Maxwell's resourcefulness these manipulations become at times instruments of expression, fetching significance out of seemingly meaningless agglomerations of symbols. It is all done through the substitution of algebraic expressions for their equivalents, rearrangements of symbols for tactical purposes.

If rhetoric is the art of saying the same thing in interestingly different ways, what we are seeing here is the creative use of mathematical rhetoric—an art in which Maxwell excels. Let us follow these steps carefully in this instance, then, as a case study in the application of mathematical rhetoric.

The beginning is with two of the equations of group (2)*, line 233:

$$p_x = k\alpha^2 - p_1$$

from (2)*

$$s = k\alpha\beta$$

These tell us how the $x$-component of stress, $p_x$, varies along the $x$-direction and how shear $s$ varies in the $y$-direction.

Further, we have the general equation (3)*, line 244, that tells us how the net force in the $x$-direction, denoted $X$, depends on these two quantities:

$$X = \frac{dp_x}{dx} + \frac{ds}{dy} \tag{3*}$$

To begin, we simply substitute from (2)* into equation (3)*, and then go to work by applying the rules of differentiation to the resulting expression. Just substituting yields

$$\frac{dp_x}{dx} = \frac{d}{dx}\left(k\alpha^2 - \frac{dp_1}{dx}\right) \tag{i}$$

$$\frac{ds}{dy} = \frac{d}{dy}(k\alpha\beta) = k\beta\frac{d\alpha}{dy} + k\alpha\frac{d\beta}{dy} \tag{ii}$$

In the second equation, we have taken the further step of applying the *product rule* of differentiation taking into account as well the fact that the constant $k$ can be factored out as the differentiation is unaffected by it. (The rules cited here are all to be found in Section III of Discussion [D1] of this chapter.)

We now proceed to apply the rule for differentiating the expression $x^n$. Combining terms and assembling these results, we see that orderly application of the rules has yielded the expression.

$$X = 2k\alpha \frac{d\alpha}{dx} + k\beta \frac{d\alpha}{dy} + k\alpha \frac{d\beta}{dy} - \frac{dp_1}{dx} \tag{iii}$$

Equation (iii) is tidy but unilluminating, as it tells us nothing with respect to the physics of electromagnetism. This is just the point at which Maxwell's mathematical rhetoric must intervene. He has noticed the possibility of a drastic rearrangement of this expression, to form these symbols into patterns that will have significance. He wants to show that this expression for the total force may be seen to be made up of distinct contributions corresponding to forces of distinct and recognizable types. To follow him in this, we may look at the first three terms of our expression for the force $X$, initially changing only the order in which they are written. We then arrange them, as the arrows suggest, into a set of patterns—those that will become Maxwell's component forces labeled (A)\*, (B)\*, and (C)\* at line 252. Just follow the arrows to see how three significant patterns are being sculpted out of the initial set of three terms, shown in Flow Chart I.

FLOW CHART I

We have left question marks as temporary place-holders at points at which the gestalt is annoyingly incomplete. From this point on, the derivation becomes a crafty business of bringing these latent patterns to the light. We may first, in Table I, list the three terms in their present, preliminary forms, giving each its letter designation. Remember that in the end each is to measure one contribution to the total force on a portion of the fluid.

TABLE I

$$\alpha\left(\frac{d}{dx}k\alpha + \frac{d}{dy}k\beta\right) \qquad \text{(A)*}$$

$$k\alpha\left(\frac{d\alpha}{dx} + ?\right) \qquad \text{(B)*}$$

$$k\beta\left(\frac{d\alpha}{dy} + ?\right) \qquad \text{(C)*}$$

The first term (A)* in Table I is already in the form in which Maxwell wants it. But in (B)* and (C)* we have to fill in the question-mark blanks. At this point Maxwell turns to a piece of overt algebraic trickery. *It is always permissible to add zero to anything,* so it is by the same token permissible to add and subtract the same expression to the mix—not the sort of thing one would do without a firm sense of purpose! Maxwell adds and subtracts such an artfully chosen expression as follows:

$$k\beta\frac{d\beta}{dx} - k\beta\frac{d\beta}{dx} \qquad \text{(iv)}$$

Since the terms (A)*, (B)*, and (C)* are all to be summed in the expression for $X$, the positive part of (iv) can be added to one term and the negative to another, as our Flow Chart II reveals, the positive term is added to (B)* and the negative to (C)*.

FLOW CHART II

Maxwell now has these terms in very nearly the form in which he wants them. (We note that a little legitimate factoring has been done in the case of (C)\*. We may tabulate the results thus far, as shown in Table II.

TABLE II

$$\alpha\left(\frac{d}{dx}k\alpha + \frac{d}{dy}k\beta\right) \tag{A)*}$$

$$\left(k\alpha\frac{d\alpha}{dx} + k\beta\frac{d\beta}{dx}\right) \tag{B)*}$$

$$k\beta\left(\frac{d\alpha}{dy} - \frac{d\beta}{dx}\right) \tag{C)*}$$

Only one finishing touch remains, this time applied to term (B)\*. Again, we reason backward—so often the most fruitful procedure! Remembering that $\alpha$ and $\beta$ represent velocities of flow and that mechanically kinetic energy is proportional to the square of velocity, Maxwell sees that (B)\* can be construed as the derivative of a kinetic energy. The trick is the rule for the differentiation of a *power* (worked out in Section III of Discussion [D1] in this chapter). The general rule is, for any $x$ and any $n$,

POWER RULE : If

$$y = a\,x^n$$

then

$$\frac{dy}{dx} = nax^{(n-1)}$$

We will also need the following

CHAIN RULE : If

$$y = f(\alpha)$$

while in turn $\alpha = g(x)$, then

$$\frac{dy}{dx} = \frac{d}{d\alpha}f(\alpha)\frac{d\alpha}{dx}$$

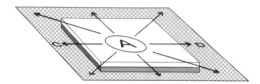

Fig. 3.30. Maxwell's fig.1, redrawn to surround A with a test box.

In words: If we have a function of α while α is itself a function of x, then the derivative of the function with respect to α must be multiplied by the derivative of α with respect to x—like most such rules, as appealing to common sense as it is provably true!

Applied to just the two terms of (B)*, we have:

$$\frac{d}{dx}\left(k\alpha^2\right) = 2k\alpha\frac{d\alpha}{dx}$$

$$\frac{d}{dx}\left(k\beta^2\right) = 2k\beta\frac{d\beta}{dx}$$

so that

$$2k\alpha\frac{d\alpha}{dx} + 2k\beta\frac{d\beta}{dx} = \frac{d}{dx}\left(k\alpha^2\right) + \frac{d}{dx}\left(k\beta^2\right) \qquad \text{(v)}$$

If we now take into account the principle that the derivative of a sum is the sum of the derivatives, we can group the last terms of (v) in a more interesting way.

$$2k\alpha\frac{d\alpha}{dx} + 2k\beta\frac{d\beta}{dx} = \frac{d}{dx}\left(k\alpha^2\right) + \frac{d}{dx}\left(k\beta^2\right)$$

$$= \frac{d}{dx}\left(k\alpha^2 + k\beta^2\right) \qquad \text{(vi)}$$

Maxwell will interpret this as the x-derivative of the *total* kinetic energy due to the velocities in the x and y directions. As energy is a scalar quantity, the two terms of (vi) sum nicely to give the total energy. Maxwell will go on to discuss each of these contributions to the force X in subsequent passages, but we can already perceive how he has in a sense brought the total kinetic energy into the light, out of the shadows of mere symbols.

The final result of all of this, with further adjustment for a factor of 2 which lurks in (vi), is the set of terms that Maxwell gives us. We have restored the equal sign and so present the result in its full form as the multitermed equation for the total force in the *x*-direction on an element of the fluid, namely equation (4)* at line 248 of the text.

$X =$

$$\alpha \left( \frac{d}{dx} k\alpha + \frac{d}{dy} k\beta \right) \qquad \text{(A)*}$$

$$+ \frac{k}{2} \frac{d}{dx} \left( \alpha^2 + \beta^2 \right) \qquad \text{(B)*}$$

$$+ k\beta \left( \frac{d\alpha}{dy} - \frac{d\beta}{dx} \right) \qquad \text{(C)*}$$

$$- \frac{dp_1}{dx} \qquad \text{(D)*}$$

Here we have not forgotten to add at the end the (D)* term, which has not come in for discussion because it called for no modification. Maxwell's next task will be to interpret in physical terms the significance of each of the terms he has so artfully prepared.

## D8. On Total Magnetic Induction from a Source [line 277]

We may look ahead to Maxwell's fig. 1, redrawing it as Fig. 3.30 so as to include a thin exploratory test box of sides $\Delta x$, $\Delta y$, and $\Delta z$ around $A$. For our present purposes, it will be useful to think of the two horizontal directions as *x* and *y*, with *z* as the vertical axis. Considering field components in only the *x*- and *y*-directions, we will compute the total number of lines passing outward through the surface of the box—by the continuity of the lines, the shape of the box will not matter. That number of lines will serve as a measure of the magnitude of the source *A* inside the box. Maxwell now claims that the expression (6)* in turn represents that number of lines, the "total amount of magnetic induction."

To see this, we can turn for a moment to an even simpler case, depicted in Fig. 3.31. Here the lines are all in the *x*-direction, though their number increases as we move from left to right—apparent evidence of hidden sources within the box! If the *rate of increase* of the *x*-component of magnetic induction $\mu\alpha$ as we move from left to right through the box is given by the derivative

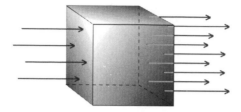

Fig. 3.31. Divergence in one dimension.

$$\frac{d(\mu\alpha)}{dx} \qquad\qquad (a)$$

then the total increase will be that rate times $\Delta x$:

$$\Delta(\mu\alpha) = \left[\frac{d(\mu\alpha)}{dx}\right]\Delta x \qquad\qquad (b)$$

and the net induction through the box will be that quantity times the area of the face of the box, $\Delta y\Delta z$. This measures the strength of the source $m$ in the box. The amount of "imaginary magnetic matter" per unit volume will be proportional to

$$\left[\frac{d(\mu\alpha)}{dx}\right]\Delta x\Delta y\Delta z \qquad\qquad (c)$$

The question of a possible constant of proportionality remains to be discussed.

We do well to remind ourselves that these boxes we so often use represent calculations carried to the limit. The relations we speak of become true only as the box shrinks to a point. Thus for example in this case it seems that we must think of the the "source" indicated by the divergent lines as rather a "sourceful point" in the field, or a point in a "sourceful" region than as a piece of magnetic matter hidden in a box. What we are actually looking at is a state of the field.

Turning turn back to Fig. 3.30, we must now take into account variation in the $y$-direction as well. To do so, we will need a second term to measure the contribution of the variation in $\beta$, the vertical component of **H:**

$$\Delta(\mu\beta) = \left[\frac{d(\mu\beta)}{dy}\right]\Delta y \qquad\qquad (d)$$

measuring an additional contribution to the strength of the source:

$$\left[\frac{d(\mu\beta)}{dy}\right]\Delta y \Delta x \Delta z \tag{e}$$

The apparent source will now be the sum of these two contributions:

$$\left[\frac{d(\mu\alpha)}{dx} + \frac{d(\mu\beta)}{dy}\right]\Delta x \Delta y \Delta z \tag{f}$$

which is the left-hand side of Maxwell's equation (6)*. The intensity of the source per unit volume will be proportional to

$$\left[\frac{d(\mu\alpha)}{dx} + \frac{d(\mu\beta)}{dy}\right] \tag{g}$$

The next step is one of *interpretation*, not simply of algebra. We know empirically that the north pole of a magnet placed in a uniform field of intensity $\alpha$ as in Fig. 3.30 will be subject to a force proportional to the product of the strength of the pole and the intensity of the field:

$$F = \alpha\, m \tag{h}$$

However, the (A)* term of our force equation (4)* falls into just that pattern!

$$\alpha\left[\frac{d(k\alpha)}{dx} + \frac{d(k\beta)}{dy}\right] \tag{A)*}$$

With the substitution $k = \mu/4\pi$, this becomes

$$\alpha\left(\frac{1}{4\pi}\right)\left[\frac{d(\mu\alpha)}{dx} + \frac{d(\mu\beta)}{dy}\right] \tag{i}$$

We now *recognize that the pattern we need lies before us:* all we have to do is to identify *m*, which we already know is *proportional* to expression (g) above, with the whole quantity

$$\left(\frac{1}{4\pi}\right)\left[\frac{d(\mu\alpha)}{dx} + \frac{d(\mu\beta)}{dy}\right] \tag{j}$$

settling the question of the constant of proportionality. Equation (i) is now exactly in the form (h)—the vortices have yielded their own account of the force on a magnetic needle!

There is a secret lurking here, and a certain genius in Maxwell's fig. 1 (our Fig. 3.30). The secret is that there *are* no magnetic poles: the magnetic lines are everywhere continuous, they have no sources, and there is no "imaginary magnetic matter" ever, anywhere. (Recall Fig. 2.12, which envisioned the continuity of the lines of force through the body of the bar magnet.) Always, when all components are taken into account, the divergence of the lines of magnetic induction at a point is zero. This has already been adumbrated and will become explicit in Chapter 4.

Yet we know that magnetic needles do point—they act as if they contained magnetic poles. That is because, as in Maxwell's figure, they look at only part of the picture. By choosing his thin section in a horizontal field, Maxwell leaves aside the question of the full reckoning, including the z-axis and all lines leaving the box. It is not possible to have a box in which magnetic lines diverge in this way from the top as well—in other words, no one has ever discovered a *magnetic monopole*. Maxwell has not tricked us, just focused our attention! The phenomenon he pictures is real and the vortex mechanism has done a spectacular job of accounting for it.

To avoid confusion, we might add that the situation is altogether different in electrostatics; there isolated sources do exist and the analogous expression for the divergence from a charged source yields a finite value.

## D9. On the Logic of Physical Speculation [line 311]

The logic of this mode of speculation seems to be this:

1. We think of a fluid capable of sustaining complex motions and stresses and fully intelligible through its obedience to the known laws of mechanical systems.
2. We compute the forces that will result, analytically.
3. We undertake to *interpret* the analytic results by associating, Maxwell proposes, relative tension along axis of the vortex system with magnetic intensity **H.**
4. Analytically, we find two quantities multiplied to form expression (A)* for one term of the force on a portion of the fluid. One of these

has the form of a divergence and the other is the magnetic intensity itself. The interpretation takes the divergence as a measure of the quantity of "magnetic matter," a concept really only borrowed from a theory of a very different kind, yet useful here.

5. We envision a situation in which these two components are made explicit—one as a field of intensity $\alpha$ and the other as a pole of strength $m$.

6. We see that this identification has neatly yielded an account of a primary phenomenon: the force on a magnetic pole placed in a magnetic field. The vortex model has passed its first test!

It is remarkable that although the interpretation uses the concept of "magnetic matter," which belongs rather to action-at-a-distance theory, it has in fact generated both magnetic poles and the forces on them entirely in continuum terms. The magnets, the fields they are in, and the forces that result are all no more than aspects of the vortical motion of Maxwell's one conceptual fluid.

# Chapter 4

## *A Dynamical Theory of the Electromagnetic Field*

Maxwell opens this third paper with reflections on the difference between his own method and that of theorists who work in a strictly mathematical mode, accepting without question the action of one body on another at a distance without the intervention of a medium. This, he says, seems "at first sight" the natural mode of explanation in electricity and magnetism. This "first sight" has given rise to nearly all of the formal, mathematical electrical theories up to Maxwell's time and, indeed, some of these have come close to giving a complete account, if one is satisfied with the action-at-a-distance mode of accounting for the phenomena. Such theories are thoroughly rational if the whole work of reason in mathematics is the elegant and accurate demonstration of consequences from clearly defined sets of first principles, in conformity with the observed phenomena of the world. Maxwell speaks with genuine respect of the work of men such as Weber who work in this mode.

Yet his own whole work seems to lie in going beyond this "first sight," to meet some further demand of reason, differently understood. What we might then call Maxwell's "second sight" has to do with meeting an additional, intuitive requirement of intelligibility. In "Physical Lines," we have seen this taking the form of a vision of a connected mechanism; at least to a first approximation, he has satisfied himself there that a literal mechanism can be envisioned that would fill the gap between two interacting centers of force. But in "Physical Lines" he may have gone much too far in the direction of sheer imagination, proposing devices for which he has no evidence. That account was an important, perhaps necessary, excursion of the mind, but it was far too *literal* to be true.

Now in this third paper we turn from speculation and imagination to scientific truth, in the most disciplined form Maxwell is able to give it. There will be no arbitrary hypotheses, no flights of fancy: the outcome will be a verifiable prediction and, as we shall see in our final section, Maxwell will turn from theorist to careful experimenter to bring that prediction to the test of truth. This paper may then serve as an exemplar of science in the strictest sense, and the empirical test to which it leads may stand as a classic instance of the *experimentum crucis,* the crucial experiment.

The term "dynamical" in Maxwell's new attempt to which we now turn refers in general to more abstract formulations in terms of systems of *energy* instead of the specific pushes and pulls of Newton's laws of motion. This approach is a relatively new historical development as Maxwell writes. His friends Thomson and Tait have set out a virtual manifesto of the new generalized mechanics in an ambitious *Treatise on Natural Philosophy.* Does this mean that had the chronology of these developments been otherwise—had the method of generalized mechanics reached him sooner—Maxwell might have skipped one huge step and gone directly from "Faraday's Lines," to the abstract dynamical account in which the machine becomes metaphor? Was there any inherent necessity that the *physical* account intervene? We may leave this to the reader as an unresolved question. However, the thought does suggest itself that for Maxwell at least, there had to be a stage in which the mind would play over its object, intensely and in detail, carrying to completion a thought which in strictness it knows must finally remain suspended. We will end, in the dynamical theory, with a level of abstraction in which force, momentum, and even mass itself become no more than mathematical metaphors; but before we admit metaphor, we may have needed concreteness, if only in the comic and virtual mode of "Physical Lines," as a springboard from which the metaphorical could take off.

[From the Royal Society Transactions, Vol. CLV]
XXV. A Dynamical Theory of the Electromagnetic Field
(Received October 27, Read December 8, 1864)

## Part I. Introductory

(1) The most obvious mechanical phenomenon in electrical and
magnetical experiments is the mutual action by which bodies in certain states set each other in motion while still at a sensible distance
5    from each other. The first step, therefore, in reducing these phenomena into scientific form, is to ascertain the magnitude and direction of
the force acting between the bodies, and when it is found that this
force depends in a certain way upon the relative position of the bodies
and on their electric or magnetic condition, it seems at first sight natu-
10    ral to explain the facts by assuming the existence of something either
at rest or in motion in each body, constituting its electric or magnetic
state, and capable of acting at a distance according to mathematical
laws.

In this way mathematical theories of statical electricity, of magnet-
15    ism, of the mechanical action between conductors carrying currents,
and of the induction of currents have been formed. In these theories
the force acting between the two bodies is treated with reference only
to the condition of the bodies and their relative position, and without
any express consideration of the surrounding medium.

20    These theories assume, more or less explicitly, the existence of substances the particles of which have the property of acting on one another
at a distance by attraction or repulsion. The most complete development
of a theory of this kind is that of M. W. Weber,* who has made the same
theory include electrostatic and electromagnetic phenomena.

25    In doing so, however, he has found it necessary to assume that the
force between two electric particles depends on their relative velocity,
as well as on their distance.

This theory, as developed by MM. W. Weber and C. Neumann,** is
exceedingly ingenious, and wonderfully comprehensive in its applica-
30    tion to the phenomena of statical electricity, electromagnetic attractions, induction of currents, and diamagnetic phenomena; and it

*"Electrodynamische Maasbestimmungen." *Leipzig Trans.* Vol. I 1849, and
Taylor's *Scientific Memoirs,* Vol. V. art. xiv.
**Explicare tentatur quomodo fiat ut lucis planum polarizationis per vires electri-
cas vel magneticus delinatur.*—Halis Saxonum, 1858.

comes to us with the more authority, as it has served to guide the speculations of one who has made so great an advance in the practical part of electric science, both by introducing a consistent system of

35   units in electrical measurement, and by actually determining electrical quantities with an accuracy hitherto unknown.

(2) The mechanical difficulties, however, which are involved in the assumption of particles acting at a distance with forces which depend on their velocities are such as to prevent me from considering this the-

40   ory as an ultimate one though it may have been, and may yet be useful in leading to the coordination of phenomena.

I have therefore preferred to seek an explanation of the fact in another direction, by supposing them to be produced by actions which go on in the surrounding medium as well as in the excited bodies, and

45   endeavouring to explain the action between distant bodies without assuming the existence of forces capable of acting directly at sensible distances.

(3) The theory I propose may therefore be called a theory of the *Electromagnetic Field,* because it has to do with the space in the

50   neighbourhood of the electric or magnetic bodies, and it may be called *Dynamical* Theory, because it assumes that in that space there is matter in motion, by which the observed electromagnetic phenomena are produced.

(4) The electromagnetic field is that part of space which contains

55   and surrounds bodies in electric or magnetic conditions.

It may be filled with any kind of matter, or we may endeavour to render it empty of all gross matter, as in the case of Geissler's tubes and other so-called vacua.

There is always, however, enough of matter left to receive and

60   transmit the undulations of light and heat, and it is because the transmission of these radiations is not greatly altered when transparent bodies of measurable density are substituted for the so-called vacuum, that we are obliged to admit that the undulations are those of an aethereal substance, and not of the gross matter, the presence of

65   which merely modifies in some way the motion of the other.

We have therefore some reason to believe, from the phenomena of light and heat, that there is an aethereal medium filling space and permeating bodies, capable of being set in motion and of transmitting that motion from one part to another, and of communicating that motion to

70   gross matter so as to heat it and affect it in various ways.

(5) Now the energy communicated to the body in heating it must have formerly existed in the moving medium, for the undulations had left the source of heat some time before they reached the body, and during that time the energy must have been half in the form of motion
75   of the medium and a half in the form of elastic resilience. From these considerations Professor W. Thomson has argued,* that the medium must have a density capable of comparison with that of gross matter, and has even assigned an inferior limit to that density.

(6) We may therefore receive, as a datum derived from a branch of
80   science independent of that with which we have to deal, the existence of a pervading medium, of small but real density, capable of being set in motion, and of transmitting motion from one part to another with great, but not infinite, velocity.

Hence the parts of this medium must be so connected that the mo-
85   tion of one part depends in some way on the motion of the rest; and at the same time these connexions must be capable of a certain kind of elastic yielding, since the communication of motion is not instantaneous, but occupies time.

The medium is therefore capable of receiving and storing up two
90   kinds of energy, namely, the "actual" energy depending on the motions of its parts, and "potential" energy, consisting of the work which the medium will do in recovering from displacement in virtue of its elasticity.

The propagation of undulations consists in the continual transforma-
95   tion of one of these forms of energy into the other alternately, and at any instant the amount of energy in the whole medium is equally divided, so that half is energy of motion, and half is elastic resilience.

(7) A medium having such a constitution may be capable of other kinds of motion and displacement than those which produce the phe-
100   nomena of light and heat, and some of these may be of such a kind that they may be evidenced to our senses by the phenomena they produce.

(8) Now we know that the luminiferous medium is in certain cases acted on by magnetism; for Faraday** discovered that when a plane
105   polarized ray traverses a transparent diamagnetic medium in the di-

*"On the Possible Density of the Luminiferous Medium, and on the Mechanical Value of a Cubic Mile of Sunlight," *Transactions of the Royal Society of Edinburgh* (1854), p. 57.
**Experimental Researches, Series XIX.*

rection of the lines of magnetic force produced by magnets or currents in the neighbourhood, the plane of polarization is caused to rotate.

This rotation is always in the direction in which positive electricity must be carried round the diamagnetic body in order to produce the
110  actual magnetization of the field.

M. Verdet* has since discovered that if a paramagnetic body, such as solution of perchloride of iron in ether, be substituted for the diamagnetic body, the rotation is in the opposite direction.

Now Professor W. Thomson** has pointed out that no distribution
115  of forces acting between the parts of a medium whose only motion is that of the luminous vibrations is sufficient to account for the phenomena, but that we must admit the existence of a motion in the medium depending on the magnetization, in addition to the vibratory motion which constitutes light.

120  It is true that the rotation by magnetism of the plane of polarization has been observed only in media of considerable density; but the properties of the magnetic field are not so much altered by the substitution of one medium for another, or for a vacuum, as to allow us to suppose that the dense medium does anything more than merely
125  modify the motion of the ether. We have therefore warrantable grounds for inquiring whether there may not be a motion of the ethereal medium going on wherever magnetic effects are observed, and we have some reason to suppose that this motion is one of rotation, having the direction of the force as its axis.

130  (9) We may now consider another phenomenon observed in the electromagnetic field. When a body is moved across the lines of magnetic force it experiences what is called an electromotive force; the two extremities of the body tend to become oppositely electrified, and an electric current tends to flow through the body. When the electro-
135  motive force is sufficiently powerful, and is made to act on certain compound bodies, it decomposes them, and causes one of their components to pass towards one extremity of the body, and the other in the opposite direction.

Here we have evidence of a force causing an electric current in
140  spite of resistance; electrifying the extremities of a body in opposite ways, a condition which is sustained only by the action of the

*Comptes Rendus* (1856, second half year, p. 529, and 1857, first half year, p. 1209).
**Proceedings of the Royal Society,* June 1856 and June 1861.

electromotive force, and which, as soon as that force is removed, tends, with an equal and opposite force, to produce a counter current through the body and to restore the original electrical state of the
145 body; and finally, if strong enough, tearing to pieces chemical compounds and carrying their components in opposite directions, while their natural tendency is to combine, and to combine with a force which can generate an electromotive force in the reverse direction.

This, then, is a force acting on a body caused by its motion through
150 the electromagnetic field, or by changes occurring in that field itself; and the effect of the force is either to produce a current and heat the body, or to decompose the body, or, when it can do neither, to put the body in a state of electric polarization, -a state of constraint in which opposite extremities are oppositely electrified, and from which the
155 body tends to relieve itself as soon as the disturbing force is removed.

(10) According to the theory which I propose to explain, this "electromotive force" is the force called into play during the communication of motion from one part of the medium to another, and it is by means of this force that the motion of one part causes motion in another part.
160 When electromotive force acts on a conducting circuit, it produces a current, which, as it meets with resistance, occasions a continual transformation of electrical energy into heat, which is incapable of being restored again to the form of electrical energy by any reversal of the process.

165 (11) But when electromotive force acts on a dielectric it produces a state of polarization of its parts similar in distribution to the polarity of the parts of a mass of iron under the influence of a magnet, and like the magnetic polarization, capable of being described as a state in which every particle has its opposite poles in opposite conditions.*
170 In a dielectric under the action of electromotive force, we may conceive that the electricity in each molecule is so displaced that one side is rendered positively and the other negatively electrical, but that the electricity remains entirely connected with the molecule, and does not pass from one molecule to another. The effect of this action on the
175 whole dielectric mass is to produce a general displacement of electricity in a certain direction. This displacement does not amount to a current, because when it has attained to a certain value it remains

*Faraday, *Experimental Researches,* Series XI.; Mossotti, *Mem. della Soc. Italiana* (Modena), Vol. XXIV. Part 2, p. 49.

constant, but it is the commencement of a current, and its variations
constitute currents in the positive or the negative direction according
180   as the displacement is increasing or decreasing. In the interior of the
dielectric there is no indication of electrification, because the electrification of the surface of any molecule is neutralized by the opposite
electrification of the surface of the molecules in contact with it; but at
the bounding surface of the dielectric, where the electrification is not
185   neutralized, we find the phenomena which indicate positive or negative electrification.

The relation between the electromotive force and the amount of
electric displacement it produces depends on the nature of the dielectric, the same electromotive force producing generally a
190   greater electric displacement in solid dielectrics, such as glass or
sulphur, than in air.

(12) Here, then, we perceive another effect of electromotive force,
namely, electric displacement, which according to our theory is a kind
of elastic yielding to the action of the force, similar to that which takes
195   place in structures and machines owing to the want of perfect rigidity
of the connexions.

\* \* \*

(15) It appears therefore that certain phenomena in electricity and
magnetism lead to the same conclusion as those of optics, namely,
200   that there is an aethereal medium pervading all bodies, and modified
only in degree by their presence; that the parts of this medium are capable of being set in motion by electric currents and magnets; that this
motion is communicated from one part of the medium to another by
forces arising from the connexions of those parts; that under the ac-
205   tion of these forces there is a certain yielding depending on the elasticity of these connexions; and that therefore energy in two different
forms may exist in the medium, the one form being the actual energy
of motion of its parts, and the other being the potential energy stored
up in the connexions, in virtue of their elasticity.
210   (16) Thus, then, we are led to the conception of a complicated
mechanism capable of a vast variety of motion, but at the same time
so connected that the motion of one part depends, according to definite relations, on the motion of other parts, these motions being communicated by forces arising from the relative displacement of the
215   connected parts, in virtue of their elasticity. Such a mechanism must

be subject to the general laws of Dynamics, and we ought to be able to work out all the consequences of its motion, provided we know the form of the relation between the motions of the parts.

(17) We know that when an electric current is established in a
220 conducting circuit, the neighbouring part of the field is charac-
terized by certain magnetic properties, and that if two circuits are in the field, the magnetic properties of the field due to the two cur-rents are combined. Thus each part of the field is in connexion with both currents, and the two currents are put in connexion with each
225 other in virtue of their connexion with the magnetization of the field. The first result of this connexion that I propose to examine, is the induction of one current by another, and by the motion of con-ductors in the field.

The second result, which is deduced from this, is the mechanical
230 action between conductors carrying currents. The phenomenon of the induction of currents has been deduced from their mechanical action by Helmholtz* and Thomson.** I have followed the reverse order, and deduced the mechanical action from the laws of induction. I have then described experimental methods of determining the quantities *L, M,*
235 *N,* on which these phenomena depend.

(18) I then apply the phenomena of induction and attraction of cur-rents to the exploration of the electromagnetic field, and the laying down systems of lines of magnetic force which indicate its magnetic properties. By exploring the same field with a magnet, I shew the dis-
240 tribution of its equipotential magnetic surfaces, cutting the lines of force at right angles.

In order to bring these results within the power of symbolical calcu-lation, I then express them in the form of the General Equations of the Electromagnetic Field. These equations express—

245    (A)   The relation between electric displacement, true conduction, and the total current, compounded of both.
      (B)   The relation between the lines of magnetic force and the in-ductive coefficients of a circuit, as already deduced from the laws of induction.

---

*"Conservation of Force," *Physical Society of Berlin,* 1847; and Taylor's *Scien-tific Memoirs,* 1853, p. 114.
**Reports of the British Association,* 1848; *Philosophical Magazine,* Dec. 1851.

250    (C)  The relation between the strength of a current and its magnetic effects, according to the electromagnetic system of
measurement.

    (D)  The value of the electromotive force in a body, as arising
from the motion of the body in the field, the alteration of the
255    field itself, and the variation of electric potential from one
part of the field to another.

    (E)  The relation between electric displacement, and the electro-
motive force which produces it.

    (F)  The relation between an electric current, and the electro-
260    motive force which produces it.

    (G)  The relation between the amount of free electricity at any
point, and the electric displacements in the neighbourhood.

    (H)  The relation between the increase or diminution of free elec-
tricity and the electric currents in the neighbourhood.

265    There are twenty of these equations in all, involving twenty variable
quantities.

(19) I then express in terms of these quantities the intrinsic energy
of the Electromagnetic Field as depending partly on its magnetic and
partly on its electric polarization at every point.

270    From this I determine the mechanical force acting, 1st, on a moveable conductor carrying an electric current; 2ndly, on a magnetic pole;
3rdly, on an electrified body.

The last result, namely, the mechanical force acting on an electrified body, gives rise to an independent method of electrical measure-
275  ment founded on its electrostatic effects. The relation between the
units employed in the two methods is shewn to depend on what I
have called the "electric elasticity" of the medium, and to be a velocity,
which has been experimentally determined by MM. Weber and
Kohlrausch.*

280    I then shew how to calculate the electrostatic capacity of a condenser, and the specific inductive capacity of a dielectric.

The case of a condenser composed of parallel layers of substances
of different electric resistances and inductive capacities is next examined, and it is shewn that the phenomenon called electric absorption

*Leipzig Transactions, Vol. v. (1857), p. 260, or Poggendorf's *Annalen,* Aug.
1856, p. 10.

285  will generally occur, that is, the condenser, when suddenly dis-
     charged, will after a short time shew signs of a *residual* charge.

     (20) The general equations are next applied to the case of a mag-
     netic disturbance propagated through a non-conducting field, and it is
     shewn that the only disturbances which can be so propagated are
290  those which are transverse to the direction of propagation, and that
     the velocity of propagation is the velocity *v,* found from experiments
     such as those of Weber, which expresses the number of electrostatic
     units of electricity which are contained in one electromagnetic unit.

     The velocity is so nearly that of light, that it seems we have strong
295  reason to conclude that light itself (including radiant heat, and other ra-
     diations if any) is an electromagnetic disturbance in the form of waves
     propagated through the electromagnetic field according to electromag-
     netic laws. If so, the agreement between the elasticity of the medium
     as calculated from the rapid alternations of luminous vibrations, and
300  as found by the slow processes of electrical experiments, shews how
     perfect and regular the elastic properties of the medium must be when
     not encumbered with any matter denser than air. If the same charac-
     ter of the elasticity is retained in dense transparent bodies, it appears
     that the square of the index of refraction is equal to the product of the
305  specific dielectric capacity and the specific magnetic capacity. Con-
     ducting media are shewn to absorb such radiations rapidly, and there-
     fore to be generally opaque.

     This conception of the propagation of transverse magnetic distur-
     bances to the exclusion of normal ones is distinctly set forth by Professor
310  Faraday* in his "Thoughts on Ray Vibrations." The electromagnetic the-
     ory of light, as proposed by him, is the same in substance as that which I
     have begun to develop in this paper, except that in 1846 there were no
     data to calculate the velocity of propagation.

     (21) The general equations are then applied to the calculation of
315  the coefficients of mutual induction of two circular currents and the co-
     efficient of self-induction in a coil. The want of uniformity of the current
     in the different parts of the section of a wire at the commencement of
     the current is investigated, I believe for the first time, and the conse-
     quent correction of the coefficient of self-induction is found.

320  These results are applied to the calculation of the self-induction of
     the coil used in the experiments of the Committee of the British Asso-

---

*Philosophical Magazine,* May 1846, or *Experimental Researches,* III. p. 447.

ciation on Standards of Electric Resistance, and the value compared
with that deduced from the experiments.

## Part II. On Electromagnetic Induction

325   Electromagnetic Momentum of a Current

(22) We may begin by considering the state of the field in the neighbourhood of an electric current. We know that magnetic forces are excited in the field, their direction and magnitude depending according
to known laws upon the form of the conductor carrying the current.
330   When the strength of the current is increased, all the magnetic effects
are increased in the same proportion. Now, if the magnetic state of
the field depends on motions of the medium, a certain force must be
exerted in order to increase or diminish these motions, and when the
motions are excited they continue, so that the effect of the connexion
335   between the current and the electromagnetic field surrounding it is to
endow the current with a kind of momentum, just as the connexion between the driving-point of a machine and a fly-wheel endows the driving-point with an additional momentum, which may be called the
momentum of the fly-wheel reduced to the driving-point. The unbal-
340   anced force acting on the driving-point increases this momentum, and
is measured by the rate of its increase.

In the case of electric currents, the resistance to sudden increase
or diminution of strength produces effects exactly like those of momentum, but the amount of this momentum depends on the shape of the
345   conductor and the relative position of its different parts.

Mutual Action of two Currents

(23) If there are two electric currents in the field, the magnetic force at
any point is that compounded of the forces due to each current separately, and since the two currents are in connexion with every point of
350   the field, they will be in connexion with each other, so that any increase or diminution of the one will produce a force acting with or contrary to the other.

Dynamical Illustration of Reduced Momentum [D1]

(24) As a dynamical illustration, let us suppose a body C so con-
355   nected with two independent driving-points A and B that its velocity
is p times that of A together with q times that of B. Let u be the

velocity of A, v that of B, and w that of C, and let $\delta x$, $\delta y$, $\delta z$ be their simultaneous displacements, then by the general equation of dynamics,*

360

$$C\left(\frac{dw}{dt}\right)\delta z = X\delta x + Y\delta y,$$

where X and Y are the forces acting at A and B.
But

$$\frac{dw}{dt} = p\frac{du}{dt} + q\frac{dv}{dt},$$

and

365

$$\delta z = p\delta x + q\delta y.$$

Substituting (see [D2]), and remembering that $\delta x$ and $\delta y$ are independent,

$$X = \frac{d}{dt}\left(Cp^2u + Cpqv\right)$$

(1)

$$Y = \frac{d}{dt}\left(Cpqu + Cq^2v\right)$$

370 We may call $Cp^2u + Cpqv$ the momentum of C referred to A, and $Cpqu + Cq^2v$ its momentum referred to B; then we may say that the effect of the force X is to increase the momentum of C referred to A, and that of Y to increase its momentum referred to B.

If there are many bodies connected with A and B in similar ways
375 but with different values of p and q, we may treat the question in the same way by assuming

$$L = \Sigma\left(Cp^2\right), \quad M = \Sigma(Cpq), \quad \text{and } N = \Sigma\left(Cq^2\right),$$

where the summation is extended to all the bodies with their proper values of C, p, and q. Then the momentum of the system referred to
380 A is

$$Lu + Mv,$$

and referred to B,

$$Mu + Nv,$$

*Lagrange, Mec. Anal. II 2 §5.

and we shall have

385

$$X = \frac{d}{dt}(Lu + Mv)$$

(2)

$$Y = \frac{d}{dt}(Mu + Nv),$$

where $X$ and $Y$ are the external forces acting on $A$ and $B$.

(25) To make the illustration more complete we have only to suppose that the motion of $A$ is resisted by a force proportional to its ve-
390  locity, which we may call $Ru$, and that of $B$ by a similar force, which we may call $Sv$, $R$ and $S$ being coefficients of resistance. Then if $\xi$ and $\eta$ are the forces on $A$ and $B$,

$$\xi = X + Ru = Ru + \frac{d}{dt}(Lu + Mv)$$

(3)

$$\eta = Y + Sv = Sv + \frac{d}{dt}(Mu + Nv).$$

395  If the velocity of $A$ be increased at the rate $du/dt$, then in order to prevent $B$ from moving a force, $\eta = \frac{d}{dt}(Mu)$ must be applied to it [D3].

This effect on $B$, due to an increase of the velocity of $A$, corresponds to the electromotive force on one circuit arising from an increase in the strength of a neighbouring circuit.

400      This dynamical illustration is to be considered merely as assisting the reader to understand what is meant in mechanics by Reduced Momentum. The facts of the induction of currents as depending on the variations of the quantity called Electromagnetic Momentum, or Electrotonic State, rest on the experiments of Faraday,* Felici,** &c.

405  Coefficients of Induction for Two Circuits

(26) In the electromagnetic field the values of $L$, $M$, $N$ depend on the distribution of the magnetic effects due to the two circuits, and this distribution depends only on the form and relative position of the circuits. Hence $L$, $M$, $N$ are quantities depending on the form and relative posi-
410  tion of the circuits, and are subject to variation with the motion of the conductors. It will be presently seen that $L$, $M$, $N$ are geometrical quantities of the nature of lines; that is, of one dimension in space; $L$

*Experimental Researches, Series I., IX.
**Annales de Chimie, ser. 3, XXIV. (1852), p. 64.

depends on the form of the first conductor, which we shall call A, N on
that of the second, which we shall call B, and M on the relative posi-
415  tion of A and B.

(27) Let $\xi$ be the electromotive force acting on A, x the strength of
the current, and R the resistance, then Rx will be the resisting force.
In steady currents, the electromotive force just balances the resisting
force, but in variable currents the resultant force $\xi - Rx$ is expended
420  to increase the "electromagnetic momentum," using the word momentum merely to express that which is generated by a force acting during a time, that is, a velocity existing in a body.

In the case of electric currents, the force in action is not ordinary mechanical force, at least we are not as yet able to measure it as common
425  force, but we call it electromotive force, and the body moved is not merely
the electricity in the conductor, but something outside the conductor, and
capable of being affected by other conductors in the neighbourhood carrying currents. In this it resembles rather the reduced momentum of a
driving-point of a machine as influenced by its mechanical connexions,
430  than that of a simple moving body like a cannon ball, or water in a tube.

### Electromagnetic Relations of two Conducting Circuits

(28) In the case of two conducting circuits, A and B, we shall assume
that the electromagnetic momentum belonging to A is

$$Lx + My$$

435  and that belonging to B,

$$Mx + Ny,$$

where L, M, N correspond to the same quantities in the dynamical illustration, except that they are supposed to be capable of variation
when the conductors A or B are moved.
440      Then the equation of the current x in A will be

$$\xi = Rx + \frac{d}{dt}(Lx + My) \tag{4}$$

and that of y in B

$$\eta = Sy + \frac{d}{dt}(Mx + Ny) \tag{5}$$

where $\xi$ and $\eta$ are the electromotive forces, x and y the currents, and
445  R and S the resistances in A and B respectively.

### Induction of One Current by Another

(29) Case 1st. Let there be no electromotive force on *B*, except that which arises from the action of *A*, and let the current of *A* increase from 0 to the value *x*, then

450
$$Sy + \frac{d}{dt}(Mx + Ny) = 0,$$

whence

$$Y = \int_0^t y\,dt = -\frac{M}{S}x, \tag{6}$$

that is, a quantity of electricity *Y*, being the total induced current, will flow through *B* when *x* rises from 0 to *x*. This is induction by variation
455 of the current in the primary conductor. When *M* is positive, the induced current due to increase of the primary current is negative.

### Induction by Motion of Conductor

(30) Case 2nd. Let *x* remain constant, and let *M* change from *M* to *M′*, then

460
$$Y = -\frac{M' - M}{S} \tag{7}$$

so that if *M* is increased, which it will be by the primary and secondary circuits approaching each other, there will be a negative induced current, the total quantity of electricity passed through *B* being *Y*.

This is induction by the relative motion of the primary and secondary conductors [D4].
465

### Equation of Work and Energy

(31) To form the equation between work done and energy produced, multiply (1) by *x* and (2) by *y*, and add

$$\xi x + \eta y = Rx^2 + Sy^2 + x\frac{d}{dt}(Lx + My) + y\frac{d}{dt}(Mx + Ny) \tag{8}$$

470 Here $\xi x$ is the work done in unit of time by the electromotive force $\xi$ acting on the current *x* and maintaining it, and $\eta y$ is the work done by the electromotive force $\eta$. Hence the left-hand side of the equation represents the work done by the electromotive forces in unit of time.

475 Heat produced by the Current

(32) On the other side of the equation we have, first,

$$Rx^2 + Sy^2 = H \qquad (9)$$

which represents the work done in overcoming the resistance of the circuits in unit of time. This is converted into heat. The remaining

480 terms represent work not converted into heat. They may be written

$$\frac{1}{2}\frac{d}{dt}\left(Lx^2 + 2Mxy + Ny^2\right) + \frac{1}{2}\frac{dL}{dt}x^2 + \frac{dM}{dt}xy + \frac{1}{2}\frac{dN}{dt}y^2$$

Intrinsic Energy of the Currents

(33) If L, M, N are constant, the whole work of the electromotive forces which is not spent against resistance will be devoted to the de-

485 velopment of the currents. The whole intrinsic energy of the currents is therefore

$$\frac{1}{2}Lx^2 + Mxy + \frac{1}{2}Ny^2 = E \qquad (10)$$

This energy exists in a form imperceptible to our senses, probably as actual motion, the seat of this motion being not merely the conducting

490 circuits, but the space surrounding them.

Mechanical Action between Conductors

(34) The remaining terms,

$$\frac{1}{2}\frac{dL}{dt}x^2 + \frac{dM}{dt}xy + \frac{1}{2}\frac{dN}{dt}y^2 = W \qquad (11)$$

represent the work done in unit of time arising from the variations of L,

495 M, and N, or, what is the same thing, alterations in the form and position of the conducting circuits A and B.

Now if work is done when a body is moved, it must arise from ordinary mechanical force acting on the body while it is moved. Hence this part of the expression shews that there is a mechanical force urg-

500 ing every part of the conductors themselves in that direction in which L, M, and N will be most increased.

The existence of the electromagnetic force between conductors carrying currents is therefore a direct consequence of the joint and independent action of each current on the electromagnetic field. If A and B

505 are allowed to approach a distance ds, so as to increase M from M to M' while the currents are x and y, then the work done will be

$$(M' - M) \, xy,$$

and the force inthe direction of *ds* will be

$$\frac{dM}{ds} xy \qquad (12)$$

510 and this will be an attraction if x and y are of the same sign, and if *M* is increased as *A* and *B* approach.

It appears, therefore, that if we admit that the unresisted part of electromotive force goes on as long as it acts, generating a self-persistent state of the current, which we may call (from mechanical analogy) its electro-
515 magnetic momentum, and that this momentum depends on circumstances external to the conductor, then both induction of currents and electromagnetic attractions may be proved by mechanical reasoning.

What I have called electromagnetic momentum is the same quantity which is called by Faraday* the electrotonic state of the circuit,
520 every change of which involves the action of an electromotive force, just as change of momentum involves the action of mechanical force.

If, therefore, the phenomena described by Faraday in the Ninth Series of his *Experimental Researches* were the only known facts about electric currents, the laws of Ampère relating to the attraction of con-
525 ductors carrying currents, as well as those of Faraday about the mutual induction of currents, might be deduced by mechanical reasoning.

In order to bring these results within the range of experimental verification, I shall next investigate the case of a single current, of two currents, and of the six currents in the electric balance, so as to enable
530 the experimenter to determine the values of *L, M, N.*

\* \* \*

Exploration of the Electromagnetic Field

(47) Let us now suppose the primary circuit *A* to be of invariable form, and let us explore the electromagnetic field by means of the secondary
535 circuit *B,* which we shall suppose to be variable in form and position.

We may begin by supposing *B* to consist of a short straight conductor with its extremities sliding on two parallel conducting rails, which are put in connexion at some distance from the sliding-piece.

Then, if sliding the moveable conductor in a given direction in-
540 creases the value of *M,* a negative electromotive force will act in the

*\*Experimental Researches,* Series I. 60, &c.

circuit $B$, tending to produce a negative current in $B$ during the motion of the sliding-piece.

If a current be kept up in the circuit $B$, then the sliding-piece will itself tend to move in that direction, which causes $M$ to increase. At
545  every point of the field there will always be a certain direction such that a conductor moved in that direction does not experience any electromotive force in whatever direction its extremities are turned. A conductor carrying a current will experience no mechanical force urging it in that direction or the opposite.
550  This direction is called the direction of the line of magnetic force through that point.

Motion of a conductor across such a line produces electromotive force in a direction perpendicular to the line and to the direction of motion, and a conductor carrying a current is urged in a direction perpen-
555  dicular to the line and to the direction of the current.

(48) We may next suppose $B$ to consist of a very small plane circuit capable of being placed in any position and of having its plane turned in any direction. The value of $M$ will be greatest when the plane of the circuit is perpendicular to the line of magnetic force. Hence if a current
560  is maintained in $B$ it will tend to set itself in this position, and will of itself indicate, like a magnet, the direction of the magnetic force.

On Lines of Magnetic Force

(49) Let any surface be drawn, cutting the lines of magnetic force, and on this surface let any system of lines be drawn at small intervals, so
565  as to lie side by side without cutting each other. Next, let any line be drawn on the surface cutting all lines, and let a second line be drawn near it, its distance from the first being such that the value of $M$ for each of the small spaces enclosed between these two lines and the lines of the first system is equal to unity.
570  In this way let more lines be drawn so as to form a second system, so that the value of $M$ for every reticulation formed by the intersection of the two systems of lines is unity.

Finally, from every point of intersection of these reticulations let a line be drawn through the field, always coinciding in direction with the
575  direction of magnetic force.

(50) In this way the whole field will be filled with lines of magnetic force at regular intervals, and the properties of the electromagnetic field will be completely expressed by them.

For, 1st, If any closed curve be drawn in the field, the value of $M$ for

580 that curve will be expressed by the number of lines of force which
pass through that closed curve.

2ndly. If this curve be a conducting circuit and be moved through
the field, an electromotive force will act in it, represented by the rate of
decrease of the number of lines passing through the curve.

585 3rdly. If a current be maintained in the circuit, the conductor will be
acted on by forces tending to move it so as to increase the number of
lines passing through it, and the amount of work done by these forces
is equal to the current in the circuit multiplied by the number of additional lines.

590 4thly. If a small plane circuit be placed in the field, and be free to
turn, it will place its plane perpendicular to the lines of force. A small
magnet will place itself with its axis in the direction of the lines of force.

5thly. If a long uniformly magnetized bar is placed in the field, each
pole will be acted on by a force in the direction of the lines of force.

595 The number of lines of force passing through unit of area is equal to
the force acting on a unit pole multiplied by a coefficient depending on
the magnetic nature of the medium, and called the coefficient of magnetic induction.

In fluids and isotropic solids the value of this coefficient $\mu$ is the

600 same in whatever direction the lines of force pass through the substance, but in crystallized, strained, and organized solids the value of
m may depend on the direction of the lines of force with respect to the
axes of crystallization, strain, or growth.

In all bodies $\mu$ is affected by temperature, and in iron it appears to

605 diminish as the intensity of the magnetization increases.

(51) If we explore the field with a uniformly magnetized bar, so long
that one of its poles is in a very weak part of the magnetic field, then
the magnetic forces will perform work on the other pole as it moves
about the field.

610 If we start from a given point, and move this pole from it to any
other point, the work performed will be independent of the path of the
pole between the two points; provided that no electric current passes
between the different paths pursued by the pole.

Hence, when there are no electric currents but only magnets in the

615 field, we may draw a series of surfaces such that the work done in
passing from one to another shall be constant whatever be the path
pursued between them, Such surfaces are called Equipotential

Surfaces, and in ordinary cases are perpendicular to the Lines of magnetic force.

620    If these surfaces are so drawn that, when a unit pole passes from any one to the next in order, unity of work is done, then the work done in any motion of a magnetic pole will be measured by the strength of the pole multiplied by the number of surfaces which it has passed through in the positive direction.

625    (52) If there are circuits carrying electric currents in the field, then there will still be equipotential surfaces in the parts of the field external to the conductors carrying the currents, but the work done on a unit pole in passing from one to another will depend on the number of times which the path of the pole circulates round any of these cur-
630    rents. Hence the potential in each surface will have a series of values in arithmetical progression, differing by the work done in passing completely round one of the currents in the field.

The equipotential surfaces will not be continuous closed surfaces, but some of them will be limited sheets, terminating in the electric circuit as
635    their common edge or boundary. The number of these will be equal to the amount of work done on a unit pole in going round the current, and this by the ordinary measurement = $4\pi\,\gamma$, where $\gamma$ is the value of the current.

These surfaces, therefore, are connected with the electric current as soap-bubbles are connected with a ring in M. Plateau's experi-
640    ments. Every current $\gamma$ has $4\pi\,\gamma$ surfaces attached to it. These surfaces have the current for their common edge, and meet it at equal angles. The form of the surfaces in other parts depends on the presence of other currents and magnets, as well as on the shape of the circuit to which they belong.

645    * * *


## Part III. General Equations of the Electromagnetic Field

(53) Let us assume three rectangular directions in space as the axes of $x$, $y$, and $z$, and let all quantities having direction be expressed by their components in these three directions.

650    * * *


Electromagnetic Momentum ($F$, $G$, $H$)

(57) Let $F$, $G$, $H$ represent the components of electromagnetic momentum at any point in the field, due to any system of magnets or currents.

Then $F$ is the total impulse of the electromotive force in the direc-
655 tion of $x$ that would be generated by the removal of these magnets or
currents from the field, that is, if $P$ be the electromotive force at any instant during the removal of the system

$$F = \int P \, dt.$$

Hence the part of the electromotive force which depends on the mo-
660 tion of magnets or currents in the field, or their alteration of intensity, is

$$P = -\frac{dF}{dt}, \quad Q = -\frac{dG}{dt}, \quad R = -\frac{dH}{dt}. \qquad (29)$$

Electromagnetic Momentum of a Circuit

(58) Let $s$ be the length of the circuit, then if we integrate

$$\int \left( F \frac{dx}{ds} + G \frac{dy}{ds} + H \frac{dz}{ds} \right) ds \qquad (30)$$

665 round the circuit, we shall get the total electromagnetic momentum of
the circuit, or the number of lines of magnetic force which pass
through it, the variations of which measure the total magnetic force in
the circuit. This electromagnetic momentum is the same thing to
which Professor Faraday has applied the name of the Electrotonic
670 State.

If the circuit be the boundary of the elementary area $dy\,dz$, then its
electromagnetic momentum is

$$\left( \frac{dH}{dy} - \frac{dG}{dz} \right) dy \, dz$$

and this is the number of lines of magnetic force which pass through
675 the area $dy\,dz$.

Magnetic Force ($\alpha$, $\beta$, $\gamma$)

(59) Let $\alpha$, $\beta$, $\gamma$ represent the force acting on a unit magnetic pole
placed at the given point resolved in the directions of $x$, $y$, and $z$.

Coefficient of Magnetic Induction ($\mu$)

680 (60) Let $\mu$ be the ratio of the magnetic induction in a given medium to
that in air under an equal magnetizing force, then the number of lines
of force in unit of area perpendicular to $x$ will be $\mu\alpha$ ($\mu$ is a quantity

depending on the nature of the medium, its temperature, the amount of magnetization already produced, and in crystalline bodies varying 685 with the direction.)

(61) Expressing the electric momentum of small circuits perpendicular to the three axes in this notation, we obtain the following

Equations of Magnetic Force

$$\mu\alpha = \frac{dH}{dy} - \frac{dG}{dz}$$

690
$$\mu\beta = \frac{dF}{dz} - \frac{dH}{dx} \tag{B}$$

$$\mu\gamma = \frac{dG}{dx} - \frac{dF}{dy}.$$

\* \* \*

Between these twenty quantities we have found twenty equations, viz.

| Three equations of | Magnetic Force | (B) |
|---|---|---|
| 695    " | Electric Currents | (C) |
| " | Electromotive Force | (D) |
| " | Electric Elasticity | (E) |
| " | Electric Resistance | (F) |
| " | Total Currents | (A) |
| 700    One equation of | Free Electricity | (G) |
| " | Continuity | (H) |

These equations are therefore sufficient to determine all the quantities which occur in them, provided we know the conditions of the problem. In many questions, however, only a few of the equations are 705 required [D5], [D6], [D7].

Intrinsic Energy of the Electromagnetic Field

(71) We have seen (33) that the intrinsic energy of any system of currents is found by multiplying half the current in each circuit into its electromagnetic momentum. This is equivalent to finding the integral

710
$$E = \frac{1}{2}\Sigma(Fp' + Gq' + Hr')dV \tag{37}$$

over all the space occupied by currents, where $p'$, $q'$, $r'$ are the components of currents, and $F$, $G$, $H$ the components of electromagnetic momentum.

Substituting the values of $p'$, $q'$, $r'$ from the equations of Currents (C), this becomes

$$\frac{1}{8\pi}\Sigma\left\{F\left(\frac{d\gamma}{dy}-\frac{d\beta}{dz}\right)+G\left(\frac{d\alpha}{dz}-\frac{d\gamma}{dx}\right)+H\left(\frac{d\beta}{dx}-\frac{d\alpha}{dy}\right)\right\}dV.$$

Integrating by parts [D8], and remembering that $\alpha$, $\beta$, $\gamma$ vanish at an infinite distance, the expression becomes

$$\frac{1}{8\pi}\Sigma\left\{\alpha\left(\frac{dH}{dy}-\frac{dG}{dz}\right)+\beta\left(\frac{dF}{dz}-\frac{dH}{dx}\right)+\gamma\left(\frac{dG}{dx}-\frac{dF}{dy}\right)\right\}dV,$$

where the integration is to be extended over all space. Referring to the equations of Magnetic Force (B) . . . , this becomes

$$E=\frac{1}{8\pi}\Sigma\{\alpha\mu\alpha+\beta\mu\beta+\gamma\mu\gamma\}dV, \tag{38}$$

where $\alpha$, $\beta$, $\gamma$ are components of magnetic intensity or the force on a unit magnetic pole, and $\mu\alpha$, $\mu\beta$, $\mu\gamma$ are the components of the quantity of magnetic induction, or the number of lines of force in unit of area. In isotropic media the value of $\mu$ is the same in all directions, and we may express the result more simply by saying that the intrinsic energy of any part of the magnetic field arising from its magnetization is

$$\frac{\mu}{8\pi}i^2$$

per unit of volume, where $I$ is the magnetic intensity.

(72) Energy may be stored up in the field in a different way, namely, by the action of electromotive force in producing electric displacement. The work done by a variable electromotive force, $P$, in producing a variable displacement, $f$, is got by integrating

$$\int P\,df$$

from $P=0$ to the given value of $P$.

Since $P=kf$ . . . , this quantity becomes

$$\int kf\,df=\frac{1}{2}kf^2=\frac{1}{2}Pf$$

Hence the intrinsic energy of any part of the field, as existing in the from of electric displacement, is

$$\frac{1}{2}\Sigma(Pf+Qg+Rh)dV.$$

The total energy existing in the field is therefore

$$E = \Sigma \left\{ \frac{1}{8\pi}(\alpha\mu\alpha + \beta\mu\beta + \gamma\mu\gamma) + \frac{1}{2}(Pf + Qg + Rh) \right\} dV. \qquad (I)$$

The first term of this expression depends on the magnetization of the field, and is explained on our theory by actual motion of some kind. The second term depends on the electric polarization of the field, and is explained on our theory by strain of some kind in an elastic medium.

(73) I have on a former occasion* attempted to describe a particular kind of motion and a particular kind of strain, so arranged as to account for the phenomena. In the present paper I avoid any hypothesis of this kind; and in using such words as electric momentum and electric elasticity in reference to the known phenomena of the induction of currents and the polarization of dielectrics, I wish merely to direct the mind of the reader to mechanical phenomena which will assist him in understanding the electrical ones. All such phrases in the present paper are to be considered as illustrative, not as explanatory.

(74) In speaking of the Energy of the field, however, I wish to be understood literally. All energy is the same as mechanical energy, whether it exists in the form of motion or in that of elasticity, or in any other form. The energy in electromagnetic phenomena is mechanical energy. The only question is, Where does it reside? On the old theories it resides in the electrified bodies, conducting circuits, and magnets, in the form of an unknown quality called potential energy, or the power of producing certain effects at a distance. On our theory it resides in the electromagnetic field, in the space surrounding the electrified and magnetic bodies, as well as in those bodies themselves, and is in two different forms, which may be described without hypothesis as magnetic polarization and electric polarization, or, according to a very probable hypothesis, as the motion and the strain of one and the same medium [D9].

(75) The conclusions arrived at in the present paper are independent of this hypothesis, being deduced from experimental facts of three kinds:

1. The induction of electric currents by the increase or diminution of neighbouring currents according to the changes in the lines of force passing through the circuit.

*"On Physical Lines of Force," *Philosophical Magazine,* 1861–62 [our Chapter 3].

775     2. The distribution of magnetic intensity according to the vari-
        ations of a magnetic potential.

        3. The induction (or influence) of statical electricity through
        dielectrics.

        We may now proceed to demonstrate from these principles the ex-
780 istence and laws of the mechanical forces which act upon electric currents, magnets, and electrified bodies placed in the electromagnetic
field.

## Part IV. Mechanical Actions in the Field

* * *

785 Mechanical Force on a Magnet

(77) In any part of the field not traversed by electric currents the distribution of magnetic intensity may be represented by the differential coefficients of a function which may be called the magnetic potential.
When there are no currents in the field, this quantity has a single
790 value for each point. When there are currents, the potential has a series of values at each point, but its differential coefficients have only
one value, namely,

$$\frac{d\varphi}{dx} = \alpha, \qquad \frac{d\varphi}{dy} = \beta, \qquad \frac{d\varphi}{dz} = \gamma.$$

Substituting these values of $\alpha$, $\beta$, $\gamma$ in the expression (equation 38) for the
795 intrinsic energy of the field, and integrating by parts, it becomes [D10]

$$-\Sigma\left\{\varphi\frac{1}{8\pi}\left(\frac{d\mu\alpha}{dx} + \frac{d\mu\beta}{dy} + \frac{d\mu\gamma}{dz}\right)\right\}dV.$$

The expression

$$\Sigma\left(\frac{d\mu\alpha}{dx} + \frac{d\mu\beta}{dy} + \frac{d\mu\gamma}{dz}\right)dV = \Sigma m\,dV \qquad (39)$$

indicates the number of lines of magnetic force which have their origin
800 within the space $V$. Now a magnetic pole is known to us only as the
origin or termination of lines of magnetic force, and a unit pole is one
which has $4\pi$ lines belonging to it, since it produces unit of magnetic
intensity at unit of distance over a sphere whose surface is $4\pi$ [D11].

Hence if $m$ is the amount of free positive magnetism in unit of volume, the above expression may be written $4\pi m$, and the expression for the energy of the field becomes

$$E = -\Sigma\left(\frac{1}{2}\varphi m\right)dV. \tag{40}$$

\* \* \*

(78) If a single magnetic pole, that is, one pole of a very long magnet, be placed in the field, the only solution of $\phi$ is

$$\varphi_1 = -\frac{m_1}{\mu}\frac{1}{r} \tag{41}$$

where $m_1$ is the strength of the pole, and $r$ the distance from it. The repulsion between two poles of strength $m_1$ and $m^2$ is

$$m_2\frac{d\varphi_1}{dr} = \frac{m_1 m_2}{\mu r^2}. \tag{42}$$

In air or any medium in which $\mu = 1$ this is simply $\frac{m_1 m_2}{r^2}$ but in other media the force acting between two given magnetic poles is inversely proportional to the coefficient of magnetic induction for the medium. This may be explained by the magnetization of the medium induced by the action of the poles.

## Mechanical Force on an Electrified Body

(79) If there is no motion or change of strength of currents or magnets in the field, the electromotive force is entirely due to variation of electric potential, and we shall have . . . .

$$P = -\frac{d\Psi}{dx}, \qquad Q = -\frac{d\Psi}{dy}, \qquad R = -\frac{d\Psi}{dz}.$$

\* \* \*

Here P, Q, and R are components of the electric field intensity, expressed as the gradient of an electric potential Ψ. Maxwell's aim is to derive Coulomb's law from the second term in the expression for the total energy of the field given by equation (I) above, as he has just done for the law of force between two magnets, derived from the first term. The argument follows exactly the same plan.

The result differs only with respect to the constant coefficient in the force law. The reason for that difference, which will turn out to be extremely significant for the electromagnetic theory of light, can be seen if we look back to our starting point in equation (I) at the close of section (72). There, the magnetic term was

$$\frac{1}{8\pi}(\alpha\mu\alpha + \beta\mu\beta + \gamma\mu\gamma)$$

while the electric term, with which we are now working, was

$$\tfrac{1}{2}\,(Pf + Qg + Rh)$$

The analogy between the terms will be clearer if we use the fact that $P = kf$, or $f = (1/k)P$, relating the displacement $f$ to the electric intensity $P$. The electric expression then takes the same form as the magnetic:

$$\frac{1}{2}\left[P\left(\frac{1}{k}\right)P + Q\left(\frac{1}{k}\right)Q + R\left(\frac{1}{k}\right)R\right]$$

To fetch out the role of the coefficients, we can divide out the $k$'s and the $\mu$'s in the two expressions:

Magnetic: $\quad \dfrac{\mu}{8\pi}\left(\alpha^2 + \beta^2 + \gamma^2\right)$

Electric: $\quad \dfrac{1}{2k}\left(P^2 + Q^2 + R^2\right)$

We can therefore confidently write the electric force law by replacing the magnetic equation with the corresponding electric terms. Where $m$ appears in the magnetic equation, we will have $4\pi/k$ in the electric:

Magnetic force law: $\quad f = \left(\dfrac{1}{\mu}\right)\dfrac{m_1 m_2}{r^2}$

[cf.equation (42)]

Electric force law: $\quad f = \left(\dfrac{k}{4\pi}\right)\dfrac{e_1 e_2}{r^2} \qquad (44)*$

## Measurement of Electrostatic Effects

(80) The quantities with which we have had to do have been hitherto expressed in terms of the Electromagnetic System of measurement,

which is founded on the mechanical action between currents [D12].
The electrostatic system of measurement is founded on the mechani-
860 cal action between electrified bodies, and is independent of, and in-
compatible with, the electromagnetic system; so that the units of the
different kinds of quantity have different values according to the system we adopt, and to pass from the one system to the other, a reduction of all the quantities is required.

865    According to the electrostatic system, the repulsion between two
small bodies charged with quantities $\eta_1$ and $\eta_2$ of electricity is

$$\frac{\eta_1 \eta_2}{r^2},$$

where $r$ is the distance between them.

Let the relation of the two systems be such that one electromag-
870 netic unit of electricity contains $v$ electrostatic units; then $\eta_1 = v e_1$ and
$\eta_2 = v e_2$, and this repulsion becomes

$$\left(v^2\right)\frac{e_1 e_2}{r^2} = \left(\frac{k}{4\pi}\right)\frac{e_1 e_2}{r^2} \quad \dots \tag{45}$$

whence $k$, the coefficient of "electric elasticity" in the medium in
which the experiments are made, i.e. common air, is related to $v$,
875 the number of electrostatic units in one electromagnetic, by the
equation

$$k = 4\pi v^2 \tag{46}$$

The quantity $v$ may be determined by experiment in several ways.
According to the experiments of MM. Weber and Kohlrausch,

880        $v = 310{,}740{,}000$ metres per second [D13].

(81) It appears from this investigation, that if we assume that the
medium which constitutes the electromagnetic field is, when dielectric, capable of receiving in every part of it an electric polarization,
in which the opposite sides of every element into which we may
885 conceive the medium divided are oppositely electrified, and if we
also assume that this polarization or electric displacement is proportional to the electromotive force which produces or maintains it,
then we can shew that electrified bodies in a dielectric medium will
act on one another with forces obeying the same laws as are estab-
890 lished by experiment.

The energy, by the expenditure of which electrical attractions and repulsions are produced, we suppose to be stored up on the dielectric medium which surrounds the electrified bodies, and not on the surface of those bodies themselves, which on our theory are merely the
895   bounding surfaces of the air or other dielectric in which the true springs of action are to be sought.

\* \* \*

## Part VI. Electromagnetic Theory of Light

(91) At the commencement of this paper we made use of the op-
900   tical hypothesis of an elastic medium through which the vibrations of light are propagated, in order to shew that we have warrantable grounds for seeking, in the same medium, the cause of other phenomena as well as those of light. We then examined electromagnetic phenomena, seeking for their explanation in the properties of
905   the field which surrounds the electrified or magnetic bodies. In this way we arrived at certain equations expressing certain properties of the electromagnetic field. We now proceed to investigate whether these properties of that which constitutes the electromagnetic field, deduced from electromagnetic phenomena alone, are
910   sufficient to explain the propagation of light through the same substance.

\* \* \*

At this point, Maxwell embarks on an analytic demonstration, in the most general terms, of the existence of an electromagnetic
915   wave propagating with the velocity $v$. We have been in the habit of substituting for Maxwell's general demonstrations other demonstrations that have been simplified by being limited to variations, where possible, in a single dimension.

Very fortunately, in this crucial instance, Maxwell has done
920   our work for us. A few years after publication of the "Dynamical Theory," in conjunction with an article reporting his own new measurements of the value of the factor $v$, he included a simplified derivation of the electromagnetic wave equation. We will do well, then, to depart from our program to incorporate at this point
925   Maxwell's own derivation, in which the field vectors align conveniently with the coordinate axes, and propagation will be

shown to occur in the $z$-direction alone. Only a few adjustments will have to be made in equation references and notation*:

930 Let the direction of propagation be taken as the axis of $z$, and let all the quantities be functions of $z$ and of $t$ the time; that is, let every portion of any plane perpendicular to $z$ be in the same condition at the same instant.

Let us also suppose that the magnetic force is in the direction of the axis of $y$, and let $\beta$ be the magnetic intensity in that direction at any
935 point.

Let [a] closed curve . . . consist of a parallelogram in the plane $yz$, two of whose sides are $b$ along the axis of $y$, and $z$ along the axis of $z$. The integral of the magnetic intensity taken round this parallelogram is $b\,(\beta_0 - \beta)$, where $\beta_0$ is the value of $\beta$ at the origin.
940 Now let $p$ be the quantity of electric current in the direction of $x$ per unit of area taken at any point, then the whole current through the parallelogram will be

$$\int_0^z bp\,dz,$$

and we have . . .

945
$$b(\beta_0 - \beta) = 4\pi \int_0^z bp\,dz.$$

If we divide by $b$ and differentiate with respect to $z$, we find

$$\frac{d\beta}{dz} = -4\pi p. \tag{14}$$

Let us next consider a parallelogram in the plane of $xz$, two of whose sides are a along the axis of $x$, and $z$ along the axis of $z$.
950 If $P$ is the electromotive force per unit of length in the direction of $x$, then the total electromotive force round this parallelogram is $a(P - P_0)$.

If $\mu$ is the coefficient of magnetic induction, then the number of lines of force embraced by this parallelogram will be

$$\int_0^z a\mu\beta\,dz,$$

*[The paper in question is "On a Method of Making a Direct Comparison of Electrostatic and Electromagnetic Force; with a Note on the Electromagnetic Theory of Light," (*SP.* ii/125). We will refer to this report again, for his account of the experiment itself; at the moment, we will quote from the appended "Note" beginning at p. 140.]

955 and since . . . the total electromotive force is equal to the rate of diminution of the number of lines in unit of time,

$$a(P - P_0) = -\frac{d}{dt}\int_0^z a\mu\beta dz.$$

Dividing by *a* and differentiating with respect to *z*, we find

$$\frac{dP}{dz} = -\mu\frac{d\beta}{dt}. \tag{15}$$

960    Let the nature of the dielectric be such that an electric displacement
*f* is produced by an electromotive force *P*,

$$P = kf, \tag{16}$$

where *k* is a quantity depending on the particular dielectric, which may
be called its "electric elasticity."

965    Finally, let the current *p*, already considered, be supposed entirely
due to the variation of *f*, the electric displacement, then

$$p = \frac{df}{dt}. \tag{17}$$

We have now four equations, (14), (15), (16), (17), between the
four quantities $\beta$, *p*, *P*, and *f*. If we eliminate *p*, *P*, and *f*, we find

970 [D14]

$$\frac{d^2\beta}{dt^2} = \left(\frac{k}{4\pi\mu}\right)\frac{d^2\beta}{dz^2}. \tag{18}$$

If we put

$$\frac{k}{4\pi\mu} = V^2, \tag{19}$$

the well-known solution of this equation is

975    $$\beta = \Phi_1(z - Vt) + \Phi_2(z + Vt), \tag{20}$$

shewing that the disturbance is propagated with the velocity *V*.]

\* \* \*

Here we return to the main text of the "Dynamical Theory,"
resuming in the course of Section 95.

980    (95) . . . This wave consists entirely of magnetic disturbances,
the direction of magnetization being in the plane of the wave. No

magnetic disturbance whose direction of magnetization is not in the plane of the wave can be propagated as a plane wave at all.

Hence magnetic disturbances propagated through the electromag-
985  netic field agree with light in this, that the disturbance at any point is transverse to the direction of propagation, and such waves may have all the properties of polarized light.

(96) The only medium in which experiments have been made to de-termine the value of $k$ is air, in which $\mu = 1$ , and therefore, by equa-
990  tion (46),

$$V = v. \tag{72}$$

By the electromagnetic experiments of MM. Weber and Kohlrausch

$$v = 310{,}740{,}000 \text{ metres per second}$$

is the number of electrostatic units in one electromagnetic unit of elec-
995  tricity, and this, according to our result, should be equal to the velocity of light in air or vacuum.

The velocity of light in air, by M. Fizeau's* . . . experiments, is

$$V = 314{,}858{,}000 \; ;$$

according to the more accurate experiments of M. Foucault** . . . ,

1000  $$V = 298{,}000{,}000.$$

The velocity of light in the space surrounding the earth, deduced from the coefficient of aberration and the received value of the radius of the earth's orbit, is

$$V = 308{,}000{,}000.$$

1005  (97) Hence the velocity of light deduced from experiment agrees sufficiently well with the value of $v$ deduced from the only set of experi-ments we as yet possess. The value of $v$ was determined by measur-ing the electromotive force with which a condenser of known capacity was charged, and then discharging the condenser through a galva-
1010  nometer, so as to measure the quantity of electricity in it in electro-magnetic measure. The only use made of light in the experiment was to see the instruments. The value of $V$ found by M. Foucault was ob-tained by determining the angle through which a revolving mirror

*Comptes Rendus, Vol. xxix. (1849), p. 90.
**Ibid. Vol. lv. (1862), (1862), pp. 501, 792.

turned, while the light reflected from it went and returned along a meas-
1015 ured course. No use whatever was made of electricity or magnetism.

The agreement of the results seems to shew that light and magnetism are affections of the same substance, and that light is an electromagnetic disturbance propagated through the field according to
electromagnetic laws.

1020 * * *

(100) The equations of the electromagnetic field, deduced from
purely experimental evidence, shew that transversal vibrations only
can be propagated. If we were to go beyond our experimental knowledge and to assign a definite density to a substance which we should
1025 call the electric fluid, and select either vitreous or resinous electricity
as the representative of that fluid, then we might have normal vibrations propagated with a velocity depending on this density. We have,
however, no evidence as to the density of electricity, as we do not
even know whether to consider vitreous electricity as a substance or
1030 as the absence of a substance.

Hence electromagnetic science leads to exactly the same conclusions as optical science with respect to the direction of the disturbances which can be propagated through the field; both affirm
the propagation of transverse vibrations, and both give the same
1035 velocity of propagation. On the other hand, both sciences are at a
loss when called on to affirm or deny the existence of normal
vibrations.

Relation between the Index of Refraction and the Electromagnetic
Character of the substance

1040 (101) The velocity of light in a medium, according to the Undulatory
Theory, is

$$(1/i)\ V_0,$$

where $i$ is the index of refraction and $V_0$ is the velocity in vacuum. The
velocity, according to the Electromagnetic Theory, is

1045
$$\sqrt{\frac{k}{4\pi\mu}}$$

where, by equations (49) and (71), $k = (1/D)\ k_0$, and

$$k_0 = 4\pi V_0^2.$$

Hence

$$D = \frac{i^2}{\mu},\tag{80}$$

1050 or the Specific Inductive Capacity is equal to the square of the index of refraction divided by the coefficient of magnetic induction.

* * *

### Absolute Values of the Electromotive and Magnetic Forces Called into Play in the Propagation of Light

1055 (108) If the equation of Propagation of light is

$$F = A \cos\left(\frac{2\pi}{\lambda}\right)(z - Vt),$$

the electromotive force will be

$$P = -A\left(\frac{2\pi}{\lambda}\right)V \sin\left(\frac{2\pi}{\lambda}\right)(z - Vt);$$

and the energy per unit of volume will be [D15]

1060
$$\frac{P^2}{8\pi\mu V^2},$$

where $P$ represents the greatest value of the electromotive force. Half of this consists of magnetic and half of electric energy.

The energy passing through a unit of area is

$$W = \frac{P^2}{8\pi\mu V};$$

1065 so that

$$P = \sqrt{(8\pi\mu VW)},$$

where $V$ is the velocity of light, and $W$ is the energy communicated to unit of area by the light in a second.

According to Pouillet's data, as calculated by Professor W. Thom-
1070 son,* the mechanical value of direct sunlight at the Earth is

---

* *Transactions of the Royal Society of Edinburgh* 1854 ("Mechanical Energies of the Solar System").

83.4 foot-pounds per second per square foot.

This gives the maximum value of *P* in direct sunlight at the Earth's distance from the Sun,

$$P = 60,000,000,$$

1075 or about 600 Daniell's cells per metre.

At the Sun's surface the value of *P* would be about
13,000 Daniell's cells per metre.
At the Earth the maximum magnetic force would be 193.*
At the Sun it would be 4'13.

1080    These electromotive and magnetic forces must be conceived to be reversed twice in every vibration of light; that is, more than a thousand million times in a second.

*The horizontal magnetic force at Kew is about 1.76 in metrical units.

# Interpretive Notes

23  "Weber"
The theory of Wilhelm Weber is indeed simple and impressive, incorporating as Maxwell says both electrostatic and electromagnetic phenomena in a single comprehensive equation. It is based on the assumption that electromagnetic effects arise from the forces between pairs of electric particles acting upon one another at a distance. Its genius is to include electromagnetic effects such as the force between currents by making these forces dependent on the motions of the particles. There is no role in it for an intervening medium.

It is expressed in a single equation, containing just three terms, the first of which is Coulomb's law. The other two involve not just the positions, but the relative *velocities* and *accelerations* of the charges:

$$f = \frac{q_1 q_2}{r^2}\left[1 + \tfrac{1}{2}c^2\left(\frac{dr}{dt}\right) + \left(\frac{1}{c^2}\right)\left(r\frac{d^2 r}{dt^2}\right)\right]$$

Here $r$ is the distance between the two particles, $q_1$ and $q_2$ are their charges, and $c$ is a constant whose significance will become apparent to us later. Thus the first term is indeed simply Coulomb's law:

$$f = \frac{q_1 q_2}{r^2}$$

The second term speaks of a force proportional to $dr/dt$, the *relative velocity* of the two charges, and the third term involves the factor $d^2 r/dt^2$—the *second derivative* (the derivative of the derivative, or the rate of the rate) of $r$ with respect to $t$. This is the derivative of the velocity, or the relative *acceleration* of the two particles. If we assume with him that electric currents consist of motions of these charged particles, Weber is able to show that this combination of terms not only accounts for all the forces that act between current-carrying conductors, but implies the phenomena of electric induction in varying magnetic fields as well.

Maxwell has serious concerns about certain implications of Weber's law, but his principal reason for turning from it is surely the simple fact that it omits the electromagnetic field and ascribes all effects to the action of charged particles upon one another over distances. However successful, it remains a theory in the mode of the *first sight*.

37  "mechanical difficulties"
Are these not *metaphysical* difficulties? Or is Maxwell making a point: that our notion of *mechanics* belongs to the domain of metaphysics?

42  "in another direction"
This choice of a different, and quite opposite, "direction," is surely a striking example of the conscious authorship of what is commonly called today, following Thomas Kuhn, a "new paradigm."

48 "The electromagnetic field"

The term, as well as the concept, is becoming increasingly explicit as Maxwell proceeds. If the field is indeed "a part of space," then we know that "space" must not mean for Maxwell a merely geometrical concept, but a substance belonging to the world. Space is an entity that includes what must be a new form of matter—for though dense and massive, it has no weight—and that matter must be in a state of intense motion. This is the foundation of the concept of the "ether." Note that thus stated, Maxwell's notion of the electromagnetic field seems quite compatible with contemporary quantum theory of the photon, and with relativistic theories of space. The photon is massive, bears energy, and is in motion with the velocity of light. But it has no rest mass and in this sense, no "weight." We might say that the photon is the answer to the riddle that Maxwell poses quite reasonably in formulating the concept of the field. And if "space" is to become a *physical* construct, we may construct it, relativistically or otherwise, as our experience of the cosmos directs.

57 "gross matter"

This does not seem to be a problem of the inadequacy of vacuum pumps but a conceptual watershed. There must be another kind of matter, which is not "gross"—that is, matter which is massive yet without proportionate weight.

67 "an aetherial medium"

The phenomena of heat referred to here are those of *radiant* heat, that is, of a wave motion with wavelengths in the infrared, longer than those of red light. The burner on an electric stove, for example, radiates in this way, heat without light. Light and radiant heat are transmitted without diminution—all the better, actually—through the most complete vacuum.

71 "must have formerly existed"

The light and radiant heat of the sun take some eight minutes to reach the earth: on absorption, of course, this radiation is primarily converted to "thermal" form—the random molecular motion of a hot body. This assures us that what is transmitted to us in radiant form is indeed a vast quantity of energy that can be used to produce mechanical effects and estimated in the common unit of energy measurement, the calorie. This energy must therefore have existed in some form during the long period of transmission in the space between sun and earth. We may know nothing with confidence concerning a mechanism, but in this sense, the *dynamical* approach to the ether, in terms of energy, seems founded on unexceptionable evidence. Space becomes an energy-containing entity, or "ether," of some kind.

80    "the existence of a pervading medium"

We see that the existence of the optical ether is a given for Maxwell, not a result of his theory. His theory, which we saw emerging from "Physical Lines," serves to *identify* the known optical ether with an electromagnetic medium. Of course as a result, we come to know much more about the formal structure of the optical ether.

If there is such a pervading medium, the question immediately arises as to our motion with respect to it. Maxwell did his own experiments to find evidence of such an ether drift, looking in fact for a shift of spectral lines when observations are taken at successive seasons of the year—that is, when the earth is moving in different directions through a supposed fixed ether. His results were negative, and were in a sense a precursor of the famous Michelson–Morley experiment in which it is established definitively that no such "absolute" measurement is possible. The absence of measurable motion with respect to space is one foundation of relativity theory. Though his own study was relatively crude, Maxwell's negative result must have left him puzzled. We know that the question was in fact on his mind and would surely have been pursued had he lived longer.

90    "'actual' energy"

The usual term is "kinetic" energy, but Maxwell is perhaps making a point. Leibniz, from whom the notion of energy in modern physics derives, acquired the term from Aristotle. Aristotle talks about the act, which most constitutes the being of a thing (we are in a sense most human when we are doing the human thing, exercising human virtue—not when we are asleep or resting on our laurels!). By contrast, for example, the acorn is only an oak tree *potentially*. The Greek term for the acorn's state got Latinized as *potential,* while the Greek for the being-in-act was retained as *en-ergeia,* "[being] at-work." Maxwell wants, it seems, to point to the sense in which the real, "actual" energy is that of motion. It is as if kinetic energy were in some way more real than potential energy. The ether, transmitting energy from the sun to the earth, is not just conveying a possibility but is in itself full of actual energy—that is, of matter in intense motion.

101    "evidenced to our senses"
This remark cuts two ways: we should, on the one hand, be alert for sensible evidence of other activities of the ether. But there may be "some of these" kinds of motion which leave no trace among the phenomena.

104    "acted on by magnetism"
This was Faraday's way of putting it, in his description of what is now called the "Faraday effect," a magnetic effect upon a ray of plane polarized light. According to the wave theory of light, the vibrations are always transverse to the direction of propagation, that is, occur in a plane perpendicular to the direction of the ray. In ordinary light, such as that produced by a flame or an incandescent lamp, these vibrations occur at random angles within that plane. In a beam that has been polarized by passage through a polarizing filter, however, these vibrations occur in only a single transverse direction.

Faraday discovered that the plane of polarization of polarized light is rotated when the light is passed through glass placed in a strong magnetic field. While one might think it more natural to say that the field acted on the glass, whose properties as a transparent medium were thereby altered, Faraday—and now Maxwell—prefer to say that the action is upon the *luminiferous* medium itself. The argument that follows from this view, which is for him an important clue, comes on the next page.

139    "Here we have evidence"
That is, mere motion of a body through "empty" space in which there is a magnetic field will cause the impressive effects Maxwell recites, those of the "electromotive force." A conducting wire,

which yields manifest currents, is only one example. Therefore space containing a magnetic field, however empty otherwise, is demonstrably full of power. Faraday was similarly impressed by this evidence from the "moving wire" that becomes an alternative means, beside the iron filings, of tracing the structure of the field.[1]

165    "But when electromotive force acts"
This is Maxwell's crucial move, which we have already observed in "Physical Lines." Despite his generous references to Faraday and to Mossotti, Maxwell is very much going out on a limb of his own, and demonstrating his deep confidence in the concept of the field, in asserting the existence of an effect for which there is no direct supporting evidence—perhaps never could be. We will soon see that the "dielectric" referred to here need not be a physical nonconductor such as glass but may be the vacuum itself. Then space itself he is claiming will be "polarized" by the electro-motive force. As we have seen in earlier discussions, what is in question is not only an understanding of the state of affairs in the vacuum, but the nature of "charge" itself.

Figure 4.1 represents perhaps the simplest form of the situation Maxwell is describing: the parallel-plate capacitor, sustaining an electric field in a dielectric that fills the space between the plates. The figure is taken from Hertz, who believes that the drawing at the top best represents Maxwell's concept of charge.[2] Hertz uses black to denote positive charge and gray, negative. In the top drawing, there is *nothing*—no "charge" of an electric substance—on the plates themselves. Neither is there any net charge in the dielectric between for, in any section such as that at *B,* adjacent charges cancel.

If the space contains no dielectric substance, but is the vacuum itself, we cannot understand the "polarization" in terms of a displacement of positive and negative electricities, yet the phe-nomenon of charge remains the same, and Maxwell retains the concepts of polarization and displacement. We have seen how he addressed this concept in "Physical Lines," where it was a ques-tion of elastic deformation of the very physical cells. Here in the context of dynamical reasoning, where our terms will be more abstract, details of the concept may be left open. In either case, there will be nothing deposited on the "charged" plates: electric

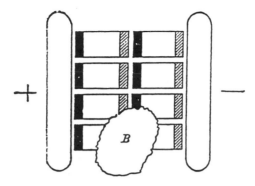

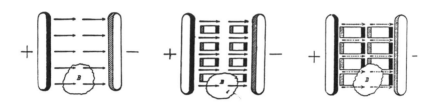

Fig. 4.1. Polarization of the vacuum in the parallel-plate capacitor: (top) Maxwell's theory, as interpreted by Hertz; (below) alternative interpretations of polarization.

"charge" will consist entirely of a state of polarization of the field between.

By contrast to Maxwell's theory, the lower set of Hertz's drawings all presuppose the existence of an electric substance deposited on the charged plates. The first represents pure action at a distance, the electricities on the opposite plates acting independently of any intervening medium. In the second, the medium conveys the force and may exert a modifying effect, while in the third, though electricity is present on the plates, the source of action is shifted entirely to the field.

178   "its variations constitute currents"

Here Maxwell draws the bottom line of this reasoning. *If the electromotive force acts in empty space to polarize it, then the variations of this polarization, at its outset and at its release, will*

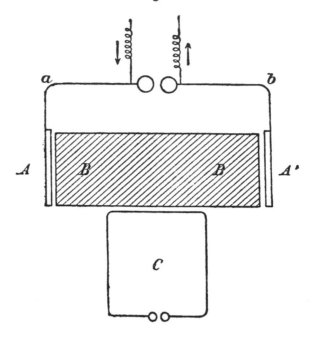

Fig. 4.2. Hertz's sketch of his own experiment to demonstrate the existence of Maxwell's displacement current.

*constitute currents.* This is not a mere thought, but immediately implies an entirely new phenomenon—for as we have remarked in Chapter 3, if there are indeed these displacement currents in empty space, there must be associated Oersted magnetic fields as well.

Think again of the space between the plates of the parallel-plate capacitor of Fig. 4.1—in a perfect vacuum if we wish. As the plates are charged, an electrostatic field arises, as we know. But according to this new proposal, the charging capacitor ought to generate a *magnetic* field as well, for as the switch is closed the sudden polarization of the field will be equivalent to a current, though in empty space. Such a transient current should have magnetic effects, even if equally transient, according to the same laws as ordinary currents. This effect of the hypothetical "displacement current" would be a phenomenon for which Maxwell has no empirical evidence—and, indeed, does not expect any.

This question would seem in principle to admit experimental testing, but calculation showed from the outset that the effects must be extremely slight. Nonetheless, in an effort to put Maxwell's theory to the test of a crucial and decisive experiment, Hermann Helmholtz in Berlin did years later propose this experimental challenge as a prize problem. The young Heinrich Hertz took up the challenge and devised an apparatus that would set up rapidly alternating electric fields in a paraffin dielectric. In Fig. 4.2, an oscillating electric potential is applied through wires connected at $a$ and $b$ to plates $A$ and $A'$. $B$ is the paraffin block, the dielectric in which the displacement currents are to arise. The loop $C$, intercepting the magnetic fields of the displacement currents, will experience induced potentials, exhibited at the spark gap at the bottom of the figure.

The outcome, however, was that the experiment quickly got out of hand, with unanticipated wave effects radiating from the apparatus throughout the room and reflecting in turn from the walls. These dramatic results of an experiment that succeeded beyond expectation became the first published demonstration of artificially produced electromagnetic propagation—the phenomenon we know as "radio" waves.[3]

205    "a certain yielding"
It was this "yielding" that constituted the elastic deformation Maxwell assigned to his vortices, in "Physical Lines" (recall Fig. 3.23).

211    "at the same time so connected"
Maxwell's visions of the connectedness of affairs in the electromagnetic field led to sketches such as those of Fig. 4.3, reproduced here from the *Treatise*.

232    "followed the reverse order"
Maxwell's reference here is to the fundamental paper in which Helmholtz introduced the breathtaking principle of the universal conservation of energy—the insight that underlies "dynamical" theory. Helmholtz and Thomson each showed that if the laws of electromagnetic forces acting mechanically on conductors are assumed to be true, then it must follow from the newly recognized principle that there will exist compensating phenomena of electromagnetic induction. The mechanical action

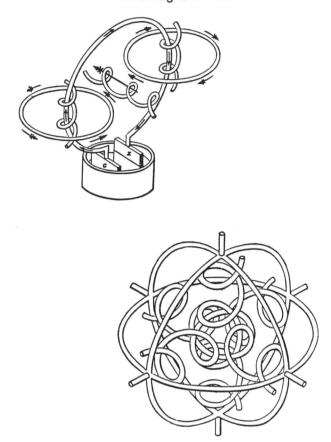

Fig. 4.3. Maxwell's sketches suggesting the connectedness of the electromagnetic field.

must be matched by some counterforce acting on the electric current itself to balance the energy bookkeeping. The precise laws of electromagnetic induction can be derived in this way without recourse to independent experimental evidence. Nature has to concur!

More generally, "Lenz's law," the "law of perversity," which says that induced electric currents will always flow in such a direction that their fields will oppose the motion that produced them, directly follows from the conservation of energy. All of this assures that we do not get something for nothing.

Maxwell, as he says, will follow the "reverse order." That is, he will begin with the phenomena of electromagnetic *induction* and their laws, and then (in Part IV) will derive the mechanical forces on interacting currents from the laws of induction together with the principle of conservation of field energy. Does it matter in which order we proceed? It seems that there is at least an important rhetorical distinction. Maxwell's method from the outset directs attention to the *field*—not as derivative, an inference from mechanical forces, but as the primary seat of energy from which mechanical effects derive.

243 "General Equations of the Electromagnetic Field"
This will be the set of first principles of electromagnetic theory, now known to the world as "Maxwell's equations." Maxwell states them here in words, but will match these words with equations as the argument develops. The letter designations he gives them here will remain constant through all subsequent reformulations in this paper. In modern vector formulation, they are usually expressed as four; Maxwell will give us twenty, summarizing them under eight headings:

> (A) Maxwell places the mysterious "displacement current" principle first: *total current* is ordinary conduction *plus* the supposed displacement current.
> (B) Circuits are coupled inductively, through an electromagnetic momentum proportional to the number of lines of magnetic induction that link them.
> (C) A current (now, *total* current) gives rise to a magnetic field that encircles the current (the Oersted effect). Expressing this now in energy terms, the total work to carry a unit charge around the current in any path is proportional to the magnitude of the current alone.
> (D) Induced emf in a body arises from one or more of three terms: (1) its motion through a magnetic field, (2) time variation of the field, and (3) electrostatic potentials.
> (E) A dielectric can be treated as an elastic medium, with electric displacement proportional to intensity of the electric field.
> (F) In conductors, current flow is proportional to electromotive force; the constant of proportionality expresses the conductivity of the medium (Ohm's law).

(G) What has been known as electric "charge" ["free elec-
    tricity"] in field terms corresponds to a divergence of the
    lines of electric displacement. (It may indeed *consist* of
    such divergence.)

(H) Electric displacement is continuous: there are no sources
    or sinks. Hence it increases or diminishes in a region
    only as currents flow in or out.

267   "intrinsic energy of the Electromagnetic Field"
      As we have seen, the dynamical theory focuses on the concept of
      *energy* as essential to our thinking about the field, in effect lifting
      the discussion above the details of physical theory. The field
      becomes the primary locus of energy, upon which bodies in the
      field, in their interactions, are drawing.

274   "an independent method of electrical measurement"
      What is perhaps unwritten here is the centrality in the dynamical
      theory of the question of units and standards of measurement of
      electrical quantities. What might be supposed to be a merely
      practical question—and, indeed, is a practical question of increas-
      ing urgency in Britain as Maxwell writes—also has much deeper
      significance. Thus, the program of measurement in the electro-
      magnetic and electrostatic systems referred to here is no mere
      corollary of a completed theory but constitutes the culmination of
      the theoretical investigation itself, making its completion possi-
      ble. It is a theoretical question, then, that leads Maxwell into a
      central project of experimental investigation. This will be the
      theme of the "Postscript" with which our study will conclude.

288   "propagated through a non-conducting field"
      An innocent remark: but the "non-conducting field" of central
      interest becomes *space itself.* What Maxwell is proposing to show
      is that the intricate assemblage of electric and magnetic phenom-
      ena his theory encompasses, and which ordinarily we associate
      with conducting wires and iron magnets, in fact occurs as well in
      a perfect vacuum—in "empty" space itself. The array of experi-
      mental apparatus and the behaviors we have associated with
      "electromagnetism" have been incidental: the intrinsic process
      goes on, prior to and apart from bodies, within the field itself.

295 "that light itself . . . is an electromagnetic process"
It is still difficult to comprehend this proposition, familiar though it is
to every schoolchild. Like a good showman (or a good teacher, if
there is a difference), Maxwell draws from his investigation below
the startling question, "What is the voltage of sunlight?" and shows
us how to compute the answer. Despite some question about his
arithmetic, the result is that an arm's length of sunlight has a voltage
several times that which comes out of a modern light socket. When,
lawyer-like, he adds the inclusive phrase "and other radiations if
any," he includes, on principle, the spectrum of possible electromagnetic radiations, which we now understand runs beyond the visible
range through the ultraviolet, x-rays, and gamma rays.

300 "how perfect and regular the elastic properties of the medium"
Maxwell will compute the velocity of the electromagnetic propagation predicted by his theory on the basis of electrical measurements made at leisure in a laboratory. Yet the vibrations that
constitute the propagation must occur at rates of "more than a
thousand million times in a second." Hence his awe at the excel-
lence of the medium, which performs in precisely the same way at
the lowest frequencies and the highest.

304 "the index of refraction"
If light is an electromagnetic process, then optical phenomena
such as refraction should have their ground and explanation in
electric and magnetic properties of materials. Maxwell sees a
good prospect that the electrically measured dielectric constant of
glass will account for its refraction of light. As Maxwell perceives, the measured properties will no longer be the same at very
low and very high frequencies. Prisms make spectra because the
dielectric constant of glass is wavelength-dependent.

305 "Conducting media are shown"
Again, if light is electromagnetism, then a good conductor, such
as silver, may short-circuit light and reflect it back on itself.

309 "set forth by Professor Faraday"
Not only the first paper, "On Faraday's Lines of Force," but the
whole series of Maxwell's investigations is here presented with

genuine grace, as the implementation of Faraday's own thought. In "Thoughts on Ray Vibrations," Faraday had said:

> The view which I am so bold as to put forth considers, therefore, radiation as a high species of vibration in the lines of force which are known to connect particles and also masses of matter together. . . . The kind of vibration which, I believe, can alone account for the wonderful, varied, and beautiful phenomena of polarization, is not the same as that which occurs on the surface of disturbed water, or the waves of sound, for the vibrations in these cases are direct, or to and from the centre of action, whereas the former are lateral. . . .
>
> The occurrence of a change at one end of a line of force easily suggests a consequent change at the other. The propagation of light, and therefore probably of all radiant action, occupies time; and, that a vibration of the line of force should account for the phenomena of radiation, it is necessary that such vibration should occupy time also. I am not aware whether there are any data by which it has been, or could be ascertained whether such a power as gravitation acts without occupying time, or whether lines of force being already in existence, such a lateral disturbance of them at one end as I have suggested above, would require time, or must of necessity be felt instantly at the other end. . . .[4]

Maxwell is perhaps thinking of that phrase, "whether there are any data . . ." as he remarks, ". . . in 1846 there were no data. . . ." But what a long road must be traveled, to make the data that now so interest Maxwell—above all, the relation of the units of charge in electromagnetic and electrostatic systems of measurement—relevant to the question, of the velocity of transmission of electromagnetic effects!

I take Maxwell to be perfectly candid in his statement that the electromagnetic theory of light, as proposed by himself, is "the same in substance" as the concepts set forth by Faraday. Since it is surely not the same in form, Maxwell must be speaking here of a shared intuitive conviction concerning the world and an accompanying belief as to the broad type of account that the sciences must give. There must in truth be a field, real and substantive, that fills that interval between sun and earth which for the mathematicians is represented merely by numbers generated by equations. This field in some substantive way bears those powers that manifest themselves throughout nature, and thus, filling all interstices, makes the universe substantively and intelligibly whole.

Such a conviction, it seems, Maxwell met adumbrated in Faraday's exploratory thoughts; he now returns it to its author, fully realized, in the form of a rounded dynamical theory.

330 "When the strength of the current"

Maxwell points to a fundamental feature of the magnetic effects of currents: increasing the current increases the magnetic field in direct proportion. We have to be reminded that things might be otherwise! If they were, and some sort of saturation were involved, any proposed medium would be "nonlinear" and the problem would be far more complicated. Real magnetic materials do present such nonlinear effects.

337 "between the driving point . . . and a flywheel"

Maxwell asks us to think—though in very general terms—of the relation between an electric current and its magnetic effects as if it were a mechanical connection, like that of one part of a system to another. We might imagine, for example, that the "driving point" in this case is the crankshaft of an automobile engine, at just the point at which it connects to the clutch. The running engine itself has a certain amount of momentum, including that of its flywheel. But when the clutch is engaged and the autombile is in motion at full speed, the engine is bound to the momentum of the entire vehicle. On a downhill, the vehicle may drive its engine rather than the other way around.

All of this is felt by the engine as a single momentum, equivalent to that of a flywheel, at the point of attachment, which we have taken as the "driving point." This single effect is the *reduced momentum*—"reduced" in the sense that its detail can be ignored and a merely formal equivalent substituted.

In the same way, if the magnetic field has the effect of a momentum-bearing system coupled to the current that produces it, then without knowing anything about the details of the field, we can speak quantitatively and accurately of its *reduced momentum* at the wire.

339 "The unbalanced force"

Maxwell is invoking Newton's second law of motion: the force on a body is proportional to the time-rate of change of its momentum. The net force is the "unbalanced" force.

Fig. 4.4. Peter Guthrie Tait, lifelong friend, who with Thomson turned Maxwell's attention to Lagrangian theory.

347    "If there are two electric currents"
The magnetic effects of two currents sum directly to yield a single result (could we imagine things otherwise?). Then we have in effect two driving points for our one connected system. Connected in this way to the same system, they are connected as well to each other. A motion of one can be expected in general to produce an effect of some sort at the other.

353    [D1] A Dynamical Illustration

358    "by the general equation of dynamics"
Maxwell here cites Lagrange, who near the end of the eighteenth century produced a work of great importance in analytic mechanics, his *Traité Analytique,* with which Maxwell has been becoming acquainted.[5] The power of Lagrange's methods lies in their generality: its symbols specify quantities, with no

more detail concerning actual existences than the algebra requires.

Thus, a *reduced momentum* corresponding to the motions of some unspecified system will do as well, for Lagrange, as a knowledge of the actual motions of the parts. Indeed, the greater ignorance will serve us better in this case, as any demand for further detail would only lead us beyond the bounds of attainable truth. These are often called *generalized* quantities.

Here, $C$ is such a generalized quantity, the *effective* mass of the system; $w$ is the corresponding generalized velocity associated with it. The "general equation" relates work and energy. On the right, each generalized force times its generalized displacement measures an increment of work; the right-hand side thus measures the total work done. On the left, the term $C\,dw/dt$ is in generalized terms the mass times acceleration of the whole system; that product equals the accelerative force applied. But that force times its displacement $\delta z$ is a quantity of very real work: namely, the work done in accelerating the system and thereafter stored in the form of its increased kinetic energy.

363  "$dw/dt = p\,du/dt + q\,dv/dt$"

We assumed that the velocity of the whole body $C$ was equal to the sum of the velocities of the drive points $A$ and $B$, each multiplied by some appropriate factor measuring its connection to the system, i.e.,

$$w = pu + qv$$

We get the corresponding statement for accelerations by computing rates, i.e., differentiating each term here with respect to time:

$$\frac{dw}{dt} = p\frac{du}{dt} + q\frac{dv}{dt}$$

$p$ and $q$ are unaffected by the time differentiation, as they are assumed to be constants. Similarly, if the velocities add, so must the displacements.

366  [D2] Derivation of Equation (1)

372  "We may call $Cp^2u + Cpqv$ the momentum of $C$"

Above, Maxwell spoke of "substituting" in the general equation to take account of the particular relations we have specified. The way in which the pair of equations numbered (1) follow is shown in Discussion [D2]. Each of these tells us what will be experienced at one of the "driving points" of our system.

Here we go back to the basic law of motion, Newton's second law, relating force and momentum. This is commonly expressed in a tidy equation $f = ma$ (force equals mass times acceleration). Such a brief statement is appropriate when the "mass" in the case is a fixed quantity, that of a cart or a stone. But we need a broader statement in the case of a complex connected system in which elements are not so rigid or neatly isolated. More generally, we may state that a force applied to a system is always equal to the time-rate of change of its momentum. If the momentum is for the moment denoted $M$, then,

$$f = \frac{d}{dt}(M)$$

If we apply this to our present system, the effect of a force applied at $A$ turns out (see Discussion [D2]) to be given by

$$x = \frac{d}{dt}\left(Cp^2u + Cpqv\right)$$

Force is always the time-rate of change of momentum, as $F = ma$ is identical to such a state of change. Thus, an experimenter applying forces at $A$ will in effect be encountering a "momentum" equal to $(Cp^2u + Cpqv)$—the "reduced momentum" of the system referred to $A$.

Each driving force, at whatever point it is applied, is acting to accelerate the same system, but is coupled to the system in a different way and yields what is in effect a different perspective. Since the momentum reduced to $A$ has two components, each in the form of a constant multiplied by a velocity—e.g., $Cp^2$ multiplied by $u$—while "momentum" was originally defined as the product of mass and velocity, we might speak of that constant as a virtual mass. Thus, $Cp^2$ would be the virtual mass referred to $A$ and associated with generalized velocity $u$. Such a virtual mass is only *metaphorically* a "mass" arising from the connectedness of the system: there is no specific body anywhere to correspond to it: nothing "has" that mass—yet the *system* does!

374 "If there are many bodies"

At this point, things really get interesting. Previously, we had two driving points, A and B, connected to a single body C. Note that there is just one velocity associated with each driving point—that is what we mean by a "driving point." We now go on to imagine that a number of bodies, say $C_1$, $C_2$, $C_3$, etc., are joined in a single overall system and that in general each driving point is connected to all of them. We extend the concept of virtual mass to collect, for example, all the mass terms referred to A and associated with its velocity $u$, summing them into a single mass-equivalent. Maxwell names this term L, and writes:

$$L = \Sigma\, (C\, p^2)$$

along with corresponding expressions for M and N.

389 "the motion of A is resisted"

This further step introduces what we might call a "dissipative force" specifically associated with just the velocity of the driving point A, namely the force $Ru$. Then the total force observed by an experimenter at point A will be the sum of forces Maxwell calls ξ, given by equation (3).

396 [D3] "Maxwell's Example of the Bell Ringers"

397 "This effect on B"

We begin to see where Maxwell is going with his paradigm of a generalized mechanical system in its application to electromagnetism. The two driving points become two circuits in the field, in effect two handles on the field, which is the connected system in our case. In the electromagnetic system, these are points at which we have something tangible to deal with—on which to make observations and take measurements. These will be the circuits composed of actual wires, which are subject to forces we can feel as we move them from place to place.

By contrast, we cannot observe "currents" flowing "in" them—we specifically want to get away from any such assumptions! We can place meters in them—voltmeters and ammeters—and the motions of their needles will yield readings, quite observable quantities associated with the instruments as electromagnetic "driving points."

Thus the "velocity" at *A* is to be the *current* there; it works formally as a velocity because it can be written as a time-rate of change—*what* is moving or changing we no longer need to ask in this very metaphorical approach. In turn, an increase in that "current" will be, equally formally, the time-rate of change of a "velocity," or an "acceleration," which will show up as an effect at driving point *B*. The *force* there will be an "electromotive force" . . . reported by a voltmeter, not a force of the sort we could feel, or in any obvious way a "force" at all. The *field* becomes the "body" composed of all the component systems we have called $C_1$, $C_2$, etc., responsible for these mysterious, but quantitatively very precise connections. Perhaps the only quantity reliably exempted from this play of mind is *energy* itself: the work done on the system and borne within it is always real and literal work. Other quantities may take literal form from time to time, as we shall see.

Lagrangian theory is a genuinely rhetorical instrument, part of the symbolic repertory of our age. These are by now genuinely metaphorical quantities—they play the roles of their namesakes. If we are not to suppose that anything is actually moving in the wire when a "current" "flows," how is a "current" a *velocity?* It is enough for the Lagrangian theory that it appears as a time rate of change of a quantity we can call a displacement. Is this some sort of rhetorical conjuring trick—or by shedding our too literal presuppositions, have we hit on the root ideas of the parameters of a *mechanical system rightly understood?*

409    "depending on the form and position"
We have been thinking of *L, M,* and *N* as measures of the inertial properties of a machine-like system, and hence have taken them as constants. Maxwell now reminds us that these coupling constants of an electromagnetic system depend entirely on the forms and positions of the circuits in it: we have now to think of these masslike coefficients as varying in time. It is as if the "machine" had become flexible, no longer a rigid system.

Since we are now speaking of moving tangible mechanical objects such as wires and coils, we meet at this point the possibility of bringing "mechanical" considerations in the literal sense into the picture. There will be ponderable Newtonian forces at work on these now movable circuit elements, forces which will act on straight wires or coils as *L, M,* and *N* change, even if the electrical variables are held constant.

419   "using the word momentum"
Once again, we make no assertion about where this "momentum" may exist; if it is "a velocity existing in a body," we have no idea where this "body" is. Surely for Maxwell, the "momentum" of the current that we say is "flowing in the wire," must rather be thought of as distributed throughout the entire field, which has no outer edge. That is, the "momentum" in question must be distributed from the wire to infinity—must be everywhere *outside* the wire!

424   "we call it electromotive force"
In this paragraph, Maxwell reminds us of the new scope we are giving to words. As Maxwell's theory stands, "force" here is "not ordinary mechanical force." We must wonder, as it is clear Maxwell does, about this precarious transformation of the most fundamental terms of the physical sciences. When we make this kind of formal or metaphorical substitution, using symbolic structures to justify our usage and carry our meanings, are we draining our terms of their significance? Is there any content to the term "momentum," when it is only a *reduced* momentum we are referring to and there is nothing we can point to that can serve as a "body" by which the so-called momentum is to be borne?

If we were to say that it is no concern of the physicist's to weigh such questions, so long as the meters read and the equations work, might we not be speaking from the modern side of a major watershed that separates the conventional wisdom of the present age from Maxwell's broader and deeper, less tethered concerns? My own suspicion is that Maxwell, like Faraday, very much does want to *know*—insofar as knowledge is humanly accessible. If that is the case, then this reshaping of words such as "force" and "momentum" is not an abandonment of the philosophical concern for their meanings but a redirection of the philosophic search. We are witnessing, in this third of Maxwell's efforts, not at all an abandonment of the search for an underlying reality but a critical recasting of the terms of the search into more valid, more potent

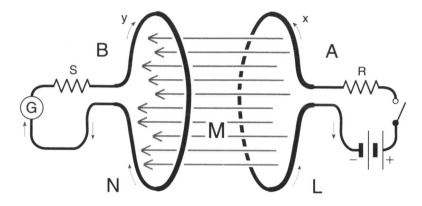

Fig. 4.5. Electromagnetic induction.

forms. We are dealing here with entities the mind can track, though the eye cannot see.

434  "$Lx + My$"
Maxwell now writes $x$ and $y$ where we have been seeing $u$ and $v$—electrical currents are playing the role of virtual velocities.

440  "the equation of the current"
The force on the current has two tasks, which we may think of sequentially: it must first overcome the dissipative resistance to the current's flow, which generates heat; then the net, remaining force will accelerate the current. It is as if, in pushing on a box, the force $Rx$ had to be exerted to overcome sliding friction, while only the remaining force were used to accelerate the box. If the push were just sufficient to overcome friction, the box would slide along without acceleration. The "friction" in the electrical case, directly proportional to the current, is electrical resistance as expressed by Ohm's law.

447  "Let there be no electromotive force"
Maxwell is speaking of two circuits electromagnetically coupled, as in Fig. 4.5. If we think of them as lying in parallel planes, their coupling will depend entirely on their proximity, while the empty space between them, with its equivalent coefficient of momentum

*M*, will play a very active role in the behavior of the system. Apart from this coupling, we have two independent loops, which we may think of as primary and secondary, with the current in each driven by an electromagnetic force. The electromotive force in the primary *A* is denoted $\xi$ and arises from the action of the battery through a resistor and switch. The secondary, on the left, for which the electromotive force is $\eta$, is passive; there is no battery but only a resistor and meter.

The only electromagnetic force on the current *y* in the secondary will thus be due to changes in the current $\xi$ in *A* by way of the electromagnetic coupling *M* between the two. We can write $\eta = 0$ and $N = 0$ in the second of equations (5) above, where the assertion that $\eta = 0$ means that there is no independent driving force on the secondary loop. Thus the only forces on the current in *B* will arise by way of the coupling *M* from the current in *A*, and the Ohm's law force resisting the flow of *y* itself, namely *Sy*. The equation of motion becomes

$$Sy + \frac{d}{dt} Mx = 0$$

Since we assume here that the circuits are rigid and only the currents vary, *M* is constant and factors out in differentiating:

$$\frac{d}{dt}(Mx) = M\left(\frac{dx}{dt}\right)$$

Then

$$Sy = -M\left(\frac{dx}{dt}\right)$$

and

$$y = -\left(\frac{M}{S}\right)\frac{dx}{dt}$$

This is already very interesting: there will be an inertial force on the current *y* in *B* due to the acceleration *dx/dt* of the current in *A*. This arises from their coupling coefficient (their "mutual inductance") *M* and demonstrates vividly that the electromagnetic momentum of empty space can have very real physical effects.

Maxwell goes on to integrate both sides of this equation with respect to time, and calls the time integral of the current $y$ the "total induced current," denoted by the symbol $Y$. If in a water-pipe analogy we think of current as the "flow" of charge, this time integral is the bucket of water or total charge that has passed in circuit $B$ due to the changing current in $A$. The negative sign expresses Lenz's law, the "law of perversity" which we have met before. Here again, Lenz's law insures that we conserve energy, and not get "something for nothing."

465    [D4] Induction by Relative Motion of the Conductors

467    "the equation between work done and energy produced"
At this point, Maxwell begins reshaping the equations of electromagnetic induction. Up to this point, they have addressed principally force and momentum; now they will speak of the relationships of work and energy. We may keep Fig. 4.5 in view and think again of the simpler case in which the circuits are rigidly mounted, so that $L$, $M$, and $N$ are constant. Forces—however metaphorical—are acting on both currents and doing work upon them. Work, we recall, is force times displacement; *power*, by definition, is *rate of doing work*. Thus power is force times time-rate of displacement, or velocity. (Maxwell refers to power as "work done in unit of time.")

Maxwell's equation (8), then, which multiplies the electromotive force in each circuit by the corresponding current, is computing *power*. It sums the total power in the two circuits. To get it, he has simply multiplied each of the equations (5), which gave the forces, by the corresponding current, and then added them up.

476    "we have, first"
Again, we are speaking of *power*, the time-rate of doing work: if $R$ and $S$ were the resistances of two light bulbs, $H$ would be the total of the power they emitted as heat and light. (Recall that power dissipated in a resistor is the product of its resistance and the square of the current flowing through it.) In the field view, the power manifested in $S$ was first present *en route* in the space between loop $A$ and loop $B$, in its transmission from primary to secondary. The energy that exists altogether in the space itself at

one moment may be used in the next by a motor at *S* to lift a heavy weight!

480 "They may be written"

What follows is an identity that results simply from using the rules of differentiation and working through the algebra. The two rules are, for any *u*, *v*, and *x* ([D1] of Chapter 3):

The product rule:

$$\frac{d}{dx}(uv) = u\frac{dv}{dx} + v\frac{du}{dx}$$

The power rule:

$$\frac{d}{dx}\left(x^n\right) = nx^{(n-1)} \quad \text{([D1] of Chapter 3).}$$

Persistent application of these rules (and taking *L*, *M*, and *N* as variables) will show the equivalence of Maxwell's new expression to equation (8).

You might well wonder why he has gone to the trouble of doing this, if the two expressions are equivalent? The striking fact is that though they are *algebraically equivalent*, they have interestingly different rhetorical force—as we have seen before, a consideration important to Maxwell. The new left-hand expression, which becomes equation (10), bears special meaning, which he goes on to discuss.

483 "If *L*, *M*, *N* are constant"

In this case, the time derivatives of *L*, *M*, and *N* vanish and, with them, the last three terms of Maxwell's equation. The remaining terms emerge as the *intrinsic energy of the currents* alone, as if the currents had been built up in rigid circuits.

489 "the seat of this motion"

This sentence is perhaps the focus of the still-incipient concept of the "field," Maxwell is contemplating the expression he has derived [equation 10] for the "intrinsic energy of the currents." It really is energy: there is nothing metaphorical about it—such as we have seen in the cases of "electromotive *force*," "electromagnetic *momentum*," or even displacement or velocity, when applied to the motions of something that is called "electricity" but is not in the wire. Since this energy is associated inherently with a motion of some sort [$x$ and $y$ in equation (10) are the time derivatives we call "currents"], the terms of the equation must be *kinetic* energies, and hence they must represent an "actual motion" which, though "imperceptible to our senses," "exists"—"in the space surrounding" the circuits. We should note that Maxwell inserts the word "probably."

The "field," then, is existent yet imperceptible, in space. What it is, is a question left completely open. The virtue of the dynamical theory—unlike the physical one—will be to bring the inquiry exactly to this point and not to go an inch further into hypothesis or speculation. An important assertion has been made about *being: space* is not an empty geometrical container but a coherent, connected physical system bearing the energy of motion. It cannot again be thought of as an Empedoclean vacuum in which atoms are flying about. Yet as it is both *imperceptible* and at the same time *physical,* the concept of the "physical" itself must be prepared to yield. Fundamental concepts of natural philosophy are shaken here. "Mass" is present, bearing energy, yet we are asked now to think of a form of mass that is radically imperceptible.

492    "The remaining terms"
In each of these terms of equation (11), work is done through the variation of one of the inductive coefficients, represented by $dL/dt$, $dM/dt$, or $dN/dt$. Since these coefficients depend on the geometrical form and relation of the circuits, their variations must depend on the motion or reshaping of a (tangible) physical body—for example, the wire constituting a circuit. And that, Maxwell says, will give rise to a (tangible) mechanical force.

504    "If $A$ and $B$ are allowed to approach
The term in equation (11) that reflects their mutual position will be $(dM/dt)$ $xy$, where $M$ measures their "mutual inductance." Remember that (11) expresses the power—the time-rate of doing work—so the actual work done will be the time integral of this term. But summing (integrating) the time-rate over the time just gives the difference in $M$ itself, so the work done is $(M' - M)xy$,

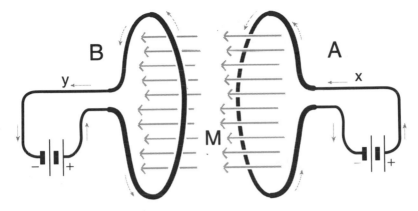

Fig. 4.6. Two circuits attracting.

where $M$ is some initial value and $M'$ a varying final value. However, since work is force times distance, force in turn is the space-rate of variation of work. In this case, with $M$ the varying quantity, the force will thus be just $(dM/ds)xy$.

510   "this will be an attraction"
How can we figure out whether we are dealing with an attraction or a repulsion in equation (12)? To go back to the beginning, keeping Fig. 4.5 in view, we initially wrote equations for the forces $\xi$ and $\eta$ acting on $A$ and $B$. These were the equations numbered (3). When we converted to equations representing the rate at which work was done [equations (8)], we were computing the work done on the circuits $A$ and $B$. Hence if the product $xy$ is positive ("$x$ and $y$ are of the same sign"), and if $dM/ds$ is such that $M$ increases when $s$ decreases (i.e., when $A$ and $B$ approach one another), work will be done on the two circuits by their mutual force whenever they approach; this represents energy being drawn from the field. This is what we mean by "attraction."

Let us apply commonsense to this situation, together with what we know about circuits and their magnetic fields. If currents flow in $A$ and $B$ as shown in Fig. 4.6 (currents flowing in the same sense), they will set up corresponding magnetic fields, indicated by the arrows along the axis in the figure. But fields of the same sort could be set up by two permanent magnets that had opposite poles adjacent. And we know that opposite poles attract! There-fore, our interpretation of the energy equation makes sense in familiar terms.

519    "the electrotonic state"
Recall the discussions of the electrotonic state in Chapter 3.

531    * * *
Maxwell now has the task of bringing these very general equa-tions to bear on specific situations upon which experiments can be done. We may safely leave these mathematical labors to him and rejoin him when it is time to enjoy their fruits.

534    "let us explore the electromagnetic field"
In "Faraday's Lines," as we saw, Maxwell began his investigation of electromagnetism by building the intuitive concept of a shaped continuum, on the model of a geometrical fluid. The result marked a new direction in physics, with an approach applicable to electric and magnetic phenomena in a number of ways; these ways did not connect, however, into a single vision of a coherent system. In "Physical Lines," he envisioned a space-filling mechanism that might in principle supply the missing connection: it was not a hypothesis, only an idea, yet even so it went too far into the domain of untestable fiction. Now, by addressing the physical system of the second paper only *in principle,* and avoiding all detailed hypotheses, he has found a way into his subject that seems coextensive with what is permitted to him to know.

Yet just at this point we must confront a new question. The idea of the connected mechanical system has left us with equations and, perhaps, dynamical ideas of a fundamentally new order. But what has happened to the idea of the field as the *continuum* of "Faraday's Lines"? Not only has the concept itself slipped away, but the method of geometrical intuition seems abandoned, yielding to equations and their symbols. This is not a situation to satisfy Maxwell. He sets out now toward a reconciliation of methods: from the equations of the dynamical theory, he seeks to fetch the continuum of "Faraday's Lines." It will be free of the aura of hypothesis that troubled the second paper, yet empowered with the clear vision of connectedness that the first paper had been unable to achieve.

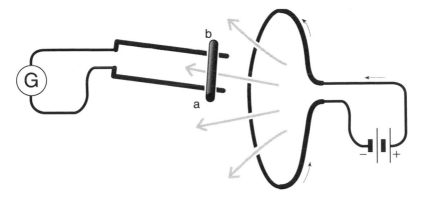

Fig. 4.7. The exploratory circuit.

As the dynamical theory is founded on the mutual inductance of two circuits, this—and not iron filings or the exploratory compass needle—must now supply the principle for tracing the shape of the field and interpreting its meaning. Figure 4.7 represents the method by which Maxwell proposes to do this. Here, the left-hand circuit takes the form of a rectangle in which one short leg *ab* slides on two rigid wires. To trace the field established by the current in the coil on the right, we move the sliding wire at a steady rate and watch the meter. If it is a sensitive galvanometer, its indications will be proportional to the time-rate at which the flux through the left-hand circuit is changing, i.e., the rate at which ab is cutting field lines. Surprisingly, a sluggish meter works even better, for such a "ballistic" indicator responds not to current but to the total effect, or time integral of the current. This means in this case that it will indicate the total change in flux cut by *ab* in a motion, however irregularly *ab* may be carried over the path.

If *ab* on the other hand is moved along the direction of the lines, the flux will not vary and the meter will not register. The slider thus becomes an indicator within dynamical theory of the direction of the magnetic field. The inverse phenomenon Maxwell describes is especially interesting: if a current is passed through *ab* and is free to slide, it will move outward to include greater flux and thus maximize the value of *M*, for we have seen at line 499 that the tendency toward the increase of electromagnetic momentum is the driving principle of all mechanical forces exerted by the field.

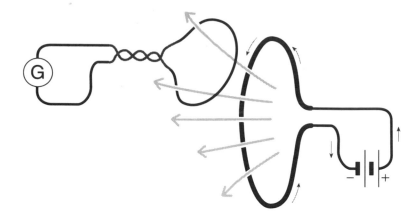

Fig. 4.8. The rigid exploratory circuit.

In turning to this new mode of tracing the magnetic field, Maxwell has by no means abandoned his allegiance to Faraday. For Faraday, too, the method of the "moving wire" provides a new kind of insight into the shape and nature of the field, and occupies a prominent place in the later section of the *Experimental Researches.*

556    "suppose *B* to consist of a very small plane circuit"
We now return to the circuit of Fig. 4.5, but replace *B* with a very small wire loop provided with a flexible connection in order to explore the field freely (Fig. 4.8). Driven by the tendency of *M* to increase, the loop will tend to turn so as to align its axis with the direction of the field, and indeed we see that in this way Maxwell has quietly brought us back to the exploring compass needle with which we began. Now, however, we see the compass needle in a new light, as an instrument for delineating the electromagnetic momentum of the field.

We now complete our return to the original method of "Faraday's Lines" by shrinking *B* to a small, movable rigid circuit. Maxwell asserts that the value of *M* will be a maximum when the plane of the circuit is perpendicular to the newly defined lines of force. Why should this be?

We know from our earlier discussion of the relation between *M* and the force on a conductor leading to Maxwell's equation (12)

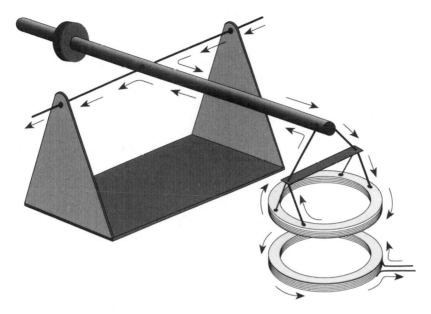

Fig. 4.9. Two repelling poles.

that under the conditions described there, there will be a force due to a motion *ds* if currents *x* and *y* are flowing and *dM/ds* is positive, i.e., if the motion tends to increase *M*. If our detecting circuit above is allowed to turn, it will do so until there is no further increase in *M*, i.e., *M* has reached a maximum. As no force acts to move it, it is aligned with the magnetic lines as Maxwell has now defined them, and the axis of our small circuit points along the direction of the field lines.

We can see that this has indeed brought us back to the beginnings of "Faraday's Lines." For the small circuit, we have seen, acts like a magnet and, thus, in turn, like the magnetic needle with which we first defined the field lines. By this route, we have returned to the small tracing needle with which it all began! And the field we are now describing is the same field with which Faraday began his study of the lines of force. Maxwell has reconciled Faraday's later method of the moving wire with the original method of the magnetic needle and iron filings.

576    "the whole field will be filled"
       In this account, we see Maxwell retracing the steps by which the
       field was developed in "Faraday's Lines." An important differ-
       ence now, however, is the understanding of the entire structure in
       terms of the values of *M*.

582    "If this curve be a conducting circuit"
       A subtle shift is occurring from the physical exploration of the
       field to the geometric understanding of its form. Above, Maxwell
       *drew* a "closed curve" and spoke of it as having an *M*! Evidently,
       we must be prepared to do as he now suggests and think of this
       merely geometric curve as replacing the physical wire as the path
       of our exploratory circuit. Since our definitions have assured that
       *M* measures the number of lines of force passing through a circuit;
       any curve, understood as a virtual circuit, will similarly catch a
       number of lines and measure *M*. As *M* now measures the number
       of lines of force and vice-versa, an increase in the number of *lines*
       intercepted immediately entails an increase in *M*, and hence a
       force acting, as Maxwell claims here.

587    "the amount of work done"
       We saw earlier that with a change in *M*, the work done on a circuit
       carrying current *y* was proportional to the value of *y* and the
       increment in *M*. If we denote with the symbol *W* the work done as
       a consequence of a change in *M*, we may write

$$W = (M' - M)\, xy = (\Delta M)\, xy$$

       where *x* and *y* are the primary and secondary currents. With the
       identification of *M* and the number of lines, the increment in the
       number of lines intercepted by it now measures the work done. In
       this way, by virtue of the dynamical theory, the lines of force are
       taking on new, direct physical meaning. We have to remember,
       however, that there is no claim as to where in the field the energy
       that yields this work resides.

596    "the force acting on a unit pole"
       It is not improper at this point to recall the simpler talk, long ago,
       of permanent magnets with "poles." We can now simulate them at
       will by means of circuits to which the dynamical theory applies.

Where once we spoke of two similar magnetic poles repelling one another at unit distance with unit force, we may now speak of two small coils—as in Fig. 4.9—each carrying current and of a mutual inductance $M$ (or should we say, a derivative $dM/ds$) so contrived that they repel one another with unit force at unit distance in the vacuum.

Now, in media other than the vacuum, they will be found to repel with a different force. For the same number of lines, the force will be less by a factor here termed the "coefficient of magnetic induction" of the medium. For iron, that factor is exceptionally large. In other words, for the same force, the number of lines is much larger: this is the sense in which iron is highly "magnetic."

611   "the work performed will be independent of the path"
As Maxwell reconstructs the magnetic field on the new, dynamical basis, the question arises whether the *potential* concept can still be used. As we have seen, a potential can be assigned to each point in a space if for some force, the work done in carrying a test body from one point to another is independent of the path taken. One of the two points can serve as reference point, so that a unique potential can be assigned to every other point in the space—the convention being to take the "point at infinity" as reference. In Chapter 1, we took gravitational potential as a model in this and assigned electric potentials in the same way.

As long as we spoke of magnets and their poles, there was no problem in assigning magnetic potentials and we did that effectively in our discussion of "Faraday's Lines." However, when currents flow in circuits, it becomes possible to carry a test pole completely around a current-carrying conductor—from a point $P$, let us say, by way of another point $Q$ back to $P$ again. Work is then done in carrying the test body "from $P$ to $P$," and evidently when that is the case, no unique value of potential can be assigned to $P$; it all depends on where you have been.

We know that a permanent magnet can be understood as the consequence of little circuits inside the iron, and here Maxwell finds a source of suggestion. Although these circuits in principle introduce the ambiguity of potential, in fact we do not have access to them and remain always outside the loop. Under these circumstances, the

problem vanishes and potentials can once again be uniquely defined. This is the principle Maxwell now adopts in general: as
long as we remain outside our circuits, promise not to pass through
them, and only look at the effects of whole circuits upon one
another, the potential concept remains tenable and valuable in the
new dynamical context. The way has been opened to look at
mutual inductances from the point of view of potential.

Here, Maxwell simply contrives to reintroduce the magnetic
test pole, keeping the inevitable second pole comfortably out of
the way. In this manner, he re-establishes magnetic potentials and
equipotential surfaces.

625    "If there are circuits carrying electric currents"
In effect, Maxwell is warning us of the ambiguity that, as we have
seen, underlies our equipotential surfaces: in principle they will
not have unique values but "a series of values," if our test pole is
allowed to trace a path that encloses a conductor. This, however,
does not prevent them from being very useful.

633    "The equipotential surfaces will not be continuous"
By a remarkable piece of geometrical insight, Maxwell is about to
show that the equipotential surface for a current flowing in a closed
loop will take a form that suggests a soap bubble attached to its ring.
In the pages of the "Dynamical Theory" that follow, but which we
must forgo here, such insights concerning the potential are put to
ingenious use in arriving at a full set of equations characterizing the
basic phenomena of electromagnetism.

One such device, taken from the *Treatise,* is illustrated in
Fig. 4.10, which represents a set of equipotential lines about a
current-carrying loop—each line is then a slice through an
equipotential surface, and the set of equipotential lines are slices
through a set of nested equipotential surfaces. They look very
much like a set of nested soap bubbles, springing from the circuit
as if attached to a common ring.

At this point we need a new geometrical concept. The *solid
angle* subtended at point *P* by the current-carrying ring is defined
as that portion of the surface of an entire sphere about *P* that is cut
by the cone of which the ring is the base, where the whole sphere
is said to contain $4\pi$ units, or *steradians.*

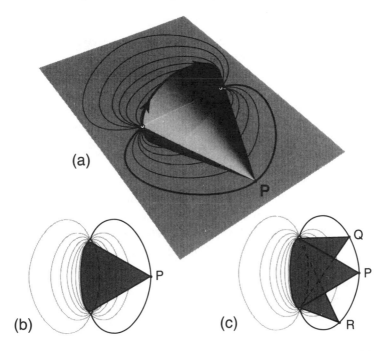

Fig. 4.10. Maxwell's "soap bubbles"—equipotential lines around a current: (a) The solid angle subtended at *P* measures the potential of the equipotential line on which *P* lies and of the equipotential surface from which it is cut; (b and c) For all points *P, Q,* and *R* on the same equipotential line and surface, the circuit subtends equal solid angles.

Maxwell's geometrical insight at this point is that an equipotential surface about a current-carrying loop is precisely a surface at all points of which the loop subtends equal solid angles. Thus the solid angles at *P, Q,* and *R* in the figure are all equal. By thus transforming a merely analytic relation into a property of a geometrical figure, Maxwell is enabled to present the field's structure to the consideration of the mind's eye. Carried through, the device becomes of great service to Maxwell in constructing drawings of fields. It is regrettable that we cannot pursue further here this line of reasoning, so close in spirit to the very concept of the field itself as a structure we can visualize as filling space.

646    "General Equations"

Maxwell now proceeds, in Part III, to reshape the dynamical principles he has established into the "General Equations of the Electromagnetic Field." They are reviewed in Discussion [D5] and expressed in vector form in [D6]. For the sake of our own goal, however—the unveiling of the electromagnetic theory of light—we must select from Maxwell's discussion only those sections that are most essential to this end.

647    "three rectangular directions in space"

To this point, we have simplified things for ourselves where possible by discussing each question with respect to one or two components rather than all three. We now look more generally at a three-dimensional case, and vector quantities "expressed," as Maxwell says, "by their components in these three." The general statement of Maxwell's equations in vector form will be the subject of Discussion [D6], below.

652    "the components of electromagnetic momentum"

Thus far in the "Dynamical Theory" we have associated this momentum with currents and conductors—the specific "driving points" of the system. Now, we formally recognize that the electromagnetic momentum is to be envisioned as distributed throughout space, and denote the momentum of the whole system as a vector with components $F$, $G$, and $H$ defined at every point throughout the field. The equivalence of *total impulse* to *change in momentum* that Maxwell goes on to assert is a general principle, one way of formulating Newton's second law of motion, which we have already applied to electromagnetic systems (see Chapter 3, Discussion [D2])

663    "if we integrate"

Maxwell is writing a general expression, applicable to a circuit of any form, in any field. We may think of the circuit as divided into segments, which in the limit, as the summation turns into an integral, become as small as we please. Each segment has its own orientation with respect to each field component. A differential term of the form $dx/ds$, for example, takes this into account by measuring the projection of the segment $ds$ on the $x$-axis, and

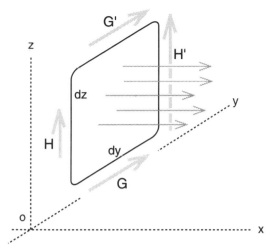

Fig. 4.11. Electromagnetic momentum and induction.

multiplying the vector component *F* by just that projected length. (*dx/ds*)*ds* is the projected length of the segment; the product of the corresponding vector component by that length becomes *F*·(*dx/ds*)*ds*. In equation (30) the same is done for each vector component and the result summed.

671    "the elementary area *dydz*"
Maxwell now very helpfully (at last!) takes a simplified case (Fig. 4.11) in which the circuit is a rectangle perpendicular to the *x*-direction. Recall that *F*, *G*, and *H* are the electromagnetic momenta *per unit length* of the circuit. As we travel around the loop in a counterclockwise direction to determine the total electromagnetic momentum, we will sum the successive products (*H' dz*), (*G' dy*), (*H dz*) and (*G dy*). Let us suppose, as the figure suggests, that in this case *H'* > *H*, while *G'* = *G*. In our reckoning, (*H' dz*) will be a positive quantity, as the vector and the path length have the same sense, while (*H dz*) will be negative, the vector and path length being of opposite senses. If *H'* and *H* were equal in magnitude, the two would cancel. However, as *H'* and *H* are unequal, there will be a net value in the summation. A nonzero value of the derivative *dH/dy*, which expresses this inequality, therefore

assures that as we travel around the loop, the two values of H will not cancel one another. (We used a formally similar mode of reasoning in Chapter 3, illustrated in Fig. 3.12.) Recall also that in Section (49) Maxwell has identified, as here, the number of lines of force embraced by a closed loop in the magnetic field with the electromagnetic momentum $M$ associated with the loop. Our integration of the electromagnetic momentum *around* the loop thus becomes at the same time a measurement of the number of lines of magnetic force *through* it.

688    "Equations of Magnetic Force"
These equations simply extend to all three components the relation asserted above for two. We will be needing them in this general form.

692    * * *
We have perforce left to Maxwell the formal task of working out the full set of equations that express the theory of the field whose dynamical foundations we have just traced. He does this in a sequence of sections, from Section 58 through Section 70, most of which we have had to omit in order to keep our present project within bounds. The result, he says, is summarized in "twenty equations"—the very relations we met in verbal form back in Section 18 (line 236). Those expressing the electrodynamic relations follow directly as interpretations of the dynamical theory we have just watched Maxwell develop.

Readers who have met "Maxwell's equations" in modern texts may well wonder why Maxwell gives twenty, when only four are included today? Considerable emphasis is often placed in fact on the elegance entailed in there being so few. For one thing, many of these are given here in sets: there are six sets of three each, followed by two individual equations. A set of three—one equation for each component—is replaced in current formulations by a single vector statement, which with one voice makes a statement pertaining to all three components. (This vector form of Maxwell's equations is developed in Discussion [D6].) We should note that Maxwell contributed greatly to the development of the formal vector calculus required.

Even with that reduction to vector form, however, there would still remain eight equations in Maxwell's theory instead of the four in modern texts. It seems possible that Maxwell is interested rather in making the *fullest* statement of the elements of his science than in presenting the mathematically most *elegant* theory. If a new equation suggests an interesting insight, it may be important to add it, even though it could in principle be derived formally from the others.

Could we express this distinction in terms of the difference between rhetoric and logic? Two statements that are logically equivalent may have very different rhetorical effects, approaching a thought from different points of view. In many respects, as we have been seeing, the field itself is in this sense a rhetorical device, visualizing in terms of lines and spatial structures what can be logically computed, perhaps more elegantly, by means of equations alone. But Maxwell is on the track of ideas, not merely of computations, and he draws upon the instruments of rhetoric wherever they seem to him to give assistance to the mind.

Maxwell's equations are reviewed more fully in Discussion [D5], and following the introduction of some principles of the vector calculus are set out in modern vector form in Discussion [D6].

705   [D5] Maxwell's Equations

705   [D6] Maxwell's Equations in Vector Terms

705   [D7] "A Contradiction in Maxwell's Equations?"

707   "The intrinsic energy of any system"
The reasoning to which Maxwell refers may be summarized as follows. At equation (10), we found the energy belonging to currents. If we apply that equation to a single conductor giving rise to a magnetic field, we have

$$E = \tfrac{1}{2}Lx^2$$

Since $Lx$ is the momentum of the current, if we translate this to the new field terms, it becomes

$$E = \tfrac{1}{2}(Lx)x = \tfrac{1}{2}\Sigma\,(F)p'dV$$

where $F$ is the momentum and $p'$ the total current, integrated over the whole field. (Note that here $V$ refers to volume, and $dV$ to a portion of space.) The expression must, however, be extended to all components of current and momentum, as in Maxwell's equation (37).

714    "Substituting the values of $p'$, $q'$, $r'$"
That is, substituting for each current its Oersted field, for example in the case of $r'$:

$$r' = \frac{1}{4\pi}\left(\frac{d\beta}{dx} - \frac{d\alpha}{dy}\right)$$

717    [D8] Integrating by Parts
Integration by parts is a widely applicable analytic device (see the Discussion), which permits us here to make substitutions of the form

$$G\frac{d\alpha}{dz} = -\alpha\frac{dG}{dz}$$

throughout. This process, with tha aid of equations (B), takes us to equation (38), expressing field *energy* entirely in terms of magnetic *intensity*.

726    "the value of $\mu$"
This is a good point to reconsider the meaning in the dynamical theory of the quantity $\mu$ and of the terms *magnetic intensity* and *magnetic induction*. Recall that, having begun with individual conducting circuits as our handles on the electromagnetic system, we introduced magnetic lines (in Section 27) by way of the exploring circuit. The number of lines passing through a circuit, delineated in Maxwell's "reticulum," determines the electromagnetic momentum in the circuit, and thus a change in their number gives rise to electromotive forces. These lines are the lines of *magnetic induction*.

Things would be very simple if it were not for the effects of iron and other magnetic materials. Maxwell introduces a "long uniformly magnetized bar" with its poles. The force acting on a pole depends on the material as well as on the number of lines of induction as determined by the exploring circuit; μ is the coefficient that measures the effect of the material. "The number of lines of force (i.e., of *induction*) passing through unit of area is equal to the force acting on a unit pole (the magnetic *intensity*) multiplied by μ."

Writing α for the *x*-component of the magnetic intensity, Maxwell thus writes μα for the magnetic induction.

727  "the intrinsic energy"

Equation (38) has been rewritten, with μ assumed uniform and therefore factored out:

$$E = \frac{\mu}{8\pi} \int \left( \alpha^2 + \beta^2 + \gamma^2 \right) dV$$

(Here we substitute the conventional integral sign for Maxwell's summation notation.)

Since α, β, and γ are the *x*-, *y*-, and *z*-components of a single vector, the Pythagorean theorem (in three dimensions) tells us that the magnitude of the total intensity vector, which Maxwell writes as *I*, is the square root of the sum of their squares, so we may write

$$E = \frac{\mu}{8\pi} I^2$$

That is mere mathematics. But it is a bold and striking claim to assert that there is *in fact* this computed energy density associated with every unit of volume of empty space in a magnetic field! The energy, once associated only with the field as a whole, is now distributed as a spatial structure, associating a specific energy density with every point of the space.

733  "The work done"

This is the same problem as the energy stored in a stretched spring. If the force is $f = kx$, where $k$ is the spring constant, then the work of stretching to any extension $x$ is

$$E = \int kx\,dx = \tfrac{1}{2}kx^2 = \tfrac{1}{2}fx$$

Maxwell's equation applies this result to electric displacement and extends it to all three components.

744    "The first term"
Equation (I), measuring the energy everywhere in the field, now becomes our guide to the electromagnetic theory of light. If we assume that the medium is uniform, so that μ and $k$ are constants, the equation takes this form:

$$E = \int\left[\frac{1}{8\pi}\mu\left(\alpha^2 + \beta^2 + \gamma^2\right) + \frac{1}{2k}\left(P^2 + Q^2 + R^2\right)\right]dV$$

Magnetization                Electric Polarization
(KINETIC ENERGY)            (POTENTIAL ENERGY)

748    "On a former occasion"
In "Physical Lines," the rotation of the vortices accounted for all of the kinetic energy, since the electric particles, though they were ultimately involved in a great deal of motion, had negligible mass and thus in themselves carried no energy. Potential energy entered only with the elastic deformation of the vortices.

756    "as illustrative, not as explanatory"
This paragraph seems to epitomize the distinction between "Physical Lines" and the present essay. What was in some sense "explanatory" there has here become "illustrative"—intended, he says above, to "direct the mind . . . to mechanical phenomena which will assist . . . in understanding the electrical ones." Is this not essentially the role of metaphor, in which by looking at one thing in some way more accessible, we come to comprehend another, more remote? "Electromagnetic momentum" is, and is not, momentum in the mechanical sense; electromotive force is only metaphorically a force. To call the one a "momentum" and the other a "force," is at the same time to understand and not to understand. Is it that we are in this way extending our grasp of the meanings of the terms themselves, which we thought of as "mechanical" only because it was in that context that we first learned to use them?

"Energy" is, by contrast, a term that has already been generalized—or was conceived more universally from the outset—so that it can be applied to heat, mechanics, physiology or electricity and still be understood as *literal*.

769 [D9] The Inertia of a Current Experiment
Does any part of the electromagnetic energy actually reside inside
bodies such as wires carrying current *at all?* That seems to be an
unresolved point. Maxwell describes in his *Treatise* a crucial
experiment to determine whether the electric current in a conductor in fact carries measurable momentum.

770 "independent of this hypothesis"
It seems clear now that Maxwell is not altogether abandoning the
basic hypotheses of kinetic energy and elastic deformation (whatever
forms these might take) of "Physical Lines." He has just called such
a hypothesis "very probable." However, the work of the present
paper—here characterized in terms of "deduction" or "demonstration"—is very different. It will serve to establish the validity of the
electromagnetic theory of light, independently of any hypothesis; in
this it is reminiscent of Ampère's *Treatise,* which claimed to "deduce" its theory "uniquely from phenomena."[6] But beyond this,
through its metaphorical vision of a connected mechanism, it is
surely guiding our thought to a concept of a new sort: that of the field.

771 "facts of three kinds"
1. *electromagnetic induction:* (a) an electric current gives rise
to a magnetic field (the Oersted effect); and (b) a changing magnetic field gives rise to an induced electromagnetic force proportional to the rate of change in the number of lines passing through
the circuit.
2. *magnetic potential:* with certain important limitations to be
discussed below, a magnetic field is similar to an electrical field in
being described by means of a potential everywhere whose gradient (space-rate of change) gives the force on a pole or magnetic
*intensity.* Then the magnetic intensity of (2) and the magnetic
induction of (1) are related by a factor $\mu$ describing the medium.
(We can visualize this gradient by going back to our old picture of
the potential function as a mountainous terrain. Then the downhill
force in each direction is proportional to the slope at that point.
Compare the "topographic" model in Chapter 1, and the extension
of that form of interpretation to figures such as those of thermal
distributions in Chapter 2.)

3. *electrostatic induction:* a charged body at one location induces charge on a body at a distance, the effect again depending
on the nature of the medium, described by the constant *k*.

From these phenomena and their laws, Maxwell proposes to
demonstrate that light is an electromagnetic phenomenon, and
thus consists of nothing but these three processes.

784    * * *

From Maxwell's derivation of the major relations of electromagnetism, we select only the most direct path to the electromagnetic
theory of light.

787    "the differential coefficients of a function"

Maxwell refers here to a *potential function.* As we have seen, the
magnetic potential is the work done on a test pole to bring it from
infinity to a given point in the field, while the force on a unit pole
at that point is the gradient (space-rate) of that function. But as we
have seen earlier, the Oersted effect seems to create a problem, if
the incoming path happens to encircle a current. Each encircle-
ment of the current will increase the potential at a point by a
constant representing the work around one whole circuit. Maxwell admits this, as we saw, but points out that it creates no
difficulty. For we are interested in intensity, which is the derivative, and this is unchanged by the addition of a constant!

795    [D10] Derivation of the Field Energy Equation (40)

795    "it becomes"

This is spelled out in Discussion [D10], but meanwhile we can get
a very good sense of the direction of the argument. Recall that
"integrating by parts" (Discussion [D8]) permitted us under certain conditions to play a simple symmetry trick when summing
over a whole space. In application to the first term here, for
example, we can make the following exchange:

$$\sum \mu\alpha\left(\frac{d\varphi}{dx}\right) = -\sum \varphi\frac{d}{dx}(\mu\alpha)$$

Applied to all three terms, this takes us to equation (39).

800 "a magnetic pole is known to us"
Here we see vividly the power of Maxwell's mathematical rhetoric at work. By reshaping his equations into the form of (39), he has unveiled an expression that has become a sure sign of the existence of a magnetic pole:

$$\frac{d}{dx}(\mu\alpha) + \frac{d}{dy}(\mu\beta) + \frac{d}{dz}(\mu\gamma)$$

which is the expression being summed on the left-hand side of equation (39). We saw the same expression earlier as a measure of the divergence of lines of induction, $\mu\alpha$ from a source ["Physical Lines," equation (6)*]. As Maxwell says here, "a magnetic pole is known to us only as the origin or termination of lines. . . ." It was Faraday's view that the lines were real and the "pole" an illusion—as if the lines were the grin and the pole the Cheshire cat: and the field view is that *there are no cats!* In the present terms, "no cats" means "no divergence."

803 [D11] The Number of Lines of a Unit Pole
This follows from the definition of a "unit pole" usually encountered (one that repels an equal, like pole at unit distance with unit force in vacuum, with air for most purposes being close enough to vacuum.)

810 "the only solution"
We will be looking for some analytic function that will determine the value of the magnetic potential $\phi$ at every point in space when a pole of strength $m$ is placed at the center of coordinates. The field must be radially symmetric, i.e., have the same behavior in all directions, and so will be a function of $r$ only. It must also be such that its derivatives everywhere will satisfy equation (39). Since there are no magnetic sources anywhere except at the center, $m$ in equation (39) is zero throughout the field. Factoring out m as constant and substituting $\alpha = d\phi/dx$, etc., equation (39) tells us that

$$\frac{d^2\phi}{dx^2} + \frac{d^2\phi}{dy^2} + \frac{d^2\phi}{dz^2} = 0$$

We will not prove it, but the function Maxwell proposes [equation (41)] uniquely possesses the second derivatives to make this true.

815    "this is simply"
We seem to have made our way back to one of the simplest relationships in the science of magnetism. It was the foundation of action-at-
a-distance theory, but the culmination of a long chain of reasoning
when the field is taken as primary. Some of the major steps:

1. Electromagnetic induction between circuits (and not mechanical forces between poles!) was the starting point.

2. The concepts of generalized force and momentum were
built upon the phenomena of induction between circuits.

3. These were then generalized as *field* phenomena, functions
defined throughout space, and thus localized apart from the conductors. Exploratory circuits were used to map the lines of magnetic induction defined in these terms.

4. Magnetic intensity was defined, related to induction through
the constant $\mu$ characteristic of the medium. Intensity was taken as
independent of the medium, and derivable from a magnetic potential function.

5. Poles were finally introduced, merely as sources of magnetic
field lines, evidenced only by their divergence. Finally, the force law
between poles was derived from the expression for energy in the
field, and the unit pole defined in electromagnetic measure.

842    "$P = kf$"
This was introduced earlier to reflect the observed effects of
dielectric media. At the same time, it suggested an analogy of the
electric field to an elastic medium—something that was explicit in
"Physical Lines," where the vortices themselves were seen as
undergoing elastic deformation. As $P$ is the analogue of force and
$f$ that of displacement, the constant $k$ is like the corresponding
coefficient in the stretching of a spring, a high value of $k$ corresponding to a stiff spring. On the other hand, solving for $f$, the
reciprocal coefficient $1/k$ corresponds to the magnetic constant $\mu$,
so that a high value of $\mu$ implies a large value of induction for a
given magnetic intensity:

$$ f \;=\; \left(\frac{1}{k}\right)P \qquad \leftrightarrow \qquad (Induction) \;=\; (\mu)\alpha $$

| Displace-ment | Electric intensity | | Magnetic induction | Magnetic intensity |
|---|---|---|---|---|

Maxwell has not given us a symbol for magnetic induction, always writing $\mu\alpha$, etc. In the *Treatise* he uses symbols that have since become conventional:

$$E = (1/k)D, B = (\mu)H$$

849 "confidently write"
Since we have identified a strict isomorphism between the two terms, one for magnetic and the other for electric energy density in the field, we can substitute corresponding values of the coefficients in the resulting force laws. The coefficient $1/\mu$, then, plays the same role in the magnetic force law as $k/4\pi$ does in the electric one.

854 "Electric force law"
We need to keep in view the fact that both of these force laws—the electric as well as the magnetic—derive from a single energy equation [equation (I)]. This means that Coulomb's law as it appears here belongs to the same system of measurements and equations as the magnetic force law. As *units of measure* are defined, then, the unit pole and the unit charge are inexorably linked. Having defined the unit pole as repelling a like pole at unit distance with unit force in air, we are not free to define the unit charge in the same way: having set $\mu = 1$ in the vacuum, we cannot set $k = 1$ in the vacuum as well. Their connection is like a wrinkle in a rug: having smoothed it in one place, we have to take what we get in another! We shall see below what form this relationship takes.

858 [D12] The Law of Force between Parallel Currents

858 "the electrostatic system"
If we stand on our heads now and take the electrostatic phenomena as fundamental, we will follow a completely distinct logic and end with an equally distinct set of units of measure. Taking the force between charged bodies as our starting point, we may indeed take unit charge as repelling its equal at unit distance in the vacuum with unit force. The rug's wrinkle will then show up in the equation for magnetic force! Let us write

Coulomb's law for the case of the vacuum in the two systems of measurement:

$$f = \left(\frac{1}{4\pi}\right)\frac{e_1 e_2}{r^2} \quad \leftrightarrow \quad f = \frac{\eta_1 \eta_2}{r^2}$$

| ELECTROMAGNETIC SYSTEM (emu) | ELECTROSTATIC SYSTEM (esu) |

869    "one electromagnetic unit contains"

What is the relation between the measures of charge in the two systems? At first blush, this might seem a question one could answer by means of a little arithmetic, but this is not the case. The two measures of any given physical charge *have no theoretical connection*—at least, insofar as the science of electromagnetism is concerned. We have, in a sense, stumbled upon one of the essential parameters of the cosmos. If we take two charges that have been determined to be unit charges, by "absolute" measurement in the electromagnetic system (emu), not only will they not be unit charges in the electrostatic system (esu) but we have no way of predicting what their esu values will be. Only experiment can determine it—an experiment that is in effect testing an inherent parameter of the physical universe. In other words, we know of no reason why it should be one number rather than another. If we denote the ratio $\eta/e$ by Maxwell's symbol $v$ (line 870), then we do know by equation (46) that

$$v^2 = \frac{k}{4\pi}$$

but this is of no help to us as we have no advance knowledge of the value of $k$. We return to the same empirical question.

Although we are putting all of this in terms of the measurement of units, it is really of course the final piece in assembling the science of electromagnetism; it is as if we had laboriously constructed a jigsaw puzzle, and had reached that marvelous point of having two major parts assembled, only waiting to be joined to complete the final image. The magnetic system is one such assemblage, with its own rational order; the electric system is another. In the physical universe, they are of course one but their juncture does not lie in our hands. If we were to hyphenate the term, "electro-magnetism," we might say, we have arrived at the hyphen. To complete the picture and capture in theory what is in fact one in nature, we have to carry out what is evidently a most crucial measurement. We have to, in effect, consult the cosmos.

880 "metres per second"

How is it that this ratio of units is given in "metres per second," *with the dimensions of a velocity?* If the value of this "velocity" strikes us as familiar, we realize that we are in the presence of a clue as telling as any murmurings brought back in an earlier era from Delphi! Those dimensions, and that number, must have struck some responsive chord in Maxwell, when he first noted them, and reported this connection back in "Physical Lines." We discuss the dimensionality of $v$ in the next commentary and Discussion [D13].

880 [D13] The Dimensionality of $v$

906 "we arrived at certain equations"

One of these equations, number (46), contained the mysterious quantity $v$, whose measured value was then given in units of *velocity*. To understand how this velocity arises, we first look at a pair of physical situations and then quote from a later paper of Maxwell's.

One of these situations, the natural foundation of the electrical system, is simply the attraction of two charged bodies, given by Coulomb's law. Is there a natural counterpart in the electromagnetic system to Coulomb's balance? For this purpose, Maxwell takes the force between two parallel conductors, though other current balances suggest themselves (Fig. 4.12).

We expressed the mechanical action between conductors in the dynamical theory in terms of their mutual electromagnetic momentum [equation (12)]. We show in Discussion [D12] that this yields a force per unit length between two parallel currents, $i$ and $i'$, at distance $r$:

$$\frac{f}{l} = 2\frac{ii'}{r}$$

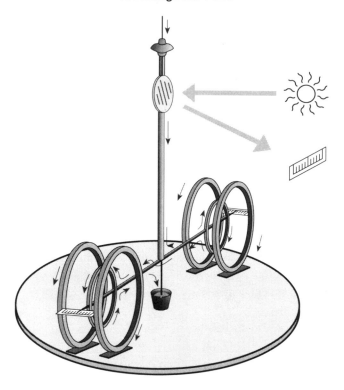

Fig. 4.12. A form of current-balance from Maxwell's *Treatise.* The coils are paired to balance out the effect of the earth's field. The suspension wire does double duty in measuring torsion and conducting the current to the arm from above; the wire dips into a mercury cup to complete the circuit at the base. By means of the mirror, a light lever sensitively indicates the deflections.

where $l$ is the length of a conducting element. With current as charge flow per unit time, the current balance becomes the electromagnetic equivalent of Coulomb's law. It serves to measure, as Maxwell explains, *unit current* (unit currents act with unit force at unit distance) and the emu *unit of charge* (unit current flowing for unit time). Maxwell's account of the dimensionality of $v$ is given in [D13].

The basic equation for the force between parallel current elements can be integrated to find the force between circular loops in parallel planes, as in Fig. 4.12, an example of a delightfully ingenious apparatus. Coils at each end of an arm suspended by a delicate wire are placed between pairs of fixed coils; the same current runs through them all, as indicated by arrows tracing the direction of flow. Note that the sense of the currents in each pair of fixed coils is such that the moving coil is in each case repelled by one coil of the pair and attracted by the other, the effects combining to deflect the arm.

In order to produce a combined rotational effect, the current runs through the moving coils at the two ends of the suspended arm in opposite senses. This has the happy consequence that any interfering effect of the earth's field is reversed at the two ends of the arm—if it aids rotation at one end, it will equally oppose it at the other. The actual measurement is made by determining the torsional force of a given angle of deflection of the calibrated suspension wire. The net effect is that the magnetic force due to the current is balanced against a precisely known mechanical force. The electric current has thus been measured in absolute, mechanical terms.

944 "we have. . ."

This is the application of Oersted's law to the present case. We wrote the law relating a current to its encircling field [equation (9)*]:

$$\frac{1}{4\pi}\left(\frac{d\beta}{dx} - \frac{d\alpha}{dy}\right) = r$$

or

$$\left(\frac{d\beta}{dx} - \frac{d\alpha}{dy}\right) = 4\pi r$$

where $r$ represented the current per unit area. Here, this current density is given as $p$; also, $\alpha = 0$, and $z$ has taken the place of $x$. Adjusting the symbols to fit the present case, we have

$$\frac{d\beta}{dz} = 4\pi p$$

and now multiplying by $b$ and integrating over the whole area to get the entire current $i$ rather than the current density:

$$b\int_0^z \left(\frac{d\beta}{dz}\right) dz = 4\pi \int_0^z bp\,dz$$

$$b(\beta_0 - \beta) = 4\pi \int_0^z bp\,dz$$

Maxwell now divides out $b$, and differentiates with respect to $z$, so that the integral on the right disappears

$$\frac{d\beta}{dz} = -4\pi p$$

What this tells us is that there is a space-rate of change of $\beta$, the magnetic field, along the $z$-axis.

951    "the total electromotive force is equal to"

We are working with the law of induction; in dynamical theory terms, the change in electromagnetic momentum is equal to the electromotive impulse around the whole circuit. Equally, since impulse is force times time, the electromotive force itself around the circuit is the time-rate of change of momentum. The electromotive force summed around the circuit is

$$a\,(P - P_0)$$

The electromagnetic momentum is equal to the magnetic induction through the circuit, which in this case is

$$\int_0^z a\mu\beta\,dz$$

Since a decrease in momentum (putting on the brakes) corresponds to a positive electromotive force, we have here

$$a(P - P_0) = -\frac{d}{dt} \int_0^z a\mu\beta\,dz$$

i.e., Force = Momentum/Time

Dividing by $a$ and differentiating with respect to $z$ as Maxwell instructs, we get his equation (15):

$$\frac{dP}{dz} = -\mu\frac{d\beta}{dt}$$

This tells us that there will be a space-rate of change of $P$ along the $z$-axis, proportional to the time-rate of change of $\beta$. Remember

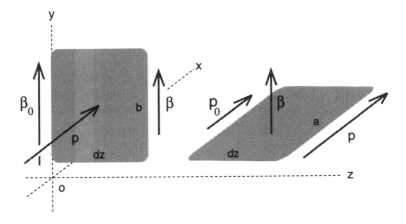

Fig. 4.13. Generation of a plane wave.

that we already have had an expression for the space-rate of change of β! Maxwell is weaving his relations into a pattern, which will be the propagating wave. A tip-off to a wave is when you have a space-rate proportional to a time-rate.

965 "entirely due to the variation of $f$"
There is no flowing current: all these processes are taking place in a dielectric medium, which we might as well think of as the total vacuum (i.e., $k = 1$).

Note that Maxwell now writes $P = kf$, eliminating the minus sign that caused the problem discussed in our note on "Physical Lines," line 931.

968 "four equations"
We have four equations, just sufficient to pin down four independent variables. Maxwell chooses to solve them together first for the magnetic field β. The result, equation (18), is in a very classic form, balancing a second derivative with respect to time against the second derivative of the same variable with respect to space. It is known to physicists as the wave equation. Derivation of equation (18) and its solution, equation (20), are the subjects of Discussion [D14].

970 [D14] The Wave Equation and Its Solution

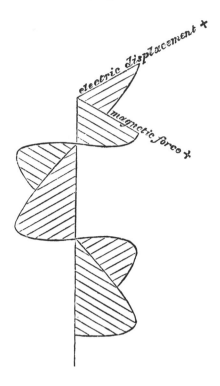

Fig. 4.14. Maxwell's illustration of an electromagnetic wave.

980    "entirely of magnetic disturbances"

Maxwell's emphasizes that all magnetic disturbances must occur in the plane of the wave—not of course that there are not other phenomena going on: there are electric displacements and displacement currents as well. Figure 4.14, from the *Treatise,* shows the relation of the electric displacement to the magnetic intensity. The fact that the electric field is everywhere at right angles to the magnetic field is evident as well in Fig. 4.13. The confinement of each field to its own plane suggests to Maxwell another crucial connection to the phenomena of light, for *polarized light* has just this property. A modern polarizing filter permits passage of light vibrating in just a single plane; when an analyzing filter is rotated at 90° to that plane, no field vibrations can get through. (We might ask, then, "What about unpolarized light?" The answer is that it is a statistical jumble of bursts of light, all polarized, but in different directions. A polarizing filter selects from that jumble just those bursts that vibrate in the filter's plane.) We can expect that other forms of electromagnetic waves will exhibit polarization as well.

992 "experiments of MM. Weber and Kohlrausch"
Maxwell is writing this in 1864. In our section below, "Maxwell as Experimenter" (1868), we will read of his own subsequent experimental determination of this value.

997 "M. Fizeau's . . . experiments"
Whittaker gives this account of Fizeau's experiment:

> In 1849 Hippolyte Louis Fizeau . . . had determined [the velocity of light in air] by rotating a toothed wheel so rapidly that a beam of light transmitted through the gap between two teeth and reflected back from a mirror was eclipsed by one of the teeth on its return journey. The velocity of light was calculated from the dimensions and angular velocity of the wheel and the distance of the mirror. . . .[7]

999 "the more accurate experiments of M. Foucault"
Whittaker continues in the same passage with this account of Foucault's experiment:

> A different experimental method was employed in 1862 by Léon Foucault . . . in this a ray from an origin O was reflected by a revolving mirror M to a fixed mirror, and so reflected back to M, and again to O. It is evident that the returning ray MO must be deviated by twice the angle through which M turns while the light passes from M to the fixed mirror and back.

1001 "deduced from the coefficient of aberration"
The phenomenon of stellar aberration arises from the observer's motion relative to the path of light from an astronomical object. In Fig. 4.15, suppose light from a star travels toward the earth, while an astronomer with his telescope travels along the line AB, arriving at B just as the light does. In what direction does his telescope point? We evidently have a simple right triangle to solve, in which the star's light traverses the hypotenuse while the observer traverses the base. Taking the earth's orbit as approximately circular,

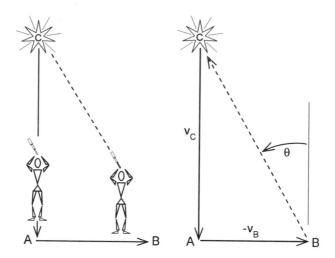

Fig. 4.15. Stellar aberration.

we can easily compute the velocity of the earth carrying the observer, $v_B$, from the radius of the orbit and the length of the year. Then the velocity of light will be

$$v = \frac{v_B}{\sin\theta}$$

Half a year later, the earth's motion at the opposite point of its orbit will be in the reverse direction and will shift the star's apparent position in the opposite sense, with the result that the annual shift in a star's position in the heavens will be $2\theta$. The first determination of the velocity of light by this method was apparently made by Bradley, at Oxford, in the early eighteenth century.[8]

1028 "we do not even know"

Franklin's rather arbitrary decision that "positive" (vitreous) charge was the *presence* of something remains for Maxwell as arbitrary as ever. Positive charge might be a substance of which negative charge is the absence, or vice-versa. But Maxwell's theory has led us to the third possibility, that neither form of charge is a substance at all.

1036  "the existence of normal vibrations"
The theoretical possibility that the electromagnetic medium might also sustain "normal" vibrations (*n*-waves), i.e., vibrations in the direction of propagation rather than transverse to it, remained a subject of conjecture after Maxwell's time. Discoveries of new phenomena, such as cathode rays, were greeted with speculations that *n*-waves might at last have manifested themselves.

1040  "The velocity of light in a medium"
Maxwell refers to a derivation developed in a section which we have omitted, including equations numbered (49) and (71); his argument is summarized in this note. Recall that light must be slowed in a refracting medium in order to account for observed optical phenomena. As Maxwell now shows [equation (80)], his electromagnetic waves will in fact be slowed in a dielectric medium such as glass, for which $\mu = 1$, while the "specific inductive capacity" $D > 1$. The theory may then be able to account for refraction. $D$ is defined as follows:

$$D = k_0/k$$

that is, $D > 1$ in a medium for which $k < k_0$. Since we had $P = kf$, a low value of $k$ means a medium that "yields" readily, so as to require a small intensity $P$ for a given displacement $f$ (like a weak spring). This is true of glass, where the electrical effect of this easy displacement is to store more charge at a given voltage, as in the apparatus shown in Fig. 4.16b.

Thus as the figure reminds us, the same yielding property of a dielectric medium gives us the optical property of bending a ray of light toward the normal as it enters a glass plate and of storing charge in a glass plate inserted in a capacitor. The electromagnetic theory of light begins to make explanatory sense!

1055  "the equation of propagation"
Here Maxwell, arbitrarily it would seem, chooses to write the equation for electromagnetic propagation in terms of the electromagnetic momentum $F$. Then the equation for electromotive force $P$, which he gives next, will be just the time derivative of the first, i.e.,

$$P = \frac{dF}{dt}$$

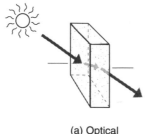

(a) Optical

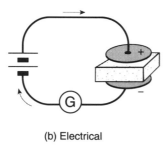

(b) Electrical

Fig. 4.16. Two effects of glass, consequences of its high inductive capacity:
(a) electromagnetic waves are slowed
in glass, and thus bend toward the
normal on entering; (b) higher energy
is stored in the field, so current flows
as glass is inserted between the
plates—$D = k_0/k$.

1059   [D15] Computation of the Voltage of Sunlight

1059   "the energy per unit of volume"
        If light, including sunlight, is an electromagnetic phenomenon,
        then it makes sense to ask about voltages and magnetic fields.
        Maxwell wants to know, "What's the voltage of sunlight?"—
        and he now has the means for arriving at an answer. The
        beginning is with the expression for the energy density in the
        electric field:

$$\tfrac{1}{2}fP$$

        where $P = kf$ relates the intensity $P$ to the displacement $f$ of the
        electric field. (By the intensity $P$ of sunlight, we mean the electro-
        motive force or the mechanical value with which it arrives.) By
        working with this in ways that are traced in Discussion [D15], and
        knowing the energy of the solar radiation as measured by collect-

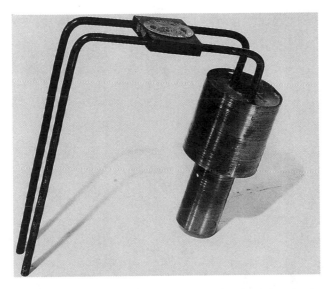

Fig. 4.17. The British Association ohm. As a legal standard, the ohm here takes the form of a physical object, a carefully encased wire, against which copies may be calibrated relatively. It in turn is calibrated absolutely through measurements of mass, length and time. As a physical quantity, it has the dimensions of velocity, in terms of which Maxwell expresses *v*. The value of the ohm is approximately 10,000,000 meters/sec.

ing it as heat in a calorimeter, we can determine the peak voltage of the oscillating electric wave in sunlight. Remembering that voltage measures electromotive force in a conductor, or around a path, we realize that *P* will be given in *volts per meter*.

Maxwell is said to have made some mistake in his result here; J. J. Thomson, as editor of an edition of the *Treatise,* corrects his result to:

$$942 \text{ volts/meter}$$

This seems an impressive figure. Why, then, do we not get shocked by sunlight? One answer might be that we *do,* but that it is the retina that takes the impact. How "shocking" is light? Well, we know better than to look at the sun!

Another answer might be that the magnitude of *P* is changing extremely rapidly (a billion times a second, Maxwell calculates here), and this computed intensity extends over extremely short

Fig. 4.18. Maxwell as experimenter: with the spinning-coil apparatus.

distances. The average value of the intensity over any reasonable length of time or space is thus zero. It all happens too fast,
and at molecular dimensions there is not enough time for us to
get shocked. Then how do we have time to see? The trick is in
the molecular structure of the eye's photosensitive substances,
which are proportioned to these wavelengths and resonate at
these frequencies. These thoughts, however, will take us to the
new domain of quantum physics, in which Maxwell's electromagnetic energies prove to be delivered in particulate form as
photons.

# Postscript: Maxwell as Experimenter

## The Problem Posed

We have seen the importance that the determination of certain electrical quantities assumes in Maxwell's theory. A secure joining of two aspects of physical reality—light and electricity, which to all appearances are distinct from one another—can be verified only if the ratio of the electrostatic to the electromagnetic units of measure takes on a certain specific value. Remarkably, as we have just seen, this otherwise rather unremarkable quantity emerges from the theory as the speed of propagation of the theoretically predicted electromagnetic waves. If, and only if, this number equals the velocity of light, will Maxwell's theory of electromagnetism have in fact unlocked the secret nature of light. Theory has thus proposed a question that can only be answered by means of exact laboratory measurements. It is a question about the constitution of the cosmos and it can only be taken to the cosmos for resolution in a crucial experiment.

Although Maxwell has said that the available experimental value of this crucial ratio "agrees sufficiently well" with recent measures of the velocity of light, it is clear that he is not prepared to rest with the one determination of Weber and Kohlrausch. His speculative work has thus led him directly to an experimental project. Fortunately, Maxwell is as much at home as an experimenter as he is as a theorist. He is perfectly ready to turn to the design of experimental apparatus, and then to roll up his sleeves and go to work in the laboratory. It must be clear, however, that we are not looking at two distinct persons here—the theorist on the one hand and the experimenter on the other—but rather seeing the same mind from a second perspective. As theorist—however outlandish some of his ideas may at times have seemed—Maxwell has always had a firm grasp of the workings of the world around him. His compulsion to see the natural world in terms of intelligible connection has some deep grounding in an understanding of how things do connect: the metaphor of the machine in "Physical Lines" is based on a sure sense of the forms well-connected machines might take.

It is striking that the demands of a new era of electrical technology—most immediately, an immense commercial undertaking to construct a cable spanning the Atlantic—converged for very different reasons on exactly Maxwell's theoretical project: the accurate determination of the fundamental electromagnetic units. In the commercial world, which knows how to get things done when important interests are at stake, this had become a

Fig. 4.19. The Great Eastern cable-laying expedition.

pressing question of law. Contracts must specify standards of performance, while performance must be certified by measurements with reference to officially recognized standards.

It happened that just as Maxwell was establishing himself at King's
College in London in 1861, the British Association for the Advancement
of Science was appointing a committee to begin work on the establishment of the needed electrical standards. In the following year, Maxwell joined that committee and, from then on, devoted a major portion of
his energies in London to their work. They were starting very nearly
from scratch. There was little agreement in England as to terminology or
standards on which to found the new electrical technology; only the
work that Gauss had begun in Germany, carried on by Weber and
Kohlrausch, provided an example. Maxwell hosted the work of the
committee in a laboratory at King's College. They readily agreed that
their first task would be to meet an urgent need that had arisen in the
process of laying the Atlantic cable: to provide sure standards for the
measurement of electrical resistance.

Measurement of resistance was crucial to the new cable technology
for two reasons. First, there was the problem of making precise, continu-

ous measurements on board the cable-laying ships during the actual process of paying-out the fragile strand. Faults or leaks in the cable would show up as anomalous electrical resistance, and determination of the magnitude of these anomalies was essential in identifying the problem or estimating as nearly as possible the location of the interruption. Second, as we shall see below, the conductivity of the copper wire going into the cable was proving to have a critical effect on signaling speeds. Thus it had to be brought under effective commercial measurement and control through standards that could be specified in enforceable contracts.

Thomson had become the leading scientific consultant to the Atlantic cable company, traveling first with a failed cable-laying expedition in 1858 over which he had been given little scientific control and, then, after gaining the confidence of the directors of the enterprise, exercising a kind of scientific mastery aboard the *Great Eastern* in the expedition that led to spectacular success in 1866. Much of the credit for this triumph would lie with Thomson's insistence on accurate measurement and continuous technical control; and for this, the researchers of the British Association Committee—whose work was undertaken largely at his instigation—were essential. As we speak of these matters, we should pause to acknowledge the enormous importance attached to the completion of this global web of cables, for they would provide the nerve system of instant communication throughout an empire, with its new order of commercial, political, and military interests. Huge human and financial investments were being made, and often lost, in these cable enterprises, of which the Atlantic cable was only the most dramatic instance.

The question of signaling speed was of particular scientific interest. It had been early recognized that signals were traveling proportionately more slowly as longer cables were laid. Stokes had written to Thomson in 1846 to ask for a mathematical account of this phenomenon. Thomson had promptly identified it in reply as a phenomenon of *diffusion,* formally analogous to the diffusion of heat, and thus fitting nicely the model he had developed with such success earlier. His equations showed that the corrective would be to keep the resistance of the cable as low as possible and as this principle came to be accepted, the conductivity of the copper that was going into the cables became an issue of extreme importance. Thomson, testing copper samples that were going into the Atlantic cable, found that their conductivities varied widely, with some being no better than iron.

Specification of legally binding levels of conductivity in contracts for the purchase of copper thus became an urgent commercial necessity:

*electrical resistance* was joining the *pound,* the *foot,* and the *second* as a foundation block of a burgeoning commercial society. It was this responsibility that had fallen upon the British Association Committee in 1861, in the interim between the two great cable expeditions. The outcome of the committee's extensive labors was the definition and evaluation of the British Association ohm, a new contribution to the technical world's body of standard measures.

Thomson, who was soon to be knighted in recognition of his contribution to this technology and would ultimately become Lord Kelvin, earned a personal fortune through his ongoing technological contributions to many fundamental aspects of the new electrical industry. Fleeming Jenkin, a member of the British Association Committee and designated as its reporter, was effectively Thomson's agent in the work. An engineer who had worked for some years with cable manufacturing and cable laying enterprises, he had in the year of the establishment of the committee gone into business in London as a consultant to cable firms. In time, he was to share with Thomson patents on many profitable devices related to telegraphic technology. This is an aspect of our story, central to any understanding of the social dialectic of the transformation of a scientific paradigm, that we cannot attempt to tell here. Whatever we might ultimately decide about an inherent relationship between the development of the new commercial society and the maturing of Maxwell's electromagnetic theory, the two surely converged dramatically in the work of the British Association Committee on Electrical Standards. The British Assciation ohm became at the same time the measuring rod of a new technology and the instrument Maxwell would bring to the cosmos in pressing the question of whether the velocity of electromagnetic propagation and of light were in truth one and the same. The committee incorporated Maxwell's investigation into its own program; the experimental determination of the ratio of the units of charge became part of the series of the committee's annual reports to the meetings of the Association.[9]

## Birth of the British Association Ohm

The committee recognized that it had two tasks. It had first to design a stable, reproducible wire coil that would serve as the new physical standard of resistance (the British Association ohm), and then to determine its value as precisely as possible by "absolute measure"—that is, by means not of comparison with other, similarly arbitrary electrical standards but by reference to experiment and calculation based on physical law and fundamental units of mass, length and time alone. Thus the value of the legal ohm was to be established by *absolute* (and hence reproducible) measurements; thereafter, replicas would be made and derivative calibrations carried out, *relatively,* using laboratory methods of comparison.

The method for making this absolute measurement, proposed by Thomson and adopted by the committee, is illustrated in Fig. 4.20; its careful design is suggested by the intricacy of this working drawing (admittedly none too clear as an explanatory diagram!). Maxwell is shown with this apparatus in Fig. 4.18. The strategy was to revolve a coil on a vertical axis in the earth's magnetic field; induced current would then flow in the coil and in turn produce a magnetic field of the coil itself. Since maximum current is induced in the rotating coil as its plane is oriented (magnetically) east–west in the magnetic field—for at that moment the rate of change of flux is greatest—this induced current will generate a magnetic field of its own in an east–west direction—i.e., perpendicular to the earth's field.

The field thus induced in the spinning coil was to be detected by a sensitive suspended magnet. The magnet initially would rest north–south in the field of the earth; then to the extent that current flowed in the coil, the detecting magnet would be deflected from its rest position. The current was limited by passage through the proposed standard ohm. By measuring the speed of rotation of the coil, observing the deflection, and making a series of other determinations, *all reducible to mass, length, and time,* the committee would be able to compute in absolute terms the values of both the induced electromotive force and the current flowing in the circuit. With these known, Ohm's law could be used to compute the actual value of the resistance of the "standard ohm." The standard ohm was after all not exactly an ohm; the point of the committee's work was to determine what its value really was!

We are meeting here the fundamental distinction between *absolute* and *relative* calibrations. The principle of the elaborate spinning coil procedure was to determine all quantities in *absolute* terms. An absolute calibration is achieved through measurements of mass, length, and time alone, without recourse to any arbitrarily introduced standard objects to represent, for example, electrical or magnetic units. In this case the physical object that was to be certified as the standard ohm, and that was

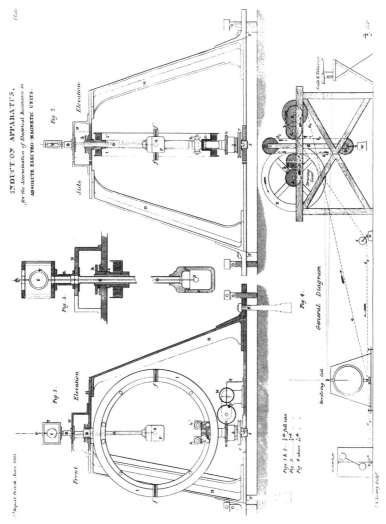

Fig. 4.20. The "spinning coil" apparatus used to determine the British Association ohm. A magnet is suspended at the center of the revolving coil; a mirror above the housing indicates its deflections.

undergoing calibration in this manner, would of course be of immediate practical value in the standards laboratory for carrying out calibrations of other instruments on which an industry was coming to depend. One would not wish to repeat procedures of the complexity of the spinning coil often; yet the physical ohm would in principle serve as only a secondary standard. If it were damaged or called into question over time, it could always be recalibrated or reproduced either by a repetition of the spinning coil procedure or by some similar absolute measurement.

A special genius of Thomson's method is that it is independent of the magnitude of the earth's field: since the same field serves to induce the current in the coil and also to restrain the deflection of the detecting magnet, the effects cancel one another—a stronger field induces more current, but also proportionately restrains the indicating magnet.[10] It is interesting to note, however, that it does depend on knowledge of the strength of the detecting magnet being deflected in the field; this determination, too, had to be carried out in absolute terms.

Once the results were announced and the new standard certified, the British Association ohm was compared carefully with previous standards, including that of Weber and Kohlrausch, and was immediately in heavy demand for replication as an industrial and laboratory standard.

## The Cosmos Speaks: Results of the Crucial Experiment

Once the ohm had been established, the time had come for Maxwell, with the help of the committee, to use it as the basis of a separate determination of the ratio of the esu unit to the emu unit, and thus to put to the test Maxwell's proposition that light is made of electricity and magnetism. Maxwell was an ingenious experimenter, as never better evidenced than in the elegance of the appparatus for this measurement, designed and executed under his direction. Later, he was to exhibit the same genius when it became his responsibility, as Cambridge University's first Professor of Physics, to design and equip the Cavendish Laboratory together with its apparatus—a project which he carried out with devotion and a very recognizable personal style.

One mark of Maxwell's manner in the design of an experiment is his tendency to prefer a "null" method: that is, an experiment in which processes are held in counterbalance, so that the whole point is that nothing happens. A nice example, from the *Treatise,* is a demonstration contrived to reveal to the viewer the law of intensity of the Oersted field

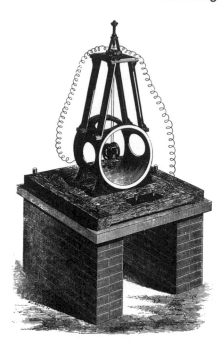

Fig. 4.21. The British Association electrodynamometer. This is the instrument used by the Committee on Electrical Standards to measure the induced current. Very much as in the current balance described earlier, the current to be measured passes through both suspended and fixed coils. The strength of their interaction is measured by the torsion on a bifilar suspension.

around a long straight conductor (Fig. 4.22). We wish to demonstrate that the intensity varies in inverse proportion to the distance from the conductor. The "experiment of illustration" sketched here reveals this nicely: the little balance merely sits at rest. It must be the case, then, that the torques of the north and south poles of each magnet are exactly equal and opposite. But we know from mechanics that the torque of a given force is proportional to the radius. Therefore, the forces must fall off exactly as the radius increases, i.e., inversely as the radius, which is the law we wished to confirm. In proof, one has only to point to the apparatus, and say, "See!"

The experiment for the comparison of units is conceived in such a "null" pattern: the general idea is given in Maxwell's sketch (Fig. 4.23). The two plates of a parallel-plate capacitor are charged alike so as to repel one another; one plate, *A*, is carried at one end of the arm *AA'* suspended by a wire from point *T*. The other plate of the capacitor is the fixed disk *C*, which is shown cut-away in the figure. (The plate *A'* at the opposite end of the suspended arm is merely a dummy, a mechanical counterpoise, to balance the arm gravitationally.) Any electrostatic field between the plates *C* and *A* will give rise to a force that will deflect the arm. This is the *electrostatic* face of the experiment.

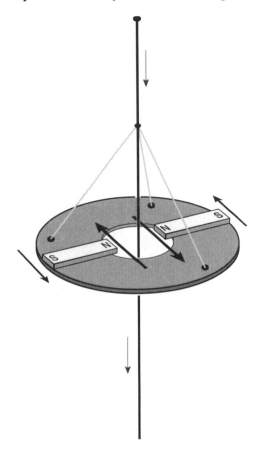

Fig. 4.22. Maxwell's null experiment to demonstrate Oersted's law (from the *Treatise*).

Attached to the same two capacitor plates is a pair of parallel coils, through which current flows in a direction so as to cause attraction: this is the *electromagnetic* aspect of the experiment. The decisive concept is that the spacing of the plates and the current through the coils are to be so contrived that the forces of the two devices exactly cancel. The electrostatic repulsion is precisely balanced by the electromagnetic attraction of the coils. Thus, when all is correctly adjusted, no deflection whatever of the arm will occur.

This physical cancellation of forces is matched by a corresponding mathematical cancellation, for in the computation of $v$, the ratio of units,

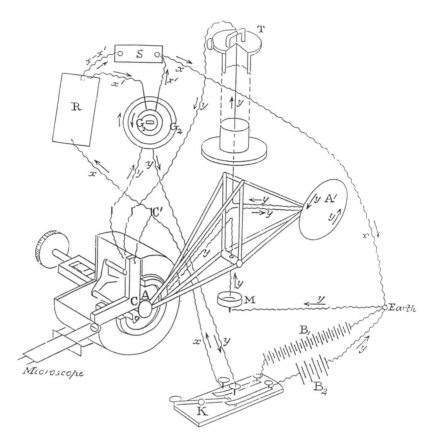

Fig. 4.23. Maxwell's sketch of the ratio-of-units experiment.

the "dimensions" cancel—length, for example, cancels length,—so that only ratios of quantities are involved. As a result, no absolute calibrations are required (the meter stick could be in error, for example: it would not affect the outcome as long as the same meter stick were used throughout) Only one absolute value must in fact be correct: none other than *the unit of resistance, ensconced in the newly minted British Association ohm.* That is, Maxwell's value for the *velocity* expressed by the ratio of units will be given in terms of the standard ohm, whose value is, as we have seen, now absolutely determined.

## How a "Resistance" Becomes a "Velocity"

How is it that an electrical resistance can become surrogate for the velocity of light? Dimensionally, it is easy to show that in the electromagnetic system of units, the units of resistance are those of velocity. We begin by writing Ohm's law in dimensional form, the square brackets indicating that we are pursuing a relationship among concepts, not numbers:

$$[R] = \frac{[E]}{[I]}$$

We may draw on the results of Discussion [D13] to evaluate [E] and [I] dimensionally. [I] as flow of charge has the dimensions of charge divided by time; substituting the emu dimensions of charge, we have

$$[I] = \left[\frac{(ML)^{\frac{1}{2}}}{T}\right] = \left[M^{\frac{1}{2}} L^{\frac{1}{2}} T^{-1}\right]$$

[E], as work per unit charge, has the dimensions of work (force multiplied by distance) divided by those of charge:

$$[E] = \left[\frac{MLT^{-2}L}{M^{\frac{1}{2}} L^{\frac{1}{2}}}\right] = \left[M^{\frac{1}{2}} L^{\frac{3}{2}} T^{-2}\right]$$

Then, dividing:

$$[R] = \frac{[E]}{[I]} = \left[\frac{M^{\frac{1}{2}} L^{\frac{3}{2}} T^{-2}}{M^{\frac{1}{2}} L^{\frac{1}{2}} T^{-1}}\right] = \left[LT^{-1}\right]$$

i.e.,

$$[R] = \left[\frac{L}{T}\right]$$

It perhaps should trouble us to be told that resistance, when evaluated in the esu system, has dimensions that are exactly the reciprocal of these, namely,

$$[R]_{esu} = \left[\frac{T}{L}\right]$$

The British Association unit is one ohm in the so-called "practical" system of units, employing the volt, the ampere, the watt, and the ohm. These are consistent units in the emu system, and the computations of the experimental results, based on the ohm, are thus in the magnetic system.

The trick of the physical measurement lies in the application of Ohm's law: the potential across the attracting plates is determined as the product of a resistance known in ohms and the current flowing through it. A second current in a known ratio to the first flows through the repelling coils. Thus in effect one current, by way of the resistance, yields both of the two counterbalancing forces. The actual equation Maxwell gives is elaborate, because it must reflect by way of theoretical calculations the details of the experiment. However, once the parameters of the apparatus have been measured and substituted, they coalesce into coefficients that are pure numbers and the equation becomes radically simple.

In such schematic form, we have:

1. Electrostatic Force (Coulomb's Law):

$$f_e = k_1 E^2 \left(\frac{a^2}{b^2}\right) = k_2 E^2$$

where $k$ is a "pure" (dimensionless) number, while $a$ and $b$ are measured lengths, the radius and spacing of the plates, respectively. Since $a/b$ is dimensionless as well, we can lump them into a second pure number $k_2$. But as we have indicated, the concept of Maxwell's measurement is that $E$ is determined as the potential across the resistance $R$ carrying current $y$:

$$f_e = k_2 (y R)^2$$

Since $y$ and $R$ are measured in practical units (amperes and ohms) in the emu system, $E$ will be as well, and we must convert to compute $f_e$. As $E$ measures potential as work per unit charge, we must convert emu charge (in the denominator) to esu by multiplying by our factor $v$:

$$f_e = k_2 E^2 = k_2 \frac{E^2}{v^2}$$

2. Electromagnetic Force (Ampère's Law): The force due to the attracting coils is computed from Ampère's law, and with pure numbers aggregated as $k_3$ takes the schematic form

$$f_m = k_3\, y^2$$

Here measurement of current in the emu system is appropriate, and no conversion is required.

3. Asserting the Balance of the Two Forces: We are now in a position to assert the equality of the two balancing forces:

$$k_2\, \frac{E^2}{v^2} = k_3\, y^2$$

$$k_2\, \frac{\left(R y^2\right)}{v^2} = k_3\, y^2$$

Once again, we may lump together the pure numbers, leaving it to the experimenter to evaluate them, and focus on the fundamental relation:

$$\frac{R^2\, y^2}{v^2} = k_4\, y^2$$

$$R^2 = k_4\, v^2$$

$$R = k\, v$$

At this point, the coefficient here designated $k$ has inherited a large aggregation of quantities, but all are pure numbers. The current, too, which as we saw has in effect (by way of the differential galvanometer) been equated between the two facets of the experiment, has canceled. Once $k$, dependent on no absolute measurement, has been evaluated, $v$ follows directly upon the determination of $R$, related to it by a dimensionless number. As a project in fundamental measurement, this is indeed a brilliant design: by virtue of null conditions and shared quantities $y$, all measurements otherwise requiring reference to authoritative standards have canceled out and vanished. The crucial universal constant $v$, the velocity of light, has been made manifest by way of that one artfully calibrated standard, the British Association ohm.

In actual execution, the experiment was delicate and in some respects, daunting. In order to achieve a sufficient force between the capacitor plates, a very large battery had to be used. Maxwell acknowledges the generosity of a Mr. Gassiot, who, he says: "with his usual liberality, placed at my disposal his magnificent battery of 2600 cells charged with corrosive sublimate, with the use of his laboratory to work in."[11]

Maxwell was thus making his delicate adjustments in the near vicinity of several thousand ill-disciplined volts! The batteries were unsteady, running down rapidly, with the result that readings had to be caught on the run. As evidence of the difficulty of governing the voltages involved, Maxwell reports further, "Another difficulty arose from the fact that when the connexions were made, but before the key was pressed, if the micrometer was touched by the hand the disk was attracted. This I have not been able satisfactorily to account for, except by leakage of electricity from the great battery through the floor."[12]

Careful precautions were thereafter taken, but it was clear that further experiments would have to be made with rearranged apparatus before a definitive result could be achieved. The results of these preliminary measurements were disappointing:

$$v = 28.798 \text{ British Association units (ohms)}$$

or

$$288,000,000 \text{ meters per second.}^{13}$$

The Weber and Kohlrausch figure had been far too high; Maxwell's is equally in error, but too low. Maxwell understands well why the German results were high: they had depended on the computed value of a reference capacitor and had not allowed for the effect of dielectric absorption that has since become clear to Maxwell. Maxwell's own result is still too preliminary to permit critical analysis—though the thought suggests itself that since the noticed leakage tended to depress the result, uncontrolled leakage might still be causing the final value to be too low as well. The measurement of the ratio of units remains as an ongoing task; Maxwell discusses several methods at length in the *Treatise*. The determined value is in any case close enough to the known velocity of light to constitute essential verification of the electromagnetic theory of light.[14]

Not long before, another experimenter had been making strenuous efforts to demonstrate a time of propagation of electromagnetic effects in a very different manner. In one of the most hopeless of experiments, Faraday was attempting to observe the "time of magnetism" across the backyard of the Royal Institution, by noting with the eye the delay in the flick of a galvanometer needle. Two galvanometers were to be compared, one near and the other far from a source of induction, marked as well by a spark of light. Some of the last pages of Faraday's diary are adorned with a profusion of sketches of alternative forms of the apparatus, which were evidently undergoing constant evaluation and rearrangement in his inimitable manner of unremitting work. At one point Faraday makes this note to himself:

> *Time.* It would appear very hopeless to find the time in Magnetic action if it at all approached to the time of light, which is about 190,000 miles in a second, or that of Electricity in copper wire, which approximates to the former. But then powers which act on interposed media are known to vary and sometimes wonderful[ly]. Thus the time of action at a distance by conduction is wonderfully different for electricity in copper, water and wax. Nor is it likely that the paramagnetic body of oxygen can exist in air and not retard the transmission of the magnetism. At least such is my hope.[15]

Without attempting to fathom the turns of thought involved in this paragraph, we can see that it remained of crucial importance to Faraday, at the end of his career, to demonstrate that "magnetic action" is not instantaneous but propagated through a medium over some elapsed time. He guesses that its velocity will approach that of light, in which case the experiment he contemplates will be "very hopeless," yet doggedly, he undertakes it. In what was literally a "backyard" experiment, behind the Royal Institution, he set out by means suggested in a sketch in the margin of the *Diary* to discover a delay between a magnetic action and its consequence. The work was as hopeless as he thought: he calculated that he might detect an effect if the velocity were 1/74 that of light. Living in retirement and increasing mental confusion by the time Maxwell's results were reported, it is not clear that Faraday would have been aware that the crucial question of the "time of magnetism" had been answered experimentally.

In a way, however unpromising this experiment—so near to being Faraday's last (he went on to look for yet another profound connection in nature, one between electricity and gravity, in an experiment that involved dropping electrified cannon balls in the Shot Tower in London)—he was on the right track in attempting to contrive an electromagnetic propagation experiment within the laboratory, as Maxwell himself did not. It seems never to have occurred to Maxwell as a practical possibility that one might initiate an electromagnetic disturbance at one point and detect it later at another, the principle that was to become the foundation of radio transmission. Evidently the extreme rapidity of the vibrations of light struck Maxwell as putting the phenomenon beyond experimental reach, for the oscillations achievable in experimental circuits would be very much slower than those of light waves.[16] Yet it was only a short while after Maxwell's death in 1879 that Hertz, as we have discussed, did stumble upon the production of the first radio waves, generated electromagnetically in the laboratory and systematically recognized as such.

For Maxwell, the electromagnetic propagation that emerged from his speculative insight, with its magic velocity $v$, was a window into the foundations of the physical universe. But when he allows for the possible existence of an electromagnetic spectrum outside the limits of the visible, he does not seem to have in mind anything like the relatively low frequency vibrations that now flood his ether with our signals of radio and television. The twentieth century, the century of the conveyance of information, remained latent in his equations.

With the conclusion of our reading of this trilogy of papers, we leave Maxwell returning to Glenlair, where he will write the immense *Treatise on Electricity and Magnetism* and pursue intensively other major lines of scientific work in which he has already established his prowess: thermodynamics, statistical mechanics, and the theory of colors and color vision. Later, he will be drawn from this retirement to return to Cambridge, where he will, as we noted before, occupy a chair new to the university as Professor of Physics, and set into operation the Cavendish Laboratory as a university facility of an entirely new kind.

Maxwell died at the height of his powers, at the age of 49 in 1879. Had he lived a little longer, he would have been among the first to know of Hertz's result and to discern its implications. I also think it is clear that he would have pursued the questions of relativity that his vivid sense of the presence of the ether were already making urgent for him.

He ends his *Treatise* with a careful evaluation of action-at-a-distance theory, and we might well close our study with a glance at these remarks. Others had been making modifications to Weber's equation to allow for *delayed* action at a distance, in order to accommodate formal theories such as Weber's to the time of transmission revealed by Maxwell's electromagnetic nature of light. It was not difficult to insert time-delay terms in the equations, but it was a matter of dispute whether a theory consistent with the laws of motion and the conservation of energy could be built in this manner. Maxwell in 1868 thought he saw an interesting conundrum arising from this concept:

From [these] . . . assumptions we may draw the conclusions, first, that action and reaction are not always equal and opposite, and, second, that apparatus may be constructed to generate any amount of work from its resources.

For let two oppositely electrified bodies *A* and *B* travel along the line joining them with equal velocities in the direction *AB*, then if either the potential or the attraction of the bodies at a given time is that due to their position at some former time (as these authors suppose), *B*, the foremost body, will attract *A* forwards more than *A* attracts *B* backwards.

Now let *A* and *B* be kept asunder by a rigid rod.

The combined system, if set in motion in the direction *AB*, will pull in that direction with a force which may either continually augment the velocity, or may be used as an inexhaustible source of energy.

I think that these remarkable deductions from the latest developments of Weber and Neumann's theory can only be avoided by recognizing the action of a medium in electrical phenomena.[17]

Yet as he closed the *Treatise,* there remained the possibility that these alternative theories, utterly excluding a medium, might by some device finally account for all the phenomena. In the final paragraph, he reflects on this prospect:

But in all of these theories the question naturally occurs: If something is transmitted from one particle to another at a distance, what is its condition after it has left the one particle and before it has reached the other? If this something is the potential energy of the two particles, as in Neumann's theory, how are we to conceive this energy as existing in a point of space, coinciding neither with the one particle nor with the other? In fact, whenever energy is transmitted from one body to another in time, there must be a medium or substance in which the energy exists after it leaves one body and before it reaches the other, for energy, as Torricelli . . . remarked, `is a quintessence of so subtile a nature that it cannot be contained in any vessel except the inmost substance of material things.' Hence all these theories lead to the conception of a medium in which the propagation takes place, and if we admit this medium as an hypothesis, I think it ought to occupy a prominent place in our investigations, and that we ought to endeavour to construct a mental representation of all the details of its action, and this has been my constant aim in this treatise.[18]

# Epilogue

We have been privileged witnesses to the construction of one of the truly great works of western science, at once a complete and coherent theory of the electromagnetic field and the unveiling of the electromagnetic theory of light. It is as if the cosmos had rewarded Maxwell for his labors in the creation of field theory with the solution of the mystery of light as a prize. In much the same way, Newton was rewarded for his general theory of the motion of bodies with the theory of universal graviation and the solution of the problem of the motions in the solar system. Maxwell's method and its endorsement by the cosmos are hardly less momentous in the history of science than was the revolution brought about by Newton. Indeed, they belong together in a way not always noticed—for Newton was never for a moment satisfied with action-at-a-distance theory as a final answer to the questions posed by natural philosophy, and, as is less often recognized, he was at least equally opposed to filling space with a Cartesian mechanism. He wrote the now-unread second book of his *Principia* specifically as a refutation of Descartes. We might almost conclude that Maxwell's theory of the field, in its final Lagrangian form, comes about as close as we could imagine to answering the overall query Newton had bequeathed to the world. And yet Maxwell's theory is, itself, as much a further query as it is an answer. It was Einstein above all who picked up this unending conversation and showed (as Maxwell was clearly already suspecting) that field theory would necessarily rock the foundations of our thinking about time, space, and matter.

Maxwell's field theory of electromagnetism, considered in itself apart from these larger consequences, has given us that complete and beautiful science known today as "classical electromagnetism," which contains within it the whole body of the phenomena of electricity and magnetism, within a range bounded by quantum theory on one side and relativity on the other. Many of these phenomena were known to Maxwell, but a cornucopia of others lay ahead whose possibilities are still being explored today. The ability to generate in the laboratory electromagnetic signals that would traverse space seemed just beyond the bounds of possibility to Maxwell. However, as we have seen it was an uncontainable consequence of his theory, which burst upon the world with Hertz's discovery very shortly after Maxwell's death.

Maxwell's new science is at the same time a new paradigm for thought about the world more generally, and in working through these three papers we have indeed been witnesses to the birth pangs of a revolution in paradigms. To the end, Maxwell is measuring the distance between his new paradigm and the old, wondering whether Weber's action-at-a-distance equations in the old mode could be stretched to take the time of propagation of the new electromagnetic waves into account. Maxwell evidently feels pretty certain that they could not, but would only wrap themselves in contradictions whose absurdity he takes no little satisfaction in pointing out.

Paradigms, once launched, know no bounds. Beyond the theory of electromagnetism itself we are indebted to Maxwell for shaping a new way of thinking about the world. It is not the notion of the field that is new so much as the demonstration that the field is a fully coherent and viable conceptual structure, an idea capable of rigorous formulation and efficient mathematical management. As we have seen in tracking the dialectical turns of thought incorporated in Maxwell's three papers, it is not so easy to have a new idea! The field is a new fundamental mode of thought in which the whole, as a fully connected, coherent system, is prior, while the parts derive their significance through their membership in that whole. Such was Maxwell's journey from the old way that took the *pole* as fundamental and attempted to build a structure as an aggregation of isolated elements, to a vision of the *field* as a system in a mathematically configured state, sensitive throughout as a whole in which a change at any one point is necessarily propagated to every other. This is not an easy idea to grasp, and it may be that our journey through these three papers has been as good a way as any to take the full measure of its richness.

We live now at a time when systems whose existence we had hardly suspected are revealing themselves to us, both in the physical and in the social worlds, while our thoughts are still far out of tune with these new realities. It is hard to know whether we should think of these new-found systems as perils or as opportunities—as closing in or opening out. As always, social thought lags far behind not only social need, but behind paradigms already long available in other areas, above all in the sciences. Maxwell, as a pioneer in mathematical physics, was at the same time breaking new ground for thought in all areas of life. As we have seen, he developed the field theory out of some driving inner conviction, identified deeply with his unswerving devotion to Faraday, concerning the nature of the world. The field thus enters Maxwell's thinking as a preconception and a passion, but emerges as a refined rational instrument, a contribution from the sciences to the liberal arts generally, not least in its power of suggestion for political vision and poetic metaphor.

It is hard to realize today what it must have meant to Maxwell first to suspect and then to demonstrate as scientific truth that light and electromagnetism are one and the same domain within the creation. This was not just a radical simplification of the sciences, though Maxwell certainly was the first to appreciate the elegance of his one set of equations from which both electromagnetism and optics could be derived. Beyond this, what Maxwell was most striving to achieve was some viable sense of the coherence and intelligibility of the system of nature as a whole. He was among the last of the great natural philosophers, for whom the philosophical questions engendered and embraced the narrower inquiries into the sciences. For Maxwell, field theory both springs from, and to a certain extent answers, questions of ontology and epistemology—the nature of the world, and grounds and limits of our ability to comprehend it.

He is looking for a new basis of intelligibility as he works out the geometry of the continuum in "Faraday's Lines," and he is striving for a sense of causality in "Physical Lines," but he subsumes and transcends them both in the third paper. What he ends with is not causality in the immediate sense of "Physical Lines," by contact and linkage, but in the larger *field* sense of which we have been speaking, in which intelligibility and causality run from the whole to the parts, from the state of the system to the contributions of the members. It does not matter if the connections are hidden and the forces generalized and metaphorical: that earlier, primitive sense of linkage has been left behind in the "Dynamical Theory." What does matter is that the connectedness, however generalized, is real, and the whole is intact. In this sense, the electromagnetic theory of light, which reaches to a whole of an entirely new order, stakes out a new claim to coherence and intelligibility in nature, which once again seems to deserve the ancient name of "cosmos"—that which fits beautifully together.

Advances since Maxwell's time in relativity and quantum physics do not seem to have left this vision behind, but point instead to the importance of considering more carefully Maxwell's sense, as philosopher, of the power of metaphoric thought to address a more subtle world. For Maxwell, discovery of the Lagrangian approach to the field suggests an answer to fundamental questions of being and truth. We can achieve a more penetrating understanding of wholeness and connection by not insisting on literal access to space, time, or causal linkage, which do not need to be literal in order to be real. This, at least, is my reading of Maxwell's parable of the bell ringers, who know their world very well—as metaphor. The reach of Maxwell's thought in the "Dynamical Theory" may yet show us the way to possibilities latent in the paradigm of the field. Understood not literally and physically, but in its power as metaphor, it becomes an invitation to a sense of membership in a world perceived as cosmos.

# Discussions

### D1. A Dynamical Illustration [line 353]

Perhaps in the same spirit in which he illustrates the electromagnetic field by means of an envisioned mechanical system in "Physical Lines," Maxwell also illustrated these mechanical principles by means of a model. On his return to Cambridge some years after completion of these papers, Maxwell undertook, as we have observed, the design of the Cavendish Laboratory. Introducing as it did manual work with actual apparatus, the new program in experimental physics constituted a very deliberate and seriously contested break with the university's tradition of purely mathematical science as training of the "mind." The new program in experimental physics was intended to teach physics with a minimum of mathematics and with emphasis instead on learning through a grasp of principles and experience in experimental work. The Cavendish at its outset had been characterized as a "research" laboratory, but it would appear that the original intention of the term was distinctly different from its use today. Maxwell seems to mean, rather literally, "re-search," the reproduction or extension of creative work done elsewhere for the sake of laboratory experience rather than altogether original research in the modern sense of the term, toward which the Cavendish subsequently led the way.

Maxwell set forth his views on "Experimental Physics" in an introductory lecture as he assumed the professorship; they are especially interesting in conjunction with these papers on electromagnetism.[19] It is clear there, for example, that the term *illustration* has special meaning for him, and that illustration of the relations expressed in Lagrange's equations by means of a model would not necessarily be an inferior but quite possibly a more insightful way of grasping the principles of a connected mechanical system.

J. J. Thomson, as editor of a later edition of the *Treatise,* describes a model Maxwell designed for the new Cavendish to illustrate Lagrange's equation, as related to an inductively coupled system (Fig. 4.24). It is an

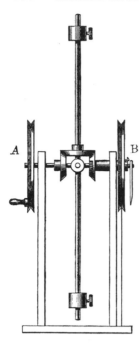

Fig. 4.24. Maxwell's model to illustrate the principles of a connected mechanical system. Here the gearing is 1:1, so $p = q = 1$ in equation (1). Is the flywheel in the center, with its weights, to be understood as the *field*, driven by the motions in "circuits" at *A* and *B*?

"illustration" that will serve us well in all of the discussion to follow of electromotive forces and electromagnetic momentum. I quote below from Thomson's note, adding my own comments in brackets. The symbols *P* and *Q* here correspond to *A* and *B* in the figure and in Section 24 above:

*P* and *Q* are two disks, the rotation of *P* represents the primary current, that of *Q* the secondary. These disks are connected together by a differential gearing [i.e, such as the "differential" which today connects the wheels of a car to the driveshaft]. The intermediate wheel carries a fly-wheel the moment of inertia of which can be altered by moving weights inwards or outwards. [A fly-wheel is a wheel specifically designed to carry high rotational momentum, to keep a rotational motion steady. We may think of "moment of inertia" as the counterpart in rotational motion of mass in linear motion.]

The resistance of the secondary [*S*, in equation (3) of Section 24] is represented by the friction of a string passing over *Q* and kept tight by an elastic band. If the disk *P* is set in rotation [a current *u* started in the primary] the disk *Q* will turn in the opposite direction [inverse current when the primary is started]. When the velocity of rotation of P becomes uniform [constant primary current], *Q* is at rest [no current in the secondary when the primary current is constant]; if the disk *P* is stopped, $Q$ commences to rotate in the direction in which $P$ was previously moving [direct current in the secondary on breaking the circuit]. The effect of an iron core in increasing the induction [i.e., the mutual induction M] can be illustrated by increasing the moment of inertia of the fly-wheel.[20]

## D2.  Derivation of Equation (1) [line 366]

This is a nice example of analytic techniques at work; let us go through this example carefully to see how these things are done. This will take a long time, because we will take care—just this once!—to make sure of each step.

We begin with

$$C\left(\frac{dw}{dt}\right)\delta z = X\delta x + Y\delta y$$

as the expression of the energy relations in a connected system composed of the elements $A, B, C$. The displacements of $A$ and $B$ are denoted as $\delta x$ and $\delta y$, the squiggly $\delta$'s indicating that these are *any* independent variations, however small: the equation must be true whatever they are. $X$ and $Y$ are the forces on $A$ and $B$, so we see that $X\delta x$ and $Y\delta y$ represent the *work* done (force times distance) by the forces on these two elements of the system. This is the work input to the system, if we think of $A$ and $B$ as the "driving points" Maxwell speaks of.

The force on $C$ is not expressed in this way as a driving force, but rather the force of acceleration, the accelerative force $F$ of Newton's second law:

$$F = ma$$

Here, the mass is denoted $C$, and the acceleration is written as the time-rate of change of $C$'s velocity, which is $w$. The second law becomes

$$F = C\left(\frac{dw}{dt}\right)$$

Then the work done in accelerating $C$ is $F\,\delta z$, and since this must equal the input work, we have the overall energy equation

$$F\delta z = C\left(\frac{dw}{dt}\right)\delta z = X\delta z + Y\delta y$$

Now, we go to work on it! The connection of the system (we might think of gearing, as in [D1] above) is expressed first in terms of the velocities: $C$'s velocity is the sum of $p$ times $A$'s plus $q$ times $B$'s:

$$w = pu + qv$$

We differentiate this with respect to time (take the time-rates) because the left-hand term in the first equation of this Discussion, namely

$$C\left(\frac{dW}{dt}\right)\delta z$$

tells us that we are interested in accelerations. Since $p$ and $q$ are constants of the gearing, they do not change with time and factor out of the derivatives:

$$\frac{dw}{dt} = p\frac{du}{dt} + q\frac{dv}{dt}$$

The same gearing assures that the displacements relate in the same way:

$$\delta z = p\,\delta x + q\,\delta y$$

Now, we follow Maxwell's advice and "substitute," namely, in the term

$$C\left(\frac{dW}{dt}\right)\delta z$$

we insert our equivalents for $dw/dt$ and $\delta z$:

$$C\left(\frac{dw}{dt}\right)\delta z = C\left(p\frac{du}{dt} + q\frac{dv}{dt}\right)(p\delta x + q\delta y)$$

This fits the general algebraic form:

$$(a + b)(c + d) = ac + ad + bc + bd$$

If you multiply this out in this form, and regroup by collecting the terms which have $\delta x$ as a common factor and then those which have $\delta y$, and do not make a mistake, you find that our $C\,dw/dt\,\delta z$ has expanded to equal the following:

$$\left(Cp^2\frac{du}{dt} + Cpq\frac{dv}{dt}\right)\delta x + \left(Cpq\frac{du}{dt} + Cq^2\frac{dv}{dt}\right)\delta y$$

We know from our first equation that $C\, dw/dt\, \delta z$ also equals $X\, \delta x + Y\, \delta y$. Therefore,

$$X\delta x + Y\delta y = \left( Cp^2 \frac{du}{dt} + Cpq \frac{dv}{dt} \right)\delta x + \left( Cpq\frac{du}{dt} + Cq^2 \frac{dv}{dt} \right)\delta y$$

Now, a very satisfying principle operates. Maxwell says, "remembering that $\delta x$ and $\delta y$ are independent. . . ." Their squiggly notation bespoke this independence. In order that these expressions *always* be ("identically") true no matter what $\delta x$ and $\delta y$ do, the coefficients of $\delta x$ must be equal to one another and so must those of $\delta y$. Thus we find that

$$X = \left( Cp^2 \frac{du}{dt} + Cpq \frac{dv}{dt} \right)$$

$$Y = \left( Cpq \frac{du}{dt} + Cq^2 \frac{dv}{dt} \right)$$

The rules of differentiation tell us that

$$\frac{d}{dt}\left( Cp^2 u + Cpqv \right) = \left( Cp^2 \frac{du}{dt} + Cpq \frac{dv}{dt} \right)$$

$$\frac{d}{dt}\left( Cpqu + Cq^2 v \right) = \left( Cpq \frac{du}{dt} + Cq^2 \frac{dv}{dt} \right)$$

The "rule" in question is that, as you might expect, the derivative of a sum is the sum of the derivatives, i.e., in general, for any $x$, $y$, $z$:

$$\frac{d}{dz}(x+y) = \frac{dx}{dz} + \frac{dy}{dz}$$

So,

$$X = \frac{d}{dt}\left( Cp^2 u + Cpqv \right)$$

$$Y = \frac{d}{dt}\left( Cpqu + Cq^2 v \right)$$

i.e., Maxwell's equations (1). If you doubt whether it was worth it, just watch what Maxwell does with this result!

### D3. "Maxwell's Example of the Bell Ringers" [line 396]

Later, in reviewing a second edition of the *Treatise on Natural Philosophy* by Thomson and Tait, Maxwell gives an illustration of the Lagrangian method. It was this Treatise, first published in 1867, and the discussions with the authors that accompanied its preparation that had guided Maxwell toward dynamical theory. His electromagnetic theory owes them much:

> The credit of breaking up the monopoly of the great masters of the spell, and making all their charms familiar in our ears as household words, belongs in great measure to Thomson and Tait. The two northern wizards were the first who, without compunction or dread, uttered in their mother tongue the true and proper names of those dynamical concepts which the magicians of old were wont to invoke only by the aid of muttered symbols and inarticulate equations. And now the feeblest among us can repeat the words of power.[21]

Here is the example he offers to help break the spell of Lagrange's analytic wizardry:

> In an ordinary belfry, each bell has one rope which comes down through a hole in the floor to the bell ringers' room. But suppose that each rope, instead of acting on one bell, contributes to the motion of many pieces of machinery, and that the motion of each piece is determined not by the motion of one rope alone, but by that of several, and suppose, further, that all this machinery is silent and utterly unknown to the men at the ropes, who only see as far as the holes in the floor above them.
>
> Supposing all this, what is the scientific duty of the men below? They have full command of the ropes, but of nothing else. They can give each rope any position and any velocity, and they can estimate its momentum by stopping all the ropes at once, and feeling what sort of tug each rope gives. If they take the trouble to ascertain how much work they have to do in order to drag the ropes down to a given set of positions, and to express this in terms of these positions, they have found the potential energy of the system in terms of the known co-ordinates. If they then find the tug on any one rope arising from a velocity equal to unity communicated to itself or to any other rope, they can express the kinetic energy in terms of the co-ordinates and velocities.

These data are sufficient to determine the motion of every one of the ropes when it and all the others are acted on by any given forces. *This is all that the men at the ropes can ever know.*[22]

### D4. Induction by Relative Motion of the Conductors [line 465]

We are working with two coupled circuits, *A* and *B* as in Fig. 4.25. We know that their relationship is expressed by the overall equation (5),

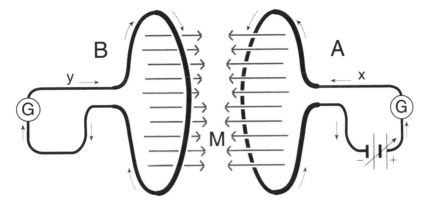

Fig. 4.25. Induction between primary and secondary coils.

which in turn had translated into electrical terms the general mechanical equation (1):

$$\xi = Rx + \frac{d}{dt}(Lx + My)$$

$$\eta = Sy + \frac{d}{dt}(Mx + Ny)$$

$\xi$ and $\eta$ are the electromotive forces acting on the two currents $x$ and $y$; $R$ and $S$ are the resistances, $L$ and $N$ are the inertial terms for the two circuits, and $M$ is their mutual inductance.

In this particular case, Maxwell proposes that we hold the current $x$ constant, keep the coils themselves rigid (so that $L$ and $N$ do not vary), but move one of the coils toward the other. $M$, then, will be the varying quantity. In this case, it is supposed that "there is no electromotive force on B except that which arises from the action of A," i.e., from the current $x$ and the mutual inductance $M$ which couples it to B. Then $\eta = 0$ (no external emf on B) and $N = 0$ (circuit B has negligible self-inductance). Maxwell makes these assumptions in order to find out the effect of varying $M$, alone. We can now write a very simple form of the equation:

$$O = Sy + \frac{d}{dt}(Mx)$$

$$Sy = -\frac{d}{dt}(Mx)$$

Since we are holding $x$ constant, it factors out of the derivative:

$$y = -\left(\frac{x}{S}\right)\frac{dm}{dt}$$

If we integrate both sides with respect to time,

$$\int y\,dt = -\left(\frac{x}{S}\right)\int dM = -\left(\frac{x}{S}\right)\Delta M$$

Note that since current can be thought of as the *flow of charge,* then summing over time on the left, we get *charge* (as summing water flow over time we get water!). A sluggish meter in the secondary would respond ("ballistically") to this total impulse. Interestingly, this is independent of the way we move the coil—fast or slow, directly or roundabout. Only the first and last positions matter! Also, we see that if we increase $M$ by bringing the coils together, we get a *negative* current—once again, Lenz's law. If it were otherwise, we would have a perpetual-motion machine!

## D5.  Maxwell's Equations [line 705]

We identify Maxwell's twenty equations by the names and letter designations that he has given them, including the letters that he assigned in Section 18 (see our note to line 243). Here, we give one representative equation from each set, reserving equation (A) for what may seem a more logical position in the sequence:

(B) MAGNETIC FORCE

$$\mu\alpha = \frac{dH}{dy} - \frac{dG}{dz}$$

Here $H$ and $G$ are components in the $z$- and $y$-directions of the electromagnetic momentum per unit length of a current flowing in a circuit. We recognize the expression on the right as the summation around one plane section through an infinitesimal box (compare Fig. 4.11). The number of magnetic lines cutting that circuit in vacuum is given by $\alpha$ and $\mu$ measures the effect of a magnetic medium: the number of lines in a medium other than the vacuum will be $\mu\alpha$.

## (C) ELECTRIC CURRENTS

$$4\pi p' = \frac{d\gamma}{dy} - \frac{d\beta}{dz}$$

Again we have drawn the infinitesimal box, this time using it to measure the work needed to carry a unit pole around a path surrounding a current-carrying conductor. If $\gamma$ and $\beta$ are the $z$- and $y$-components of the magnetic intensity at a point in the field, then the right-hand side measures the work to carry a pole around a conductor carrying current $p'$ in the $x$-direction. This is the Oersted effect.

## (D) ELECTROMOTIVE FORCE

$$P = \mu\left(\gamma\frac{dy}{dt} - \beta\frac{dz}{dt}\right) - \frac{dF}{dt} - \frac{d\psi}{dx}$$

This has three parts. The expression in parentheses measures the electromotive force induced in a circuit moving in a magnetic field. The circuit has sides $dy$ and $dz$, moving with velocities $dy/dt$ and $dz/dt$ in a magnetic field whose components are $\beta$ and $\gamma$ in the $y$- and $z$-directions. The second term represents the electromotive force in the same circuit due to an electromagnetic momentum with $x$-component $F$ changing at the rate $dF/dt$. The third, $d\Psi/dx$, symbolizes the electromotive force in an electrostatic field with a spatial gradient. All three elements combine to act with an electromotive force $P$ in the $x$-direction.

## (E) ELECTRIC ELASTICITY

$$P = kf$$

In a simple, isotropic dielectric, the polarization $f$ and the electric intensity $P$ are related in direct proportion.

## (F) ELECTRIC RESISTANCE

$$P = -rp$$

This is just Ohm's law, where $r$ is the circuit's resistivity and $p$ is the current density.

## (A) TOTAL CURRENTS

$$p' = p + df/dt$$

This expresses the total current $p'$, adding the rate of change of the electric displacement (the "displacement current") to the conventional current density $p$. Note that it is this total current $p'$ which was related to the magnetic field in equation (C) above.

## (G) FREE ELECTRICITY

Maxwell gives an equation that we can write slightly differently, in this way:

$$\frac{df}{dx} + \frac{dg}{dy} + \frac{dh}{dz} = e$$

in which $f$, $g$, $h$ are the components of electric displacement. The whole expression on the left represents the divergence of the electric field—the space increase of $f$ in its own direction, which is the $x$-direction; of $g$ in its $y$-direction; and of $h$ in its $z$-direction. What we are coming to mean by the term "free electricity" is apparently no more or less than the fact of this divergence of the field (compare Discussion [D8] of Chapter 3).

## (H) CONTINUITY

If we have true currents flowing, with component current densities $p$, $q$, $r$, then the free electricity which we have just referred to must change accordingly. The divergence of the current is equal to the decrease in the free electricity.

## D6. Maxwell's Equations in Vector Terms [line 705]

### I. The Terms of Vector Analysis

We have avoided the complications of the vector statement of Maxwell's equations in order to concentrate on the underlying physical concepts and the unfolding of the theory itself, tasks which have certainly kept us busy! Wherever possible, we have isolated a simplified two-dimensional picture from the more general three-dimensional case; this did no harm, as the physics is the same. We have tended to focus on just one component of a three-component vector and to write only one equation, where in general three vector components and a set of three analogous equations are required to express the general case. Our simplification has usually meant just a commonsense orientation of our coordinate system to line it up neatly with the direction of the field lines or the plane of an exploratory loop.

Now, however, as we review the completed theory, the time has come to put his equations in their most general form. Only with this reformulation can their inherent symmetry fully emerge. Not only will we now write the complete set of three equations where before we had written only one, but to make these more manageable we will express these in vector form, and introduce as well certain vector operators that compress in symbols the patterns of derivatives with which we have become familiar. Vector notation, and this symbolic expression of operations of the calculus of vectors, or *vector analysis,* was emerging as Maxwell wrote. He was contributing, even in this third paper, to its development, and his rhetorical genius played a role in the choice of such a term as "curl," the vector operation whose name is suggested by the Scottish game. Among his manuscript papers is an intriguing scratch list of possible names for operations we are about to describe.

Incidentally, as we do this we may as well adopt the modern convention, alluded to in our Discussion [D1] of Cbapter 3, of using curly symbols for partial derivatives.

## II. Vector Quantities

Many of the quantities we have been dealing with are vectors with components in the $x$-, $y$-, and $z$-directions. One common practice, which we will follow here, is to use boldface characters to denote vector quantities. We may list here some of the principal quantities we have been working with, giving first the components by which Maxwell has designated them, but now assigning a modern vector symbol as well as shown in Table I.

Each of the quantities shown in Table I is a vector with three components. Thus for example the magnetic force **H** is the vector sum of three components, namely components of magnitude $\alpha$ in the $x$-direction, $\beta$ in the $y$-direction, and $\gamma$ in the $z$-direction.

There are highly satisfying rules for the algebra of vector quantities, specifying their summation and more than one form of multiplication, etc., but for these we must refer the interested reader to the Bibliography. We may permit ourselves just one convenient device, the *unit vector.* Three unit vectors are defined with unit length and direction along their

Table I

| Maxwell's name | Components | Vector symbol |
| --- | --- | --- |
| Electromagnetic momentum | F, G, H | **A** |
| Electromotive force | F, Q, R | **E** |
| Electric displacement | f, g, h, | **D** |
| Magnetic force in vacuo | a, b, g | **H** |
| Magnetic force in medium μ | μα, μβ, μγ | **B** |
| Current of free electricity | p, q, r | **i** |

respective axes, namely **i, j,** and **k** in the *x*-, *y*-, and *z*-directions. Using these, we may write for example:

$$\mathbf{H} = \alpha\mathbf{i} + \beta\mathbf{j} + \gamma\mathbf{k}$$

We should remind ourselves that there are other, scalar quantities, which have a value at every point in space but no directionality. Such are the electric potential *y* and the density of free electricity *r*. To remind ourselves that they are functions of position, we may write *y(x,y,z)*, etc., and of course our vectors are equally functions of position and may be written **H**(*x,y,z*).

Let us now rewrite the equations Maxwell has given us, in vector terms, taking them in the order of Maxwell's list.

III. Maxwell's Equations in Vector Terms

EQUATION (B) AND THE OPERATION CURL: Displayed in its full set of three components, and introducing now the modern notation to denote partial differentiation, Maxwell's "equation of magnetic force" (line 690) for the three components of **B** looks like this:

$$\mu\alpha = \frac{\partial H}{\partial y} - \frac{\partial G}{\partial z}$$

$$\mu\beta = \frac{\partial F}{\partial z} - \frac{\partial H}{\partial x} \qquad \text{(B)}$$

$$\mu\gamma = \frac{\partial G}{\partial x} - \frac{\partial F}{\partial y}$$

The cyclic pattern of the symbols here reflects the structure of the three-dimensional space. We might say that isotropic space is boring: from the first equation, the others can be written down by symmetry—they contain no new information, which is the reason that vector notation can be so concise!

We worked with the first equation, setting up a $yz$-plane and generating a component of the magnetic field parallel to the $x$-axis (Fig. 4.11). We chose our coordinates then so that **B** would have no $y$- or $z$-component, but if it does, we must set up exploratory rectangles in the same way parallel to the $xz$- and again to the $xy$-planes. The physics in each case is the same, and is simply iterated to generate the three components, which add vectorially.

We asserted then that by this process we were finding the curl of the momentum. We now see that we were in fact finding just one *component* of the curl, and that curl is actually itself a vector quantity. *The entire set of three equations can be expressed in a single breathtaking statement of the vector calculus.* Recalling that **B** = μ**H,** we may write simply

$$\mathbf{B} = \text{curl } \mathbf{A} \qquad (\text{B}')$$

We see that curl is an operator, which operates on one vector **A** in this special way to produce another vector **B**.

EQUATION (C) IN TERMS OF CURL: We have used the same mathematical method in expressing Oersted's relation between a current and the corresponding magnetic field. We may now put Maxwell's "equation of electric currents" in terms of the operator curl. We need not bother to spell out the three component equations, but rather will go directly to the result:

$$4\pi\mathbf{i}' = \text{curl } \mathbf{H} \qquad (\text{C}')$$

EQUATION (D): THE OPERATION GRAD AND THE CROSS PRODUCT: Maxwell's equation for the electromotive force has three terms, which we may address one by one. The full vector equation is:

$$P = \mu\left(\gamma\frac{\partial y}{\partial t} - \beta\frac{\partial z}{\partial t}\right) - \frac{\partial F}{\partial t} - \frac{\partial \psi}{\partial x}$$

$$Q = \mu\left(\alpha\frac{\partial z}{\partial t} - \gamma\frac{\partial x}{\partial t}\right) - \frac{\partial G}{\partial t} - \frac{\partial \psi}{\partial y} \qquad (\text{D})$$

$$R = \mu\left(\beta\frac{\partial x}{\partial t} - \alpha\frac{\partial y}{\partial t}\right) - \frac{\partial H}{\partial t} - \frac{\partial \psi}{\partial z}$$

Dealing initially with just the first term, however, we have:

$$P = \left(\mu\gamma\frac{\partial y}{\partial t} - \mu\beta\frac{\partial z}{\partial t}\right)$$

$$Q = \left(\mu\alpha\frac{\partial z}{\partial t} - \mu\gamma\frac{\partial x}{\partial t}\right)$$

$$R = \left(\mu\beta\frac{\partial x}{\partial t} - \mu\alpha\frac{\partial y}{\partial t}\right)$$

We recognize this as the expression for the electromotive force **E** induced in a conductor moving in a field of magnetic flux **B** with a velocity **v** whose components are

$$\frac{\partial x}{\partial t}, \quad \frac{\partial y}{\partial t}, \quad \frac{\partial z}{\partial t}$$

If we write the components of **E** and **B** as:

$$E_x, E_y, E_z$$

$$B_x, B_y, B_z$$

the pattern clarifies appreciably:

$$E_x = (B_z v_y - B_y v_z)$$

$$E_y = (B_x v_z - B_z v_x)$$

$$E_z = (B_y v_x - B_x v_y)$$

This is the defining pattern for the *cross product* of two vectors:

$$\mathbf{E} = \mathbf{v} \times \mathbf{B}$$

As simple as that!

The second term, taken alone, is very simple. It represents the time variation of the vector **B**: formally, the vector is being differentiated with respect to the scalar *t*:

$$\mathbf{E} = -\frac{\partial \mathbf{B}}{\partial t}$$

These have both been dynamic terms. The third term represents the electrostatic field that arises as the gradient of the electric potential field. We have seen this along one axis; now we extend it, as the potential may vary with respect to all three axes, each of which generates its corresponding component of the electric field. We write this third term in component notation:

$$E_x = -\frac{\partial \psi}{\partial x}$$

$$E_y = -\frac{\partial \psi}{\partial y}$$

$$E_z = -\frac{\partial \psi}{\partial z}$$

This procedure for generating a vector by taking the space derivatives of a scalar is the operation termed gradient, and written grad:

$$\mathbf{E} = -\text{ grad } Y$$

In the new notation, then, Maxwell's equation (D) becomes

$$\mathbf{E} = (\mathbf{v} \times \mathbf{B}) - \frac{\partial \mathbf{B}}{\partial t} - \text{grad} \psi$$

EQUATION (E): ELECTRIC ELASTICITY: Maxwell understands **D** as a displacement in the direction of **E** with a proportionality constant $k$: this is now easily written in a single equation:

$$\mathbf{E} = k \, \mathbf{D} \qquad \qquad (\text{E}')$$

EQUATION (F): ELECTRIC RESISTANCE: Although we normally think of Ohm's law operating in confined circuits, Maxwell is writing it for the density of current flow in a resistive space:

$$\mathbf{E} = -r \, \mathbf{i} \qquad \qquad (\text{F}')$$

EQUATION (G): THE EQUATION OF FREE ELECTRICITY: Here we come to the crunch mentioned earlier: Maxwell has introduced a certain perversity of sign, with a minus sign where we should expect a plus. If we override his equation as explained in the preceding discussion, we will have as a measure of free electricity:

$$e = \frac{\partial E_x}{\partial x} + \frac{\partial E_y}{\partial y} + \frac{\partial E_z}{\partial z}$$

This is the operation we have identified as divergence and illustrated generally in Fig. 3.9, and specifically in the case of divergence in just one dimension in Fig. 3.31. Here it extends to divergence in all three dimensions and becomes the vector divergence operator div:

$$e = \text{div } \mathbf{D} \tag{G'}$$

EQUATION(A): TOTAL CURRENT: Here, the displacement current is added vectorially to the current of free electricity:

$$\mathbf{i}' = \mathbf{i} + \frac{\partial \mathbf{D}}{\partial t} \tag{A'}$$

EQUATION (H): THE EQUATION OF CONTINUITY: The time-rate of the decrease of free electricity in an element of volume is equal to the vector current flowing out of that element. That is, the decrease of the density of free electricity at a point is equal to the vector current density at that point:

$$\text{div}\,\mathbf{i} = -\frac{\partial e}{\partial t} \tag{H'}$$

IV. Summary: Maxwell's Equations in Vector Form
We may collect Maxwell's equations (which we might call "Maxwell's own equations") in vector form, so that we can most easily go on to compare them with the standard account of Maxwell's equations:

$$\mathbf{B} = \text{curl } \mathbf{A} \tag{B'}$$

$$4\pi\mathbf{i}' = \text{curl } \mathbf{H} \tag{C'}$$

$$\mathbf{E} = (\mathbf{v} \times \mathbf{B}) - \frac{\partial \mathbf{B}}{\partial t} - \text{grad}\,\psi \tag{D'}$$

$$\mathbf{E} = k\mathbf{D} \tag{E'}$$

$$\mathbf{E} = -r\,\mathbf{i} \tag{F'}$$

$$e = \text{div } \mathbf{D} \tag{G'}$$

$$\mathbf{i}' = \mathbf{i} + \frac{\partial \mathbf{D}}{\partial t} \tag{A'}$$

THE EQUATIONS IN FREE SPACE: If we look simply at the case of the field in a segment of free space, containing no conductors or dielectric media—other than the vacuum itself!—we get a still more simplified set of equations. There will be no $r$ and no $i$; we will have $\mu = 1$, so $\mathbf{B} = \mathbf{H}$ (we may write just $\mathbf{B}$ for each), and $k = 1$, so $\mathbf{E} = \mathbf{D}$ (we may write just $\mathbf{E}$):

$$\mathbf{B} = \text{curl } \mathbf{A} \tag{B$'$}$$

$$4\pi\left(\frac{\partial \mathbf{E}}{\partial t}\right) = \text{curl}\,\mathbf{B} \tag{C$''$}$$

$$\mathbf{E} = -\frac{\partial \mathbf{B}}{\partial t} - \text{grad}\,\psi \tag{D$''$}$$

$$e = \text{div } \mathbf{D} \tag{G$''$}$$

Apart from certain constant factors that will become important to us later in this chapter, these formulations do not differ greatly from typical modern presentations of "Maxwell's equations" except in the prominence given here to the vector $\mathbf{A}$. In modern texts $\mathbf{A}$ is normally not included in the fundamental list, but is treated separately as the "vector potential." Logically, it is not essential as a first principle. For Maxwell, on the other hand, $\mathbf{A}$ appears at the outset, since it represents for him the momentum stored in the resting field, manifested in various ways in all electromagnetic phenomena. For moderns, it is logically unnecessary and its inclusion among the basic premises would be inelegant; for Maxwell, it is conceptually at the foundation of the entire theory and it takes first place among the elements of the science!

In place of (B$'$), one sees today the equation

$$\text{div}\mathbf{B} = 0 \tag{B$''$}$$

asserting the fundamental fact that there are no isolated magnetic poles—the magnetic "unipole" does not occur. (B$'$) and (B$''$) are equivalent, for it is provable in the vector calculus that for any vector $\mathbf{v}$, the divergence of its curl is identically zero:

$$\text{div}(\text{curl } \mathbf{v}) = 0$$

The distinction between $\mathbf{B}$ and $\mathbf{H}$, and that between $\mathbf{D}$ and $\mathbf{E}$, which reflect the magnetic and dielectric properties of materials expressed in the parameters $\mu$ and $k$, may be ignored in writing the fundamental equations today. The $\mu$ and $k$ are perceived as macroscopic quantities, consequences of known molecular configurations and thus disappear on the atomic level. The fundamental set thus looks like Maxwell's free-space equations—in that sense, on the atomic level all space is free!

In general, formulations by moderns, who normally agree with Hertz's dry judgment that "Maxwell's theory is Maxwell's equations," tend to be tighter. Maxwell offers more variables and correspondingly more equations than formally necessary, guided by a different muse.

## D7.  A Contradiction in Maxwell's Equations [line 705]

### I. Demonstration of the Contradiction
Maxwell writes, as the equation of free electricity, equation (G):

$$e + \frac{df}{dx} + \frac{dg}{dy} + \frac{dh}{dz} = 0$$

i.e.,

$$\frac{df}{dx} + \frac{dg}{dy} + \frac{dh}{dz} = -e \tag{G}$$

This is demonstrably in contradiction with his equation (C), as we shall now show. This well-known error, Maxwell's minus-sign mistake, has received a great deal of serious discussion in the critical literature. It might be nice to say that it was a simple error, but indeed although it entailed some bad arithmetic in "Physical Lines," it appears to be deep-rooted in Maxwell's own struggle to reconcile the concept of electric charge to the field theory, a question addressed later in this discussion.

The contradiction may be demonstrated as follows (we take the liberty of using the vector notation introduced in Discussion [D6], above).

$$\text{div } \mathbf{D} = -e \tag{G'}$$

We have also the equation of total current:

$$\mathbf{i}' = \mathbf{i} + \frac{\partial \mathbf{D}}{\partial t} \tag{A'}$$

the equation of continuity:

$$\text{div } \mathbf{i} = -\frac{\partial e}{\partial t} \tag{H'}$$

and the equation of currents:

$$4p\,\mathbf{i}' = \text{curl } \mathbf{H} \tag{C'}$$

Substituting in (C′)

$$4\pi\left(\mathbf{i} + \frac{\partial \mathbf{D}}{\partial t}\right) = \text{curl}\mathbf{H} \tag{a}$$

For any vector **v**, it is always the case that

$$\text{div (curl } \mathbf{v}) = 0$$

Thus we have a sequence of steps:

$$\text{div}\left[4\pi\left(\mathbf{i} + \frac{\partial D}{\partial t}\right)\right] = \text{div curl } H = 0$$

$$\left[4\pi\left(\text{div}\,\mathbf{i} + \text{div}\,\frac{\partial \mathbf{D}}{\partial t}\right)\right] = 0$$

$$\left(\text{div}\,\mathbf{i} + \text{div}\,\frac{\partial \mathbf{D}}{\partial t}\right) = 0$$

or simply

$$\text{div}\,\mathbf{i} = -\,\text{div}\,\frac{\partial \mathbf{D}}{\partial t} \tag{b}$$

In the steps above we have used the fact that a constant factor (here $4\pi$) factors out of a differentiation and that the derivative of a sum is the sum of the derivatives (here in the form of div).

The crunch now comes when we use equation (H′),

$$\text{div}\,\mathbf{i} = -\frac{\partial e}{\partial t} \tag{H'}$$

to get

$$\frac{\partial e}{\partial t} = \text{div}\,\frac{\partial \mathbf{D}}{\partial t} \tag{c}$$

On the right-hand side, we can use the principle that the order of differentiation (here, with respect to space and time) may be interchanged:

$$\frac{\partial e}{\partial t} = \frac{\partial}{\partial t}(\text{div}\mathbf{D}) \tag{d}$$

Integrating both sides with respect to time:

$$e = \text{div }\mathbf{D} \tag{e}$$

to be contrasted with the original equation of free electricity:

$$-e = \text{div }\mathbf{D} \tag{G'}$$

II. Maxwell's Resolution of the Contradiction

The problem remains unresolved in the "Dynamical Theory" paper, but in the *Treatise on Electricity and Magnetism* a few years later, Maxwell writes equation (G) with the sign corrected, that is, in the form (e). The real trouble, however, is not algebraic but substantive. It arises from Maxwell's continuing discomfort with the concept of free electricity as an entity in the cosmos. In his theory the energy we call electrostatic resides in the field, and he can find no real evidence of any substance to be called electric charge accumulating on what we think of as a "charged conductor." Our note on line 165 of this paper discussed this question, and Fig. 4.1 showed how free electricity might be understood as a phenomenon of the field alone. As we shall see in Discussion [D9], whether despite of or because of these doubts Maxwell searched for experimental evidence of inertial effects due to any such fluid supposedly "flowing" in a wire when what we call a "current" exists—and found none.

The problem returns, then, to the difficult concept of *polarization,* which we discussed and illustrated in relation to the previous paper. It appears that Maxwell's resolution of the problem is as follows: In the case of the charged capacitor of Fig. 4.1, nothing has accumulated on the plates. Current flow is always continuous; there are no "open circuits." Rather, an apparent break in a circuit such as the dielectric gap of Fig. 4.2 is actually a sharp discontinuity in the elasticity of the medium (compare Fig. 3.23). On one side, the conductor supports no elastic stress; on the other, we have the electric stiffness of air or vacuum. We know that wherever a discontinuity exists (as between glass and air) apparent charges will be found, and Maxwell seems to be concluding that all charge is only apparent.

Today we know about electrons, as Maxwell did not—though we might regard his experiment depicted in Fig. 4.26 as seeking them: they proved simply too small for his apparatus to detect. The electron is an elementary body bearing, we say, mass and "charge." It is not popular today to pursue the "metaphysical" question of the meaning of this statement—that is the thrust of Hertz's petulant remark, "Maxwell's theory is Maxwell's equations." But those of us who, with Maxwell, persist in wondering about such things may point out that the fact that *electrons* demonstrably accumulate on plates does not mean that charge resides there. Whitehead, thinking the field concept view through to its implications, spoke of each electron, by way of its field, as "ingredient" in the cosmos.

## D8. Integrating by Parts [line 717]

This is part of what we might call the repertory of mathematical rhetoric, which has the effect of transforming a mathematical phrase, often to reveal new significance The concept of integration by parts is at once simple and powerful. It is based on the rule for the differentiation of a product:

$$\frac{d}{dz}(xy) = x\frac{dy}{dz} + y\frac{dx}{dz}$$

If we integrate both sides with respect to $z$—since integration is the operation inverse to differentiation—we undo the derivatives and get

$$xy = \int x\,dy + \int y\,dx$$

or equally,

$$\int y\,dx = xy - \int x\,dy$$

If the integration is between fixed limits—summing from one fixed value to another—then we must evaluate the definite integral at the limits:

$$\int_{x_1}^{x_2} y\,dx = xy\big]_{x_1}^{x_2} - \int_{x_1}^{x_2} x\,dy$$

Now, Maxwell is interested in the special case in which the limits are infinite—summing over the whole field, or all of space, we are integrating "from infinity to infinity" (something a natural philosopher can do with remarkable ease!):

$$\int_{-\infty}^{\infty} y\,dx = xy \ ]_{-\infty}^{\infty} - \int_{-\infty}^{\infty} x\,dy$$

The trick is this: $x$ and $y$ may be supposed to be functions that, whatever values they have elsewhere have the value zero at infinity. Then under this condition, their term vanishes when the limits are substituted, and the definite integral becomes:

$$\int_{-\infty}^{\infty} y\,dx = [0 - 0] - \int_{-\infty}^{\infty} x\,dy$$

$$\int_{-\infty}^{\infty} y\,dx = - \int_{-\infty}^{\infty} x\,dy$$

which is what we need for our purposes. Let us add, to acknowledge to our mathematical friends the informality of our argument (here, as elsewhere), "under reasonable conditions"!

### D9. The Inertia of a Current Experiment [line 769]

Probing, as it seems, all possibilities with respect to the substance of electricity, Maxwell describes in the *Treatise* an experiment to detect any momentum of current in the conducting wire itself—that is, to determine whether "electricity" is something massive flowing in the wire (Fig. 4.26). A strong current flows in a delicately suspended coil. When the circuit is suddenly interrupted and the current brought to an abrupt stop, a reaction should deflect the reflected light beam—for total momentum must be conserved and the momentum of the electricity must, or at least might, appear as an impulse imparted to the wire and its supporting frame. Modern experiments have detected such effects but the momentum of what has emerged as the "electron" is vanishingly small with respect to the momentum and energy relations in the field. Maxwell's experiment was not destined to detect such a small effect, and within the limits of sensitivity of the apparatus he seems right to conclude that there is no evidence for a momentum-bearing current.

### D10. Derivation of the Field Energy Equation (40) [line 795]

We begin with equation (38) for the energy in the magnetic field:

$$E = \frac{1}{8\pi} \Sigma \left( \alpha \cdot \mu\alpha + \beta \cdot \mu\beta + \gamma \cdot \mu\gamma \right) dV \tag{38}$$

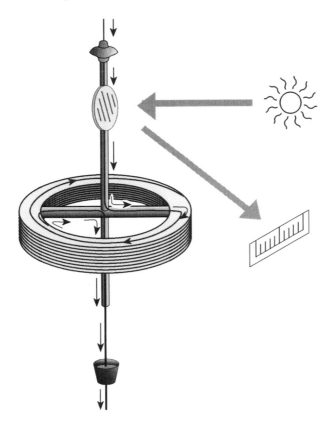

Fig. 4.26. Maxwell's design of for an experiment to detect the inertia of a current.

representing a summation over the whole field of the product of the field intensity ($\alpha$ . . .) and the magnetic induction ($\mu\alpha$ . . . .) at every point. Our purpose now is to rework this into an expression that involves the magnetic potential function $\phi$. The strategy is to arrive at an equation for the force between magnetic poles and then, in the same way, to get Coulomb's law for the force between charges. The potential function is, then, a way station to an expression for force.

We know, however, that we can write the magnetic intensity as the gradient of a potential, that is, the space-rate of change of the potential. Each component of the intensity is the space-rate of the potential with respect to its own axis:

390 Maxwell on the Electromagnetic Field

$$\frac{d\varphi}{dx} = \alpha, \qquad \frac{d\varphi}{dy} = \beta, \qquad \frac{d\varphi}{dz} = \gamma$$

Substituting, we get

$$E = \frac{1}{8\pi} \int (\mu\alpha \frac{d\varphi}{dx} + \mu\beta \frac{d\varphi}{dy} + \mu\gamma \frac{d\varphi}{dz}) dV$$

Integration by parts ([D8], above) permits interchange of the roles of φ and μα, with a change of sign:

$$E = -\frac{1}{8\pi} \int \left( \varphi \frac{d\mu\alpha}{dx} + \varphi \frac{d\mu\beta}{dy} + \varphi \frac{d\mu\gamma}{dz} \right) dV$$

The φ can now be factored out since it appears in each term:

$$E = -\frac{1}{8\pi} \int \varphi \left( \frac{d\mu\alpha}{dx} + \frac{d\mu\beta}{dy} + \frac{d\mu\gamma}{dz} \right) dV$$

From this, Maxwell takes the expression

$$\left( \frac{d\mu\alpha}{dx} + \frac{d\mu\beta}{dy} + \frac{d\mu\gamma}{dz} \right)$$

This is the "divergence" of the lines of induction, which becomes the sign indicating the presence of a "pole"—we might say, the *signature* of the pole. The "pole," on the other hand, no longer appears as a *thing* but as this specific configuration of the field.

Thus, by this restructuring of the energy equation, Maxwell has shaped it to reveal the presence of field sources, if any. A unit source yields $4\pi$ lines, so if $m$ is the strength of the source, we have

$$\left( \frac{d\mu\alpha}{dx} + \frac{d\mu\beta}{dy} + \frac{d\mu\gamma}{dz} \right) = 4\pi m$$

The energy equation is now in the form of "poles" distributed in a potential field φ. Maxwell is ready to derive the law of force between poles.

D11. The Number of Lines of a Unit Pole [line 803]

We may turn back to Fig. 2.10, representing flow from a point source, and reason quite simply from it to the conclusion that a total of $4\pi$ lines emerge from a unit pole—or more generally, the total number of lines of magnetic intensity from a pole of strength $m$ is $4\pi m$.

The *intensity* at any point in a field has been defined as the force per unit pole and, at the same time, this becomes the number of lines crossing unit area. (These are at the same time lines of induction if $\mu = 1$). But a pole of unit strength is by definition one that produces unit intensity at unit distance.

We surround such a unit pole with a unit sphere, whose area, in general $4\pi r^2$, becomes in this case just $4\pi(1)^2 = 4\pi$. Knowing that the unit pole yields just one line per unit area, we have the total number of lines in turn equal to $4\pi$. A pole of strength $m$ can be thought of as $m$ unit poles, corresponding to $4\pi\, m$ lines.

We are not to be deterred in any of this reasoning if it turns out that "poles" may not exist. For our purposes, the pole is quite strictly defined as a structural form—a pattern of divergence—in the field. We may continue to speak in terms of poles—such borrowing of concepts from unacceptable foundations is not without precedent. Navigators fully persuaded of the truth of Copernican astronomy have found their ways for centuries by means of the earth-centered celestial sphere of Ptolemy!

## D12. The Law of Force between Parallel Currents [line 858]

We can easily derive a law for the force acting between parallel conductors from the basic principle we arrived at when we applied the concept of mutual inductance $M$ to find the mechanical action between conductors. Equation (12) told us that as the coils approached through the interval $ds$ and their linkage $M$ changed by the amount $dM$, the force in the direction of $s$ would be

$$F = \frac{dM}{ds}xy \tag{12}$$

We know now that when magnetic field lines are defined by the method of the exploring coil, we have an electromagnetic momentum through a circuit equal to the total number of lines that pass through: that is, the induction (lines per unit area) times the area.

Suppose then that we have a conveniently rectangular circuit $PQRS$, carrying current $y$, placed in a uniform magnetic field $B = \mu\beta$ perpendicular to its plane (Fig. 4.27). We can think of our circuit as a secondary circuit and the field $B$ as due to a primary circuit carrying current $x$ and

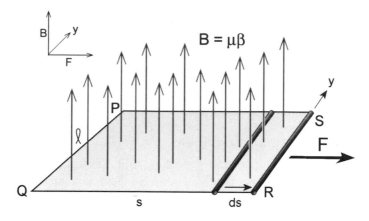

Fig. 4.27. Force on a conductor.

coupled by a mutual inductance $M$. Then the electromagnetic momentum of $x$ at the secondary will be $Mx$. This will equal the number of lines of induction through the circuit, or

$$M\,x = B(\text{area}) = B\,\ell\,s$$

Now if we let the side $RS$ slide parallel to itself the distance $ds$, the change in momentum in the secondary will be

$$x\,dM = B\ell\,ds$$

assuming that we keep $x$ and therefore also $B$ constant. In this process, we will meet a force in moving the segment of conductor, and the amount of that force will be given by (12) above:

$$F = \frac{dM}{ds}xy = \left[\frac{(x\,dM)}{ds}\right]y = \left[\frac{(B\ell\,ds)}{ds}\right]y = B\ell\,y$$

i.e.,

$$F = B\,\ell\,y$$

A conductor carrying current in a magnetic field is connected to a system possessing momentum and energy, and encounters a force tending to move it so as to minimize that energy. We see that the force F acts perpendicularly to both the field $B$ and the conductor carrying current $y$.

If the primary circuit is very large, so that $B$ is effectively the field at distance $r$ from a long straight conductor, then that field will be given by the Oersted relation. This tells us that the work to carry a unit pole around the primary current $x$ will be

$$W = 4\pi x$$

But this work to move the unit pole is just the field intensity times the distance. The distance around the circle at radius $r$ is $2\pi r$, so that if $b$ is the intensity, we have

$$\beta(2\pi r) = 4\pi x$$

The field is then just

$$\beta = \frac{2x}{r}$$

and $B = \beta$ when $\mu = 1$. Inserting that into our force equation, we find that, for the force between parallel conductors, the rule is

$$F = B\ell y = \left(\frac{2x}{r}\right)\ell y$$

This will be better expressed as force per unit length of the conductor, or

$$\frac{F}{\ell} = \frac{2xy}{r}$$

If the medium is of $\mu > 1$, then $B = \mu\beta$, and the force is proportionately increased.

   This is a good point at which to recall that we are speaking of a particular instance of the connected system with two driving points, with which we began. The two points are the two parallel currents; their interconnection is such that energy stored in the field will be released if they move toward one another.

## D13. The Dimensionality of $v$ [line 880]

We may keep in view two physical situations, both familiar to us now. Each is the natural foundation of a system of units. The first is Coulomb's balance, founded on his equation for the force between two charges; it serves as physical foundation for the electrostatic system of

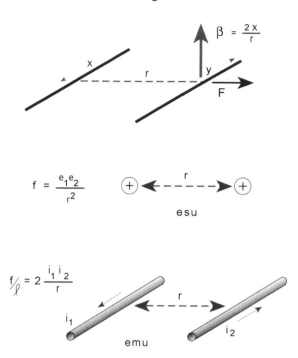

Fig. 4.28. Force between parallel conductors.

units (esu). The second is a balance, actually used by Ampère, which measures the force between two parallel conductors; it is founded on the force law we derived in the Discussion [D12] above, and serves as a natural foundation for the electromagnetic system of measurements (emu) (Fig. 4.28). Maxwell builds his explanation of the dimensionality of $v$ on these two equations:

$$f = \frac{e_1 e_2}{r^2} \qquad \frac{f}{\ell} = 2\frac{i_1 i_2}{r}$$

$$\text{(esu)} \qquad\qquad \text{(emu)}$$

In the electrostatic system we have a force equal to the product of two quantities of electricity divided by the square of the distance. The unit of electricity will therefore vary directly as the unit of length, and as the square root of the unit of force.

In the electromagnetic system we have a force equal to the product of two currents multiplied by the ratio of two lines [i.e., solving for $f$ above, we will have *l/r* on the right]. The unit of current in this system therefore varies as the square root of the unit of force; and the unit of electrical quantity, which is that which is transmitted by the unit current in unit of time, varies as the unit of time and as the square root of the unit of force.

The ratio of the electromagnetic unit to the electrostatic unit is therefore that of a certain distance to a certain time, or, in other words, this ratio is a *velocity;* and this velocity will be of the same absolute magnitude, whatever standards of length.[23]

Maxwell's line of thought here is representative of an elegant and very powerful mode of reasoning in which he pioneered, termed *dimensional analysis.* In this remarkable algebra, symbols are used to represent not numbers but the *concepts* of which the numbers are the measures. In this sense, dimensional analysis penetrates to the very elements of a physical system and is close in spirit to the very serious attention Maxwell regularly gave to the *elementary* foundations of each branch of physics he addressed. The "dimensions" to which the name refers seem to be the very parameters of the cosmos itself, as human science has at any point been able to grasp them. In this sense, too, dimensional analysis is close to the Kantian concern with the elements of human intuition of the physical world, a topic of Hamilton's lectures in metaphysics at Edinburgh that remained with Maxwell in all of his work.

In dimensional analysis, then, literal symbols represent concepts, not numbers: the things counted rather than the count itself. Thus $L$ denotes length, $T$ denotes time, $M$ denotes mass—the foundations, we might say, of our belief system at any time concerning the ultimate elements of the physical world. It is useful to follow Maxwell in distinguishing dimensional from numerical equations by writing them in square brackets. Velocity then becomes $[L/T]$ and acceleration is $[L/T^2]$, or writing the same expressions by means of negative exponents, $[LT^{-1}]$ and $[LT^{-2}]$. Force as mass times acceleration becomes $[MLT^{-2}]$. When new concepts are introduced, such as electric charge, they may be expressed in terms of the basic concepts by which we know them. Maxwell's argument above concerning the measures of electric charge may thus be expressed in these symbols:

1. By Coulomb's law we have in dimensional terms:

$$\left[ MLT^{-2} \right] = \left[ \frac{e_s \cdot e_s}{L^2} \right]$$

or, solving for $e_s$

$$[e_s] = [ML^3T^{-2}]^{\frac{1}{2}} = [M^{\frac{1}{2}}L^{\frac{3}{2}}T^{-1}] \cdot [L]$$

2. For the force between parallel currents we have:

$$\frac{[MLT^{-2}]}{[L]} = \frac{\left[\frac{e_m}{T}\right] \cdot \left[\frac{e_m}{T}\right]}{[L]} = \frac{\left[\frac{e_m}{T}\right]^2}{[L]}$$

$$\left[\frac{e_m}{T}\right]^2 = \frac{[MLT^{-2}]}{[L]} \cdot [L] = [MLT^{-2}]$$

$$[e_m]^2 = [MLT^{-2}] \cdot [T]^2$$

$$[e_m] = [M^{\frac{1}{2}}L^{\frac{1}{2}}T^{-1}] \cdot T$$

3. For the ratio of the units of charge $e_s/e_m$:

$$\left[\frac{e_s}{e_m}\right] = \frac{[M^{\frac{1}{2}}L^{\frac{3}{2}}T^{-1}] \cdot [L]}{[M^{\frac{1}{2}}L^{\frac{1}{2}}T^{-1}] \cdot [T]} = \left[\frac{L}{T}\right]$$

Here, we have spelled out the dimensional analysis fully, to see how the method works. Maxwell's verbal account leaves force unanalyzed, as it appears, as we see, equally in both expressions and cancels out.

D14. The Wave Equation and its Solution [line 970]

I. Verifying the Proposed Solution
Maxwell has arrived at the equation

$$\frac{d^2\beta}{dt^2} = \left(\frac{k}{4\pi\mu}\right)\frac{d^2\beta}{dz^2}$$

which he rewrites by simplifying the constant coefficient:

$$\frac{k}{4\pi\mu} = v^2$$

$$\frac{d\beta^2}{dt^2} = v^2 \frac{d\beta^2}{dz^2}$$

The second derivative of the magnetic intensity with respect to time is proportional to its second derivative with respect to distance. The quantity $\beta$ *is some function of distance and time for which this will be true*—for which the two derivatives will relate in this simple way. Just to give it a name, let us call this function $\varphi$ (nothing to do with any other function of the same name!), i.e., $\varphi$ is some function of $z$ and $t$:

$$\varphi = \varphi(z,t)$$

Maxwell now tells us that the "well-known" solution to our equation, i.e., the particular form $\varphi(z,t)$ that will have the right derivatives, is

$$\beta = \varphi_1(z - Vt) + \varphi_2(z + Vt) \tag{20}$$

We can lessen our labors if we just take the first form, calling it simply $\varphi$. It will be one solution; Maxwell's is merely more general in a way that we will see. We assume, as a proposed solution,

$$\beta = \varphi(z - Vt)$$

Let us designate $(z - Vt)$ as $g$, so

$$\beta = \varphi(g)$$

The *chain rule* tells us that in a case like this

$$\frac{d}{dz}\varphi(g) = \left[\frac{d}{dg}\varphi(g)\right]\left(\frac{dg}{dz}\right)$$

Equally,

$$\frac{d}{dt}\varphi(g) = \left[\frac{d}{dg}\varphi(g)\right]\left(\frac{dg}{dt}\right)$$

We do not need to know what $\varphi$ is at this point; it can be anything within reason. Any "disturbance" will work. But we do need to catalogue the derivatives of our $g$. Luckily, they are easy:

$$g = (z - Vt)$$

$$\frac{dg}{dz} = \frac{dz}{dz} = 1 \quad (t \text{ constant})$$

$$\frac{dg}{dt} = \frac{d}{dt}(-Vt) = -V \quad (z \text{ constant})$$

These are applied to φ:

$$\frac{d}{dz}\varphi(g) = \left[\frac{d}{dg}\varphi(g)\right]\left(\frac{dg}{dz}\right) = \left[\frac{d}{dg}\varphi(g)\right] = \varphi'(g)$$

$$\frac{d}{dt}\varphi(g) = \left[\frac{d}{dg}\varphi(g)\right]\left(\frac{dg}{dt}\right) = (-V)\left[\frac{d}{dg}\varphi(g)\right] = (-V)\varphi'(g)$$

We have used the notation φ′ (g) to denote whatever function the derivative turns out to be. To get the second derivatives, we just do it again—but this time, differentiating the result we got above. The chain rule operates in the same way.

$$\frac{d^2}{dz^2}\varphi(g) = \frac{d}{dz}\left[\frac{d}{dz}\varphi(g)\right] = \frac{d}{dz}\varphi'(g)$$

$$= \frac{d}{dg}\varphi'(g)\frac{dg}{dz} = \varphi''(g)$$

$$\frac{d^2}{dt^2}\varphi(g) = \frac{d}{dt}\left[\frac{d}{dt}\varphi(g)\right] = \frac{d}{dt}\left[(-V)\varphi'(g)\right]$$

$$= (-V)\frac{d}{dg}\varphi'(g)\frac{dg}{dt} = V^2\varphi''(g)$$

Have we achieved anything? Yes! If we take these two results and insert them in the original equation, we see that they do indeed satisfy it:

ORIGINAL EQUATION:

$$\frac{d^2}{dz^2}\varphi(g) = \varphi''(g) \tag{A}$$

PROPOSED SOLUTION:

$$\beta = \varphi(z - Vt) \tag{B}$$

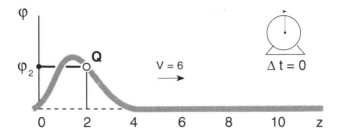

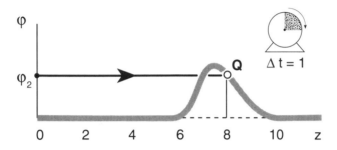

Fig. 4.29. An advancing wave.

WHAT WE HAVE FOUND:

$$\frac{d^2}{dz^2}\varphi(g) = \varphi''(g) \tag{C}$$

$$\frac{d^2}{dt^2}\varphi(g) = V^2\varphi''(g) \tag{D}$$

SUBSTITUTING IN (A):

$$V^2 \varphi''(g) = V^2 \varphi''(g)$$

Our proposed solution makes the original differential equation identically true and hence is a verified solution.

## II. Showing That It Is a "Wave" Equation

If we have a function $\varphi(z - Vt)$ and give it initial values for all $z$ at $t = 0$, it might look something like the upper image in Fig. 4.29. If, for example, we think of $z$ as the length of a stretched string, then our function at

$t = 0$ might represent a pluck. With $t = 0$, the equation of the pluck will be just $\varphi(z)$. Let us watch one particular point $Q$ on the pluck, corresponding to $\varphi(2)$.

We might ask, "Where will $Q$ be at some later time, say when $t=1$ ?" We know that $Q$ corresponds to $\varphi(2)$ . Initially,

$$t = 0, z_0 = 2: \qquad \varphi_0 = \varphi(z - Vt) = \varphi(2)$$

Now to locate $Q$ at $t = 1$ we must again have

$$(z - Vt) = (z - V) = 2$$

i.e.,

$$z_1 = 2 + V$$

*Thus z has advanced the distance V in unit time.* The whole pluck will advance in the same way, or move forward with the velocity $V$ (bottom image in Fig. 4.29). That is what we mean by a "wave" equation—the pattern, whatever it is, advances as a whole with a definite velocity. Often, we show waves as sinusoidal and Maxwell makes that assumption below. But that is just a special case.

D15.  Computation of the Voltage of Sunlight [line 1059]

Earlier, Maxwell computed the energy per unit volume of the electric field; its value is

$$\frac{1}{2}kf^2$$

Since here we are thinking in terms of electric intensity $P$, it would be useful to substitute $P = kf$ or $f = P/k$, giving

$$\frac{1}{2}\frac{P^2}{k}$$

This expresses the *electric* energy per unit volume at any point at which the electric intensity has the instantaneous value $P$. Now, however, Maxwell rather inconsiderately changes the meaning of $P$, which is to represent the maximum value of $P$ ($P$ is after all oscillating all the while). Let us call this maximum value $P_m$. What we want now is the *average* value of the energy over a volume, within which $P$ at any moment has all values from $+P_m$ to $-P_m$. Now it can be shown that the average value of the square of the function that represents $P$, namely, the *square of the sine function,* is one-half of its maximum value. Now the maximum value of electric energy per unit volume is

$$\frac{1}{2}\frac{P_m^2}{k}$$

and the *average* energy per unit volume in the electric field will be

$$\frac{1}{4}\frac{P_m^2}{k}$$

However, what Maxwell really wants is the *total* (average) energy of the propagating wave. And that is half electric, half magnetic. So we go back and multiply by 2 after all! The result is now, for *total average energy* per unit volume,

$$\frac{1}{2}\frac{P_m^2}{k}$$

This is equivalent to the expression Maxwell writes at line 1060, if we take into account the fact that

$$V^2 = \frac{k}{4\pi\mu}$$

i.e.,

$$\frac{P_m^2}{8\pi\mu V^2} = \frac{1}{2}\frac{P_m^2}{k}$$

Maxwell's form is well adapted to the expression of the conclusion he is reaching, namely the *flux* of energy when electromagnetic energy impinges on a surface (thinking of the sun's energy warming the earth's surface). For that purpose, all we have to do is multiply the average energy contained in unit volume by the rate of advance of the wave, *V*. Maxwell denotes this flux *W*, and writes

$$W = \left(\frac{P_m^2}{8\pi\mu V^2}\right)(V) = \frac{P^2}{8\pi\mu V}$$

Solving for *P*, Maxwell arrives at the number about which he has been so curious. He knows approximately the energy flux on earth from the sun, which he gives as

83.4 foot-pounds per second per square foot

Fig. 4.30. The Cavendish laboratory.

Maxwell really wants to ask, "What is the voltage?" If the solar radiation is an electrical phenomenon, he seeks to show us that it makes sense to ask *the electric intensity of sunlight.* There seems to be some dispute about Maxwell's numbers at this point. He arrives at 600 Daniell cells per meter (the potential of a single Daniell cell being a little more than 1 volt), and he repeats the same figure in the *Treatise.* There, J. J. Thomson as editor of the Third Edition says he cannot confirm Maxwell's numbers and gets instead

942 volts per meter

The reader is invited to work this out independently. Either way, it is an impressive value. Over the minute span of optical wavelengths, the actual voltages will be small but the intensity is real. Maxwell, it seems, cannot resist using the inverse square law to arrive at the corresponding figure for "the sun's surface" as well—that is, to estimate the electric intensity of sunlight at its source. If light is an electromagnetic phenomenon, then the sun is revealed as a dynamo emitting electrical energy at intensities of the order of 13,000 volts per meter, and as Maxwell goes on to emphasize, reversing the polarity of this field at the rate of a billion times per second.

# Notes

Biographical notices of scientists referred to are kept very brief, as fuller information is readily available in sources such as the *Encyclopaedia Britannica*. Articles in the *Dictionary of Scientific Biography,* found in most libraries, are generally excellent. Bibliographical references are given here in short form; the full reference is to be found in the Bibliography below.

## Chapter 1

1. Isaac Newton (1643–1727). His *Principia Mathematica Philosophiae Naturalis* (*Mathematical Principles of Natural Philosophy*) (1687) constitutes a complete theory of the motion of bodies, based on definitions, laws, and axioms—a theory that will occupy the center of our attention in Chapter 3 of the present study. Newton deliberately adopts a geometrical mode of argument, while the Continental tradition tends to be strongly analytic.

2. Laplace (1749–1827) reformulated celestial mechanics in a thoroughly analytic mode, while Joseph-Louis Lagrange (1736–1813) in his *Mécanique Analytique* recast the foundations of mechanics in terms that Maxwell will discover, with a sense of excitement, only in the third paper in this study, the "Dynamical Theory."

3. From a letter of Faraday to Ampère in 1822, to be found in de Launay, *Correspondence du Grand Ampère,* vol. 3, p. 911.

4. Jones, *Life and Letters of Faraday,* vol. 2, p. 10 (italics added).

5. Faraday, *Experimental Researches,* vol. 3, p. 530–531. (Throughout this text, reference to the *Experimental Researches* will be given in short form: [XR 3:531] or, where appropriate, in terms of Faraday's paragraph numbers, which run sequentially through the three volumes.)

6. Faraday to Maxwell, November 13, 1857, in Campbell and Garnett, *Life of James Clerk Maxwell,* p. 290. Hereafter cited simply as the *Life;* page references are to the Second Edition.

7. Maxwell to Faraday, October 19, 1861; *Life,* p. xxii.

8. Benjamin Franklin (1706–1790) was widely respected as the scientist who laid the foundations of the science of electricity; his *Experiments and Observations on Electricity* (1751) had been translated into French, German, and Italian. Joseph Priestley (1733–1804) was an experimentalist whose work in electricity was collected in his *History and Present State of Electricity* (1767) with the special encouragement of Franklin. Following attack in England for his liberal political views, he moved to the United States, where he was befriended by Jefferson and Adams; he continued his scientific work in this country, and died in Northumburland, Pennsylvania.

9. Carl Friedrich Gauss (1777–1855), Professor of Astronomy at Göttingen. Apart from the studies of magnetism of which we will speak here, which formed a relatively minor part of his distinguished career, he is known for his fundamental work in mathematics. He approached questions of magnetism with the same insight he brought to the foundations of geometry. He embarked on his magnetic work in 1833, and six annual volumes of its "Results" were published from 1837 to 1843. Note that Gauss has died, at the age of 77, early in the year in which Maxwell is beginning his reading in electricity. Wilhelm Weber (1804–1891), twenty-seven years Gauss's junior, was first his protégé, and later his partner, in a project of systematic magnetic measurements. The crux of the method, as we shall see, was to base these measurements on a rational system of magnetic units. In our year of 1855, Weber is teaching at the University of Leipzig, and has gone on to formulate an inclusive law of electromagnetic interactions based on the principle of action at a distance, which Maxwell is engaged in studying.

10. René Descartes (1596–1650). Though the point is commonly confused, Descartes' physics, as set out in his *Principles of Philosophy,* is strictly mechanical—in the sense that all actions are the consequence of contacts between parts of a plenum—while Newton's, cast in action-at-a-distance mode, is not. Descartes envisions a space-filling, mathematical medium ("matter") whose configurations and intelligible interactions account for all physical substances and their motions.

11. Charles-Augustin de Coulomb (1736–1806). As we shall see, he developed a delicate torsion balance to establish the inverse-square law of force between charged bodies, which follows the form of Newton's law of gravitation in the *Principia* exactly.

12. Ampère, *Théorie mathématiques,* p. 2.

13. Jones, *Life of Faraday,* vol. 2, p. 354.

14. This and the quotations that follow on induction are from the opening paragraphs of the Eleventh Series of Faraday's *Researches,* XR §1162–1174.

15. Siméon-Denis Poisson (1781–1840), building on the law electrostatic attraction established by Coulomb, together with the mechanics of

Laplace and Lagrange, had established a complete analytic theory of electrostatics.

16. The 1845 discovery is reported in Series 20, from which the preceding quotation is taken. The quotations which follow are selected from Series 26.

17. XR, §3271. The learned friend was apparently William Whewell, who was Master of Trinity College in Maxwell's time. By this route Faraday is coining many terms destined (unlike "sphondyloid") to become most fundamental to the science of electricity, among them anode, cathode, electrode, and ion.

18. XR §3300–3301.

19. George Simon Ohm (1789–1854), German physicist, working by analogy to the theory of flow of heat, had established the law of flow of current through a resisting wire.

20. Hans Christian Oersted (1777–1851), reportedly by surprise during a lecture in Copenhagen, had discovered the basic relation between an electric current and an associated magnetic field.

21. Joseph Henry (1797–1878), working in Albany, apparently preceded Faraday in this discovery, but America was not sufficiently included in the general scientific conversation for this work in upstate New York to be taken into account.

22. André-Marie Ampère (1775–1836), a brilliant French mathematician who in 1827 had claimed to have solved, in his *Théorie Mathématique,* the seemingly impossible problem of using Newtonian linear forces to account for the circular motions discovered by Oersted. He was memorialized on his death as "France's answer to the genius of Newton."

## Chapter 2

1. Maxwell sent this paper to Faraday, who replied: "I was at first almost frightened when I saw so much mathematical force made to bear upon the subject, and then wondered to see that the subject stood it so well." Faraday to Maxwell, March 25, 1857. Campbell, *Life,* p. 200. In a later letter, Faraday elaborated:

> I hang on to your words because they are to me weighty. . . . There is one thing I would be glad to ask you. When a mathematician engaged in investigating physical actions and results has arrived at his conclusions, may they not be expressed in common language as fully, clearly, and definitely as in mathematical formulae? If so, would it not be a great boon to such as I to express them so?—translating them out of their hieroglyphics . . . I have always found that you could convey to me a perfectly clear idea of your conclusions . . . neither above nor below the truth, and so clear in character that I can think and work from them. [Faraday to Maxwell, November 13, 1857. *Life,* p. 206]

The entire correspondence as reproduced by Campbell is most interesting.

2. Descartes, *Rules for the Direction of the Mind.* Rule xiv, pp. 28ff. But better: *Principles of Philosophy,* Principle iv: "That the nature of body consists not in weight, nor in hardness, nor colour and so on, but in extension alone." [Descartes, *Philosophical Works,* vol.ii, p. 255]

3. See Tricker, *Early Electrodynamics,* in the Bibliography.

4. This was preserved by Campbell in the *Life,* pp. 347ff.

5. Discussion of certain principles of analytic mathematics and guidance to texts will be a concern of Chapter 3.

## Chapter 3

1. [XR iii:438]

2. [*XR*.i:265ff.] Much later, Series III of the *Experimental Researches* was devoted to this question of the "identity of electricities."

3. In "Maxwell and the Direct Experimental Test of His Electromagnetic Theory" (see the Bibliography) I examine the available evidence on this point.

4. Eight pages of difficult physical argument, entangled in the physics of elastic media, intervene before he is able to make this claim! Maxwell, *Scientific Papers,* vol. 1, p. 500 (hereafter cited as [SP i/500]).

5. Newton's laws, which appear replete with important corollaries and interpretive scholia, are set out at the beginning of the *Principia.* See the helpful discussion in Densmore's edition, hereafter cited as Newton, *Principia.*

## Chapter 4

1. In summarizing a long series of investigations of the lines of force, Faraday concludes: "The general conclusions are, that the magnetic lines of force may be easily recognized and taken account of by the moving wire, both as to *direction* and *intensity,* within metals, iron or magnets, as well as in the space around; and that the wire sums up the action of many lines in one result" [XR 3:405–6]. Here, as often, one feels that Maxwell has one eye on the *Experimental Researches* as he writes.

2. Hertz, *Electric Waves,* p. 26.

3. Hertz, *op.cit.,* "Introduction," pp. 1–14. T. K. Simpson, "Maxwell and the Direct Experimental Test of His Electromagnetic Theory," pp. 416ff.

4. [XR 3:451]

5. The analytic approach to mechanics, following in the Cartesian tradition, had been highly developed in France. It is difficult to appreciate that methods developed there in the eighteenth and early nineteenth centuries, belonging to the era of the American Revolution, remain powerful tools in quantum physics, beyond the ambition of most basic courses in physics or the calculus in our own day. One example is the method of Fourier analysis, which we met in Chapter 2, Discussion [D4]; others are the method of

generalized coordinates, and the Lagrangian equations which are the foundation of Maxwell's third paper to be discussed here.

A very different spirit prevailed in England, where Newton remained the model. When Thomson embraced French methods, he was able to make striking contributions largely because each side of the Channel was so little familiar with work being done on the other.

Maxwell is inspired to write this third paper by Thomson and P. G. Tait, who in their *Treatise on Natural Philosophy* are at this moment dramatically introducing these methods on a grand scale. See Everitt, *op.cit.,* pp. 105ff.

6. The full title of Ampère's *Treatise,* in its extended cadence, epitomizes a severe theory of science: *Mathematical Theory of Electro-dynamic Phenomena Uniquely Deduced from Experiment.* Maxwell and Faraday would seem to hold a viewpoint directly opposed to that of Ampère, who with true Cartesian confidence is willing to confess that he never actually got around to doing some of the principal experiments he describes. Yet we have to wonder whether there is not a certain curious alliance occurring in the more austere environment of Maxwell's "Dynamical Theory"? The connected mechanical system of Lagrange's equations is something like a pure idea of reason; much can be "deduced" from it, with minimal cues from nature.

7. Whittaker, *History,* vol.1, p. 253.

8. James Bradley (1693–1762). Astronomer Royal, noted for the care and precision of his observations; he is said to have arrived at the interpretation of stellar aberration described here while observing the relations of wind and water when he was boating on the Thames.

9. We may note especially the classic report of 1863, which includes a description by Maxwell of the experimental apparatus and the results of the "spinning coil" method for the absolute determination of the British Association ohm, described below. Also, we note a dazzling essay by Maxwell and Fleeming Jenkin, "On the Elementary Relations between Electrical Measurements," in which it seems that the entire domain of electrical units and measurements is surveyed from the point of view of its conceptual foundation.

10. The delicacy of the apparatus with which Maxwell was working in these experiments and the particular importance of eliminating the effect of variations in the local magnetic field where possible are suggested by a remark of Maxwell's in connection with his description of the "spinning coil" experiment. The amplitude of oscillations of the detecting magnet was affected by "the passage of iron steamers in the Thames . . ."! *loc.cit.* p. 173.

11. [SP. 2:127] The Mr. Gassiot to whom Maxwell refers was a wealthy wine merchant who devoted his personal and financial resources to work in science. Some of Faraday's last diary entries (in 1858) refer to work with the great battery in Gassiot's laboratory, where experiments were done on electrical discharges in evacuated tubes.

12. [SP 2:134]

13. The British Association ohm was designed to equal a velocity of 10,000,000 meters/second, and Maxwell here assumes this value to be correct. Actually, the original ohm proved to have been appreciably lower, so presumably this would contribute to the low value of Maxwell's *v*. (See Thomson's note at Maxwell's *Treatise,* 2:436—though he does not seem to draw this conclusion.)

14. Chapter 19 of the *Treatise* is devoted to this topic. Thomson's value had come out even a little lower than his own, so that he has no really satisfactory figure to report. The values of the velocity of light he lists by comparison, though Foucault's in fact very accurate figure is at the end of the list, appear equally disparate. Maxwell is able to conclude only:

> It is manifest that the velocity of light and the ratio of the units are quantities of the same order of magnitude. Neither of them can be said to be determined as yet with such a degree of accuracy as to enable us to assert that the one is greater or less than the other. . . . Our theory . . . is certainly not contradicted by the comparison of these results such as they are. [*Treatise* 2:436]

In 1879, the year of Maxwell's death, Ayrton and Perry achieved a value in excellent agreement with Foucault's determination of the velocity of light (J.J. Thomson's note, *ibid.*).

15. Faraday, *Diary,* 7:311.

16. Such is the conclusion I once reached in trying to track down this question: T.K.Simpson, "Maxwell and the Direct Experimental Test," cited earlier. The missing link appears to have been the lack of any adequate appreciation of the possibilities of resonance—between an inductance and a capacitance—as providing rapid oscillations, and responding to them as a mode of sensitive detection. The possibility had not come into focus, and when, as it happened, radio transmission was quite clearly demonstrated to Maxwell's scientific friends shortly after his death, they saw nothing particularly interesting in it. Is it entirely certain that Maxwell would have?

17. [SP 2:138]

18. *Treatise* 2:493

19. [SP 2:241]

20. *Treatise* 2:228

21. [SP 2:782]

22. [SP 2:783] (italics mine)

23. [SP 2:125–6]

# Selected Readings

This list of readings is divided by chapter. For full citations of works mentioned, see the systematic Bibliography that follows.

## Chapter 1

### Biography and Background

Biographies of Maxwell
The *Life of James Clerk Maxwell* by Lewis Campbell, who had been his lifelong friend, remains the first reading for anyone interested in knowing Maxwell better. William Garnett, the coauthor who gives a very limited account of Maxwell's scientific work, was his assistant ("Demonstrator") at the Cavendish Laboratory. An excellent appreciation of his thought and scientific work is the biography by C. W. F. Everitt in the *Dictionary of Scientific Biography,* A fuller study is Martin Goldman's *The Demon in the Aether,* while an additional brief biography is that by Ivan Tolstoy, *James Clerk Maxwell.* Others of interest will be readily found by anyone who begins to explore the extensive literature on Maxwell.

A work of special interest in relation to Maxwell's two educations, Scottish and English, is George Elder Davie's *The Democratic Intellect: Scotland and her Universities in the Nineteenth Century.* For Davie, Maxwell serves as exemplar of a mind bridging those two very different intellectual worlds.

Biographies of Others
In general, the *Dictionary of Scientific Biography* is an excellent source of brief biographies of the scientists we will be encountering. To avoid interruption of our text, very short notices are included at appropriate points in our notes, In the case of Faraday there is both a classic life by H. Bence Jones and a modern biography by L. Pearce Williams. Of special interest to us must be Maxwell's two brief accounts of Faraday, one an article written

for *Nature,* and the other the biographical entry for the 9th Edition of the *Encyclopaedia Britannica.* These are included in Maxwell's *Scientific Papers* [hereafter, SP] Vol. 2, p. 355 and 786, respectively). Maxwell, we may note, was the scientific editor of the 9th Edition and contributed several of the articles himself; however, he had only reached "H" at the time of his death.

The standard biography of William Thomson (Lord Kelvin) is the *Life of Lord Kelvin* by Sylvanus Thompson; see also D. B. Wilson, *Kelvin and Stokes;* George Stokes was a close scientific friend of Maxwell's. A reader interested in having a look at those of Thomson's papers that Maxwell was studying will find examples in the *Reprint of Papers on Electrostatics and Magnetism.* The classic biography of P. G. Tait is that by C. G. Knott.

## Principles of Electricity

First to be noted is the *Elementary Treatise on Electricity* by Maxwell himself; he was working on this book at the time of his death, and though the published work contains only the beginnings of his project (the rest was filled out by the editors from the major *Treatise on Electricity and Magnetism*), his approach is of great interest; Maxwell attached special importance to the *elements* of any subject. One basic book which will be of special value throughout our study is Hugh Skilling, *Fundamentals of Electric Waves.*

The best way to meet Faraday is surely through an exploration of portions of his enormous *Experimental Researches in Electricity.* No paraphrase or summary can substitute for the unique modes of Faraday's mind, revealed at work in the *Experimental Researches* [hereafter XR]. One good place to investigate is Series XI.

Two books pertain most directly to our present endeavor, both by R. A. R. Tricker: *The Constributions of Faraday and Maxwell to Electrical Science* and *Early Electrodynamics.* The latter is especially helpful in developing Ampère's theory, in spirit perhaps the antithesis of the approach taken by Faraday and Maxwell. Another historical insight is found in the article "The Development of the Concept of Electric Charge," by Roller and Roller.

For anyone who can contrive access to the work, de la Rive's *Treatise on Electricity* (1853–1858) gives a very thorough account of electromagnetic theory as it was perceived at the time Maxwell is going to work.

Beyond these, each reader may perhaps best browse a good library and select a text that approaches the subject at what seems the most comfortable level and style. A delightful book for beginners is Eric Rogers, *Physics for the Inquiring Mind.* For a reader with some calculus, an account of classical physics that is to be recommended as making concepts especially clear is Lindsay and Margenau, *Foundations of Physics;* see Chapter VI for electromagnetism. For the no doubt rare reader prepared to undertake the subject in full mathematical form, there is George Owen's *Introduction to Electromagnetic*

Theory, while Feynman's *Lectures on Physics* is an exemplary text that places electricity and magnetism in the context of modern physics as a whole; classical electromagnetic theory appears in Feynman's Volume II.

# Chapter 2

## Theories of Light

On the theories of light to which Maxwell is referring at lines 58ff., see Strong's *Concepts of Classical Optics,* and Chapter I of Whittaker's *History.* The theory of Descartes is found in his *Principles of Philosophy,* and that of Huygens in his *Treatise on Light.* Thomas Young's paper on interference theory is reproduced with commentary in Shamos, *Great Experiments in Physics,* pp. 93ff.

## Fourier's Theory of Heat and the Analogy to Electrostatics

Fourier will not make rewarding reading for most of our readers, but the curious might consult his *Analytical Theory of Heat.* The analogy to electrostatics is no doubt in Maxwell's mind when he articulates his fascinating Apostle's Club essay, "Are There Real Analogies in Nature?," reprinted by Campbell in the *Life,* p. 347.

## Readings in Faraday

Maxwell's footnotes guide us to the sections in Faraday's *Experimental Researches* he has primarily in mind.

Placed in their sequence as they appear in the *Experimental Researches,* Maxwell's interests at this point focus on these areas of Faraday's work: Series XI (on electrostatic induction); Series XXVI (on magnetic conducting power); Series XXVIII (on the magnetic lines of force); and the 1852 article (without "series" designation) "On the Physical Character of the Lines of Magnetic Force," XR 3:407.

## General

### Mathematics

It is not presupposed here that readers have any special skill or preparation in mathematics, and suggestions will be made as we go along for help with new mathematical concepts as they appear. Maxwell is building his own geometry of fluids from scratch here, so we have only to follow him! But anyone who meets difficulty with the algebra that creeps in, primarily in our commentary, may turn to Levi's *Elements of Algebra* or Moore's *Algebra* as basic guides. Other such texts abound from which the reader may want to select.

Criticism of Maxwell's Way
Behind the surface of civility, it appears to me that there is a strong and
interesting polarization between England and the Continent with respect to
the standards of scientific reason. In particular, Pierre Duhem is a fierce
French critic of fuzzy ways of visualization by means of things like lines of
force. For readers who know French, the full measure of his biting critique
is to be found in *Les théories électriques de J. Clerk Maxwell;* but a
representative sample is included in *The Aim and Structure of Physical
Theory.* Ampère's highly mathematical approach, contrasting with Faraday's mental imagery, is well described in Tricker's *Early Electrodynamics.*
There is a fascinating correspondence between Ampère and Faraday, reproduced in Launay's *Correspondence.*

# Chapter 3

## Mathematics

An excellent introduction to the basic mathematical concepts Maxwell will be
using is Chapter VI of Courant and Robbins, *What Is Mathematics?,* titled
"Functions and Limits." A more practical text to fill gaps in a grasp of algebra
is Levi's *Algebra.* For the calculus, a most practical text is Niven's *Calculus,* in
which the first five chapters most directly serve our ends. Basic concepts are
very well developed in Chapter VIII of Courant and Robbins, "The Calculus,"
(skip §6 for our present purposes, but §7 is particularly pertinent).

## Physics

### General
Maxwell wrote a short text in mechanics, titled *Matter and Motion,* for the
working people he liked to teach (it was originally published by the Society
for Promoting Christian Knowledge). Needless to say, it is recommended as
first reading in the foundations of mechanics for our purposes. Another good
background text is Norman Feather's *Mass, Length and Time.* A fine text in
basic mechanics is Slater and Frank, *Mechanics,* in which Chapters I and III
bear most directly on the fundamental equations of motion. In Feynman's
*Lectures,* see Vol. I, Chapters 2, 4, 7–14. The opening chapters of Lindsay
and Margenau, *Foundations of Physics,* are to be highly recommended, and
the first sections of Chapter III, "The Foundations of Mechanics," take a
larger view of the concepts we will be using most.

### On Stress
Beginning at line 67, Maxwell makes considerable use of the concept of
stress and its analysis, to which Discussion [D5] is devoted. Background

reading on this can be found in Long, *Mechanics of Solids and Fluids;* there the complexities of Chapter I need not be mastered in order to make good use of Chapter II, "Stress in a Continuous Medium." In Slater and Frank, see Chapter XII, "Strains, Strains, and Vibrations of an Elastic Solid," See also Jaeger, *Elasticity, Fracture and Flow.*

A penetrating analysis and critique of Maxwell's concept of the vortices and the displacement current is to be found in Siegel's *Innovation in Maxwell's Electromagnetic Theory.*

# Chapter 4

## Physics

### Generalized Coordinates and Variational Mechanics

Maxwell's formulation in this paper is inspired by an approach to mechanics in some ways the opposite of that of Newton's laws of motion, that is, the mechanics of Lagrange in which the state of the system as a whole is taken as primary rather the summation of individual actions. Because tendencies are measured by evaluating variations from a given state, it is often referred to as *variational mechanics.* The method has its conceptual roots in Leibniz and was developed by Lagrange at the end of the eighteenth century, but it commanded new attention in England with the publication of the *Treatise on Natural Philosophy* by Maxwell's two friends, Thomson and Tait. His review of their book, in SP 2:776, gives a good sense of the conceptual importance he ascribes to it. It is given even greater prominence in Maxwell's *Treatise,* where it is the subject of Chapter V of vol. ii.

Apart from Maxwell's own discussion, see Chapter IV in Slater and Frank, while a very fine text on this is Lanczos, *Variational Mechanics.* See especially a "special lecture" on "The Principle of Least Action" by Feynman, *Lectures,* vol. 2, chap. 19. This is also a good time to turn to the second half of Chapter III in Lindsay and Margenau on "The Foundations of Mechanics."

### Electromagnetic Theories Based on Action at a Distance

Again, Maxwell's own accounts of these alternative theories—of which he was very much aware—in the *Treatise* may serve us best. His own account of Ampère's theory constitutes a chapter beginning at Vol. ii, p. 158 (there Maxwell pulls the neat trick of beginning with Ampère's action-at-a-distance theory and working it around until, at the end of the chapter, it has turned into Faraday's field theory—perhaps a sly Maxwellian mathematical joke. Other action-at-a-distance theories, alternatives to his own field theory, are discussed in his final chapter, *Treatise,* vol. ii, pp. 480ff.

Hertz's own interpretation of Maxwell's theory and his account of his experiment to put it to the test, in which radio waves were discovered, are found in the first two chapters of his *Electric Waves*.

Maxwell's own reflections on the possible significance of the field theory are met in his *Encyclopaedia Britannica* article on the "Ether," SP 2:763, and in his articles on "Action at a Distance," and "Attraction," SP 2:311 and 485.

A stimulating, rather testy critique of the foundations of electromagnetic theory, emphasizing action-at-a-distance alternatives to field theory, is O'Rahilly's *Electromagnetic Theory*. I review the issue of the plenum and action at a distance as Descartes and Newton may have seen it—namely, as the alternative between *mechanism* and *spirit*—in an essay "Science as Mystery: An Alternative Reading of Newton's *Principia*." West's *In The Mind's Eye* is a stimulating reflection on the metaphor of the field in terms of "vision."

# Electromagnetic Theory

## Mathematics

The mathematics of vectors, with its application to electromagnetism always in view, is well explained in Chapter II of Skilling's *Fundamentals of Electric Waves*. The chapters that follow (through Chapter IX) constitute a fine transcription into systematic vector terms of Maxwell' theory. A thorough text is Wade's *Algebra of Vectors and Matrices*. For the history of vector methods, and Maxwell's involvement with it, see Crowe, *A History of Vector Analysis*.

Maxwell's own thinking about the nature of vectors, and the relation of mathematics to physics, is found in a paper "On the Mathematical Classification of Physical Quantities," in *Scientific Papers,* ii, 257ff.

## Maxwell on the Role of Experiment

In relation to Maxwell's very special concept of the nature and role of experiment, see his "Introductory Lecture on Experimental Physics," given on his assumption of the first professorship of this subject at Cambridge [SP 2:241].

# Bibliography

Ampère, André-Marie. *Correspondence du Grand Ampère.* de Launay, L., ed. 3 vols., Paris: Gauthier-Villars, 1936–1943.

———. *Théorie mathématique des phénomènes électro-dynamiques uniquement déduite de l'expérience.* Paris: Librarie Scientifique Albert Blanchard, 1958. [original edition Paris, 1826.]

Campbell, Lewis, and William Garnett. *The Life of James Clerk Maxwell,* 2d ed. London: Macmillan, 1884; rev. ed. Robert Kargon, ed. New York: Johnson Reprint Corporation, 1970.

Conant, James B., and Duane Roller. *Harvard Case Histories in Experimental Science.* Cambridge: Harvard University Press, 1957.

Courant, Richard, and Herbert Robbins. *What is Mathematics?* New York: Oxford University Press, 1979.

Crowe, Michael J. *A History of Vector Analysis.* New York: Dover Publications, 1985.

Davie, George Elder. *The Democratic Intellect: Scotland and her Universities in the Nineteenth Century.* Edinburgh: Aldine Press, 1961.

de la Rive, Auguste. *A Treatise on Electricity.* Trans. Charles V. Walker. 3 vols. London: 1853–56.

Descartes, René. *Le Monde (The World).* French with translation by Michael Sean Mahoney. New York: Abaris Books, 1979.

———. *Principles of Philosophy.* In *The Philosophical Works of Descartes.* Trans. E. S. Haldane and G. R. T. Ross. New York: Dover Publications, 1955.

———, *Rules for the Direction of the Mind.* In *Great Books of the Western World, vol.31, Bacon, Descartes, Spinoza.* Chicago: Encyclopaedia Britannica, 1952.

*Dictionary of Scientific Biography.* C. C. Gillispie, ed. 16 vols. New York: Charles Scribner's Sons, under the auspices of the American Council of Learned Societies, 1970–1980.

Duhem, Pierre. *The Aim and Structure of Physical Theory.* Trans. Philip Wiener. Princeton: Princeton University Press, 1991.

———. *Les théories électriques de J. Clerk Maxwell.* Paris: Hermann, 1902.

Everitt, Francis W. "James Clerk Maxwell." In *Dictionary of Scientific Biography,* q.v.

Faraday, Michael. *Faraday's Diary.* Thomas Martin, ed. 8 vols. London: G. Bell and Sons, 1932–1936.

———. *Experimental Researches in Electricity.* 3 vols. London: Taylor & Francis, 1839, 1844, 1855; reprint *Great Books of the Western World, vol. 45, Lavoisier, Faraday.* Chicago: Encyclopaedia Britannica, 1952.

Feather, Norman. *Mass, Length and Time.* Baltimore: Penguin Books, 1961.

Feynman, Richard P., R. B. Leighton, and Matthew Sands. *The Feynman Lectures on Physics.* 3 vols. Reading, Mass: Addison-Wesley, 1970.

Fourier, Joseph. *Analytic Theory of Heat.* Trans. Alexander Freeman. New York: Dover Publications, 1955.

Goldman, Martin. *The Demon in the Aether.* Edinburgh: Paul Harris Publishing, 1983.

Hertz, Heinrich. *Electric Waves.* Trans. D.E. Jones. London: Macmillan, 1983; reprint New York: Dover Publications, 1962.

Hesse, Mary. *Forces and Fields . . . The Concept of Action at a Distance in the History of Physics.* London: Thomas Nelson, 1961.

Huygens, Christiaan. *Treatise on Light.* Trans. S. P. Thompson. In *Great Books of the Western World, vol.34. Newton, Huygens.* Chicago: Encyclopaedia Britannica, 1952.

Jaeger, J.C. *Elasticity, Fracture and Flow.* 2d ed., London: Methuen and Co., 1962.

Jones, H. Bence. *The Life and Letters of Faraday.* 2 vols. London: Longmans, Green, 1870.

Knott, C. G. *Life and Scientific Work of Peter Guthrie Tait.* Cambridge: Cambridge University Press, 1911.

Kuhn, Thomas. *The Structure of Scientific Revolutions.* 2d ed., Chicago: University of Chicago Press, 1970.

Lanczos, Cornelius. *The Variational Principles of Mechanics.* Toronto: University of Toronto Press, 1949; reprint New York: Dover Publications, 1986.

Levi, Howard. *Elements of Algebra.* New York: Chelsea Publishing, 1990.

Lindsay, Robert, and Henry Margenau. *Foundations of Physics.* 2d ed., Woodbridge, Conn.: Ox Bow Press, 1981.

Long, Robert R. *Mechanics of Solids and Fluids.* Englewood Cliffs, N.J.: Prentice-Hall, 1961.

Mach, Ernst. *Mechanics.* Trans. Thomas J. McCormack. 5th ed., La Salle, Ill.: Open Court Publishing Co., 1942.

Maxwell, James Clerk. *A Treatise on Electricity and Magnetism,* 2 vols., 1st ed., Oxford: Clarendon Press, 1873; 2d ed. W. D. Niven, ed. Oxford,

1881; 3d ed. J. J. Thomson, ed. Oxford, 1892; reprint New York: Dover Publications, 1953.

Maxwell, James Clerk. *An Elementary Treatise on Electricity.* ed. William Garnett. Oxford: Clarendon Press, 1891.

———. *Matter and Motion.* London: Society for Promoting Christian Knowledge, 1876; reprint New York: Dover Publications, 1952, and *Great Ideas Today: 1986.* Chicago: Encyclopaedia Britannica, 1986, pp. 348ff.

———. *The Scientific Letters and Papers of James Clerk Maxwell,* P. M. Harmon, ed. In progress; vol. 1 (1846–1862); vol. 2 (1862–1873). Cambridge: Cambridge University Press, 1990–1995.

———. Niven, W. D., ed. *The Scientific Papers of James Clerk Maxwell,* 2 vols., reprint Cambridge: Cambridge University Press, 1890. New York: Dover Publications, 1952.

———. *The Theory of Heat.* Textbooks of Science, Adapted for the Use of Artisans. London: Longmans, Green, 1871; reprint Westport, Conn.: Greenwood Publishing Group, 1970.

———. *A Treatise on Electricity and Magnetism.* 2 vols. Oxford: Clarendon Press, 1873; 2d. ed. W. D. Niven, ed., 1881; 3d ed. J. J. Thomson, ed., 1892; 3d ed. reprint New York: Dover Publications, 1953.

——— and Fleeming Jenkin. "On the Elementary Relations between Electrical Measurements," (*British Association Report: 1863,* London: British Association for the Advancement of Science, 1863, pp. 130ff.)

Moore, Gerald E. *Algebra.* New York: Harper Collins, 1971.

Newton, Isaac. *Newton's `Principia': The Central Argument.* Trans. William Donahue. Notes and Expanded Proofs by Dana Densmore. Santa Fe, New Mexico: Green Lion Press, 1995.

Niven, Ivan. *Calculus: An Introductory Approach.* Princeton, N.J.: D. Van Nostrand, 1961.

O'Rahilly, Alfred. *Electromagnetic Theory: A Critical Examination of Fundamentals.* 2 vols. New York: Dover Publications, 1965.

Owen, George E. *Introduction to Electromagnetic Theory.* Boston: Allen & Bacon, 1963.

Phillips, H. B. *Vector Analysis.* New York: John Wiley & Sons, 1959.

Rogers, Eric. *Physics for the Inquiring Mind.* Princeton: Princeton University Press, 1960.

Roller, Duane, and Duane H.D.Roller. "The Development of the Concept of Electric Charge," In *Great Experiments in Physics,* James Bryant Conant and Leonard K. Nash, eds., 2 vols. Cambridge, Mass: Harvard University Press, 1964. vol 2., pp. 543ff.

Shamos, Morris H., ed. *Great Experiments in Physics.* New York: Dover Publications, 1987.

Siegel, Daniel M. *Innovation in Maxwell's Electromagnetic Theory: Molecular Vortices, Displacement Current, and Light.* Cambridge: Cambridge University Press, 1991.

Simpson, Thomas K. "Maxwell and the Direct Experimental Test of his Electromagnetic Theory." *Isis, 57:* 423 ff., 1966.

———— "Maxwell's *Treatise* and the Restoration of the Cosmos" In *The Great Ideas Today: 1986,* Chicago: Encyclopaedia Britannica, 1986. pp. 218ff.

————. "Science as Mystery: A Speculative Reading of Newton's *Principia*" In *The Great Ideas Today: 1992,* Chicago: Encyclopaedia Britannica, 1992. pp. 97ff.

Skilling, Hugh H. *Fundamentals of Electric Waves,* 2d ed., New York: John Wiley & Sons, 1948.

Slater, John C., and Nathaniel H. Frank. *Mechanics.* Westport: Greenwood Publishing, 1983.

Strong, John. *Concepts of Classical Optics.* San Francisco: W. H. Freeman, 1958.

Thompson, Silvanus P., *Calculus Made Easy.* London: Macmillan, 1969.

————., *The Life of William Thomson, Baron Kelvin of Largs.* 2 vols. London: Macmillan & Co., 1910; reprint New York: Chelsea Publishing, 1977.

Thomson, William (Lord Kelvin). *Reprint of Papers on Electrostatics and Magnetism.* 2d ed. London: Macmillan and Co., 1884.

Thomson, William, and P. G. Tait. *Treatise on Natural Philosophy.* 2d ed. Oxford: Clarendon Press, 1878; reprinted as *Principles of Mechanics and Dynamics.* 2 vols. New York: Dover Publications, 1962.

Tolstoy, Ivan. *James Clerk Maxwell.* Chicago: University of Chicago Press, 1983.

Tricker, R. A. R. *Early Electrodynamics: the First Law of Circulation.* Elkins Park: Franklin Publishing Co., 1965.

————. *The Contributions of Faraday & Maxwell to Electrical Science.* Elkins Park: Franklin Publishing Co., 1966.

Wade, Thomas L. *The Algebra of Vectors and Matrices.* Reading, Mass.: Addison-Wesley Publishing Co., 1951.

Whittaker, Sir Edmund. *A History of the Theories of Aether and Electricity: Volume I: The Classical Theories.* Woodbury: American Institute of Physics, 1987.

West, Thomas G. *In the Mind's Eye.* Buffalo: Prometheus Books, 1991.

Williams, L. Pearce. *Michael Faraday.* New York: Da Capo Press, 1987.

Wilson, D. B. *Kelvin and Stokes: A Comparative Study in Victorian Physics.* Philadelphia: I O P Publishing, 1987.

# Illustration Credits

I am most grateful for the help of the following institutions and individuals in furnishing illustrations from their collections and granting permission for their use:

*Cover photograph (and Figure 4.18).* Courtesy of the Institution of Electrical Engineers (London).

*Figure 1.1.* Courtesy of the Masters and Fellows of Trinity College Cambridge.

*Figure 1.3.* Courtesy of the Maxwell Foundation (Edinburgh).

*Figure 1.7 (right).* Courtesy of the Hunterian Museum, University of Glasgow.

*Figures 1.8, 1.9, 1.29, and 1.38.* Courtesy of the Royal Institution of Great Britain (London).

*Figures 1.7 (left), 1.18, 1.22, 1.36, 3.20, and 4.17.* Courtesy of the Smithsonian Institution.

*Figures 1.2, 1.4, 2.28, and 4.30.* Photographs by Eric Simpson.

Original illustrations for this volume have been patiently and ingeniously created by Anne B. Farrell, Santa Fe, New Mexico. Certain of these illustrations are based upon or redrawn from figures in Maxwell's *Treatise on Electricity and Magnetism:*

*Figure 2.6.* Vol. i, plates, Fig. I.

*Figures 2.7, 2.8, and 2.25.* Vol. i, plates, Fig. XII.

*Figures 2.16 and 3.10.* Vol. ii, plates, Fig. XVIII.

*Figure 2.17.* Vol. ii, plates, Fig. XVII.

*Figure 2.18.* Vol. ii, plates, Fig. XIX.

*Figure 4.9.* Vol. ii, p. 371, Fig. 55.

*Figure 4.12.* Vol. ii, p. 372, Fig. 56.

*Figure 4.22.* Vol. ii, p. 139, Fig. 21.
*Figure 4.26.* Vol. ii, p. 217, Fig. 33.

A number of illustrations have been reproduced directly from the works of Faraday, Hertz, and Maxwell:

*Figure 1.11.* Faraday, *Experimental Researches,* vol. iii, Plate III.
*Figure 1.31.* Faraday, *Experimental Researches,* vol. iii, §2821, Fig. 3.
*Figure 4.1.* Hertz, *Electric Waves,* pp. 23–26, Figs. 2–5.
*Figure 4.2.* Hertz, *Electric Waves,* p. 5, Fig. 1.
*Figure 4.3.* Maxwell, *Treatise on Electricity and Magnetism,* vol. ii, pp. 154–155, Figs. 24 and 25.
*Figure 4.14.* Maxwell, *Treatise,* vol. ii, p. 439, Fig. 67.
*Figure 4.21.* Maxwell, *Treatise,* vol. ii, p. 360, Fig. 21.
*Figure 4.23.* Maxwell, *Scientific Papers,* vol. ii, p. 129.

One drawing is reproduced from Campbell's *Life:*

*Figure 1.5.* Campbell, *Life,* p. 27.

Full bibliographic reference to the works cited above is to be found in the Bibliography.

Studio work in the preparation of illustrations for reproduction was ably provided by Stock Studies, Saratoga Springs, New York.

# Index

acceleration: defined, 222; illustrated, 222*fig*3.26. *See also* motion

action-at-a-distance, 5, 21, 26, 115, 139, 175, 196, 293, 332, 362, 364; Maxwell on, 143; Weber's theory, 21. *See also* attraction

additive. *See* superposition

Airy, Sir George, 32

algebra, 32, 79, 82–83, 87, 134, 140, 185, 217, 220, 242, 311; differentiation and integration of forms, 134, 224–225, 311; of dimensional analysis, 395; symbols of, versus geometrical ideas, 60, 79, 133, 302, 386

ammeter, 43–44, 44*fig*1.34, 109, 140, 208, 305, 370

ampere, 43, 358

Ampère, André-Marie: electrodynamic law of, 43, 269, 358, 537; electrodynamic theory of, 18, 49–51, 110; experiments of, 394; Faraday letter to, 13; on formal scientific method, 27, 116, 409n4.6; immortalized as "answer to the genius of Newton," 407n1.22; Maxwell on, 53, 74–75, 78, 95, 109–110; method of, 18, 51, 75, 119, 239; theory of permanent magnetism of, 50

Ampère's law. *See* electrodynamic force

analogies of fluid flow: to conduction of electricity, 43, 72–73, 107–109,

407n1.19; to electric potential, 73; to electrodynamic force, 74–76; to electromagnetic induction, 76–78; to electromotive force, 73–74; to electrostatics, 68–70, 73, 94*fig*2.7, 95*fig*2.8, 99–105, 129, 130*fig*2.25, 134–138, 135*fig*2.27; to force laws in general, 85; to heat flow 91–92; to magnetic lines of force, 84*fig*2.3; which does not work, 123–124, 123*fig*2.22. *See also* capacitor, parallel-plate

analogy: and theory, 78; as embodied form of mathematical ideas, 77; measurement as universal analogy, 81; as a method, 56–57, 78–79, 82, 118–119; mixed equation expressing, 99–100; partial, 117. *See also* physical analogy; Thomson, William

analytic mathematics, 7, 28, 58, 81–83, 87, 96, 129, 134, 140, 182, 217–228, 281, 321, 369; advanced development in France, 408n4.5; Faraday's appeal from, to Maxwell, 407n2.1; interpretation of results, 186–196; Maxwell's ability in, 15. *See also* calculus; differentiation; gradient; limit; superposition

analytic methods: attraction of subtleties, 55; notations, 83; and speculative methods, 87; and synthesis, 91, 133. *See also* dimensional analysis

Apostle's Club, 4, 122
appropriate ideas, 57
arc lamp, 18*fig*1.13, 19
Archimedes, 126
Aristophanes, 140
Aristotle, 85, 91, 290; as figure in motion, 126; laws of motion of, 91, 116
astronomy, 391. *See also* stellar aberration
attraction, 25–26, 31, 34–35, 56–57, 59, 81, 120–121, 131, 173, 254, 263; and electromotive force, 49, 55, 59, 68, 74–75, 78, 261, 269, 313, 355; electrostatic, 55, 68–70, 99–100, 102–104, 136, 281, 363; Faraday on, 39, 116; gravitational, 26, 29–30, 34, 81, 120, 173, 179; and laws of force, 145–146, 173, 179; magnetic, 38, 68, 71, 115, 144–146, 173, 179; as mere appearance, 43, 116. *See also* action-at-a-distance; gravitational force

balance: Ampère's, for force between conductors, 394; Coulomb's, for electrostatic force, 28*fig*1.23; to demonstrate repelling currents, 317*fig*4.9; Maxwell's current balance, 336–337, 336*fig*4.12
bar magnet, 35, 40, 75, 90, 105–106, 116, 122, 132, 135, 250, 271, 326; field of, 13*fig*1.10; fields of groupings of, 14*fig*1.11; lines of force through, 105*fig*2.12
battery. *See* voltaic battery
British Association for the Advancement of Science: electrodynamometer of, used in determination of the ohm, 354*fig*4.21; spinning coil calibration of the ohm, 350–353, 352*fig*4.20, 409n4.9; standard ohm, 345, 350–351, 410n4.13

cable, telegraphic, 8, 20, 45, 347–350; cable galvanometer, 9*fig*1.7; Great Eastern cable-laying expedition, 348*fig*4.19

cage, Faraday's electrically charged, 35, 95, 135
calculus, 2, 7, 30, 101, 134, 140–141; chain rule, 223; differentiation of sums and products, 224; fundamental principles, 218–228, 219*fig*3.24, 221*fig*3.25, 222; integration, 225–226; Leibniz's notation, 221; partial differentiation, 224; product rule, 222, 243; vector calculus, 94, 141, 201, 324–325, 377–386
calibration: absolute, 208–209, 351, 353, 356; of electrical instruments, 351, 353, 357; of flow model, 101–103, 136–138; thermal, of electrical instruments, 208–209, 209*fig*3.21
caloric, 91–92, 116
calorimeter: in electrical calibration, 208–209, 209*fig*3.21
Cambridge Philosophical Society, 9, 55, 160
Cambridge University, 1–6; curriculum, 203, 367; Maxwell's fellowship, 1, 4; Maxwell's professorship, 353, 362, 367; Trinity College, 1, 3*fig*1.2, 4; Tripos examinations, 3–4. *See also* Apostle's Club
capacitor, equipotential lines for, 94*fig*2.7; 130*fig*2.25; parallel-plate, 129, 135*fig*2.27, 261–262, 292–294, 343, 354–356, 359–360, 386. *See also* capacity; Leyden jar; specific inductive capacity
capacity, 261, 284
cathode rays, 107, 343
cause, 59, 66, 76, 133, 146, 156–157, 196–197, 206, 215–216, 257–258; concept of causality, 39–40, 49, 59, 139–140, 281, 291, 355
Cavendish Laboratory, 353, 362, 367
cell, galvanic, 41, 73, 199, 213, 359; Daniell cell, 42*fig*1.32, 43, 287, 402; and electromotive force, 73–74
cells: in fluid flow, 65; magnetic, 105, 214; and vortices, 166, 170–172, 214, 292. *See also* unit cells

centrifugal force, 182; derivation of Maxwell's result for, in Maxwell's vortex theory, 233–237; derivation of Newtonian equation for, 231–233, 232*fig*3.28; of vortices, 146, 160, 169, 179, 182, 195. *See also* vortices

circuit, as driving point of mechanical system, 305; electric, 43–44, 44*fig*1.34, 73–74, 107, 109–111, 199, 208; electrodynamic force on, 76, 258, 266, 268, 306, 312–313, 318; electromotive force in, 310, 338, 351; electrotonic state of, 269; energy of, 311–312; equivalence to dipole, 110, 110*fig*2.14, 318–319; exploratory, 269–272, 316–317, 315*fig*4.7, 332; as geometric path, 318; induction in, 164, 265, 276, 310, 329; magnetic field of, 112, 159, 260, 319; magnetic potential of, 320, 330; mutual induction of circuits, 265–268, 297, 306, 308–309, 315, 332; oscillations in, 361. *See also* electric current; open circuits; parallel connection; primary circuit; secondary circuit; series circuit

coefficient of magnetic induction (μ), 271, 273, 278, 286, 319. *See also* magnetic induction

coefficient of mutual induction, 262, 265, 308–309, 369; measurement of, 317–318. *See also* magnetic induction

components: chemical, 258; of a mechanical system, 306, 322–323; of vectors, 63, 91, 93–94, 100, 257, 272, 274–275, 278, 304, 322–324

conducting power of a medium: dielectric, 70, 292; magnetic, 38–40, 70–71, 106, 190; thermal, 56–57, 83, 120, 130

conduction of electric current: conducting powers of substances, 25, 41–42, 73, 108, 167–168, 171, 203, 262; conductivity of copper wire, 349; electrolytic, 41, 73; fluid flow analogy to, 43–44, 72–73, 107–109; super-

conductivity, 209; theories of, 55, 72–74, 107, 109, 170, 297, 388; time in, 354. *See also* electric current; electromotive force; Ohm's law; resistance

connectedness, 96, 230, 314; of the electromagnetic field, 295, 296*fig*4.24; Lagrangian, 252, 256, 263–264, 295, 304–305, 312, 329, 347, 365–369, 393; of a mechanical system, 304, 366, 368*fig*4.24. *See also* contiguity

conduction of heat. *See* analogy; Fourier; Thomson, William

conduction of lines of force, 136, 40*fig*1.31

contiguity, 15, 57, 158–159; Faraday on, 33; of vortices, 198*fig*3.14

continuous rotation, 46. *See also* motor, electric

continuum, 83, 89, 234, 251, 314, 366

contour lines, 122, 131, 173, 191

cosmos, 138, 289, 334–335, 347, 350, 353, 366–367, 395

Coulomb, Simeon-Denis, 26; analytic theory of electrostatics, 406n1.15; electric balance of, 28*fig*1.23, 335, 393

Coulomb's law, 26, 278, 287–288, 333, 335, 358, 389, 395; Faraday on, 32–33; in fluid flow analogy, 66–67, 85, 96–98, 97*fig*2.10; in Weber's theory, 51. *See also* electrostatics

curl, in vector calculus: "curl meter," 201; defined, 377–379; as measure of current, 191*fig*3.12, 192, 199–201, 272–273, 375

current elements: in Ampère's theory, 49–51, 75, 110

current. *See* electric current

cut, as mental act, 60–62, 89

cylinder, 88, 131; in Euclid and Maxwell, 88*fig*2.4. *See also* geometrical ideas; line of sight; parallels in geometry

Daniell cell. *See* cell, galvanic

definition: of number of lines of force, 318; of properties of fluid models,

definition (*continued*)
59, 79, 85–87, 138; role in Maxwell's method, 85–87; of standard units, 41, 351; of unit charge, 333; of unit pole, 331, 391

delta (Δ), as symbol of difference, 92, 127, 218

derivative. *See* differentiation, in calculus

Descartes, René, 25; on matter as extension, 125, 408n2.2; Maxwell and, 125, 364; mechanical physics of, 406n1.10; refraction theory of, 80*fig*2.1, 117

diagrams: Maxwell's interest in, 53

diamagnetism, 39*fig*1.30, 36–40, 154, 166, 256; Faraday's interpretation of, 40*fig*1.31; in fluid flow analogy, 40, 71–72, 106–107; and magnetic inductive capacity, 190; and rotation of plane of polarized light, 256–257; Tyndall's theory of, 107; Weber's theory of, 51, 254. *See also* conduction of electric current

dielectric absorption, 360

dielectric constant: as elasticity of figure, in Maxwell's vortex theory, 214*fig*3.23; of glass, and refraction, 299

dielectrics, 33, 70, 73, 104; in fluid flow analogy, 70, 73, 104; glass as dielectric, 136; in Hertz's experiment, 295; and optical refraction, 299, 344; vacuum as dielectric, 292, 339, 383. *See also* specific inductive capacity

differential, in calculus, 97, 151, 192, 195; complete, 154, 192, 221, 240; notation 100–101

differential equation, 226, 235, 399

differential gear, 368

differentiation, in calculus, 134, 150, 185, 218–225; illustrated in theory of motion, 219*fig*3.24, 221*fig*3.25, 222*fig*3.26

diffusion, 212, 349

dimensional analysis, 357, 394–396

direction cosines: defined, 183–184, 184*fig*3.5

displacement current, 171, 215, 258–259, 293–295, 297, 339–340, 376, 382, 384; Hertz's experiment on, 294*fig*4.2, 295

divergence, in calculus: as measure of source strength, 187, 188*fig*3.9, 247–251, 331, 376; in one dimension, 248*fig*3.31

dynamical reasoning, 253, 255, 263–265, 295–297, 312, 366–369, 372

ecology, 86

elastic force: and problem of fluid model, 123, 123*fig*2.22

elastic medium, 78, 171–172, 212, 215, 256; energy in, 256, 329; as optical medium, 281; regularity of properties, 299; vacuum as, 262; waves in, 56, 117–118, 216, 259, 276, 292, 295, 297, 328–329, 332

electric charge, 25–26, 58; absolute charge, 34–35; energy of, 103; existence of, 25, 33–34, 70, 99, 294, 387; in fluid analogy, 68–70, 99, 134–136; implied by distribution over cutting surface, 90*fig*2.5; as induction, 31–32, 34; inertia of, 107, 330, 387, 389; interpretations of, by Hertz, 144, 292–293, 293*fig*4.1; and lines of force, 90; in "Physical Lines," 170, 213; in Weber's theory, 51

electric current, 22, 41–44, 55, 59, 68, 72, 76, 107–108, 110, 144, 155–156, 171, 254, 270; conventional direction of, 73; and electromagnetic field, 193, 259–260, 264, 376, 380; electromagnetic momentum of, 264, 270, 274, 302, 308–310, 323, 375; heat produced by, 269; inertia of, experiment, 330, 387, 389, 389*fig*4.26; intensity of, 41–42, 74, 109; intrinsic energy of, 269, 275, 312; in Maxwell's vortex theory, 156–158, 166–170, 194, 198, 198*fig*3.14, 205,

208, 210, 212; measurement of, 45, 109, 208; as motion of charge, 74, 107, 199, 208–209, 213, 261, 289, 336–337, 375, 377, 383–385; and nature of electricity, 197–198, 306; relation to static electricity, 73; unit of, 73, 75, 336, 395–396; as velocity, in dynamical theory, 306–307, 313, 370. *See also* circuit, electric; displacement current; electromagnetic induction; Oersted effect; Ohm's law

electric lines of force, 32–34, 58, 104, 213

electricity. *See* electric charge

electrodynamic force, 50*fig*1.39, 48–51, 101; in Ampère's theory, 49, 94, 141–142; in balance of repelling currents, 317*fig*4.9; between attracting circuits, 313*fig*4.6; between parallel currents, 394*fig*4.26; on current in magnetic field, 194*fig*3.13, 392*fig*4.27; in fluid flow analogy, 74–76; in Weber's theory, 56

electrodynamometer: of British Association for the Advancement of Science, 354*fig*4.21

electrolytic cell: in measuring current, 45

electromagnet: Faraday's coil, 49*fig*1.38; of Royal Institution, 35, 38*fig*1.29

electromagnetic force, 22, 44–48, 45*fig*1.35, 50, 78, 192, 294, 297, 326, 329–330, 337; equivalence of current to magnet in producing, 110*fig*2.14; on iron in field of current, 190*fig*3.11; Maxwell's null demonstration of law of, 355*fig*4.22. *See also* Oersted effect

electromagnetic induction, 47, 51, 68, 76–78, 106, 156, 203, 254, 260–269, 276, 295–297, 310, 326–333, 338, 360, 372; coil, used by Faraday, 49*fig*1.38; in fluid flow analogy, 76–78; illustrated, 48*fig*1.37, 308*fig*4.5, 373*fig*4.25; in Maxwell's vortex theory, 203*fig*3.17, 205*fig*3.19; possible discovery by Henry, 407n1.21

electromagnetic momentum, 116, 201–202, 263, 265–266, 269, 272–274, 311, 315–316, 322–326, 328, 332, 335, 338, 343, 368, 374–375, 377, 383, 391–392; and induction, illustrated, 323*fig*4.11; measurement of, 324, 326–327

electromagnetic theory of light, 16, 172, 216–217, 262, 279, 281–287, 299–300, 321–322, 329–367; refraction in, 344*fig*4.18; in Maxwell's vortex theory, 204*fig*3.18. *See also* light

electromagnetic waves: advancing wave, 399*fig*4.29; generation of, 399*fig*4.13; Maxwell's dismissal of the prospect of producing, 215, 285, 362, 410n4.16; Maxwell's sketch of, 340*fig*4.14; production of, by Hertz, 294*fig*4.2, 295, 362

electrometer: pith ball, 27*fig*1.22; Thomson's quadrant, 9*fig*1.7

electromotive force: around a loop, 200*fig*3.15; and electric current, 171, 198, 257–258, 261, 307, 375, 380; in Maxwell's vortex theory, 168; measurement of, 109

electron, 107, 199, 210, 386–388

electroplating, 23*fig*1.18

electrostatic force, 25*fig*1.21, 90*fig*2.5; Coulomb's analytic theory of, 406n1.15; Coulomb's balance for measuring, 38*fig*1.23; inverse square law of, 97*fig*2.10. *See also* potential

electrostatic induction, 31–34, 31*fig*1.26, 36–37, 39, 70, 73, 104–105, 171, 277, 288, 330; in curved lines, 33, 34*fig*1.27; in fluid flow analogy, 133

electrostatics, 25–35; apparatus for, 25*fig*1.21; in fluid flow analogy, 68–70; and laws of charge distribution, 101. *See also* electrostatic force; Leyden jar

electrotonic state, 86, 144, 162, 165, 175, 201–205, 265, 269, 273, 314;

electrotonic state (*continued*)
Maxwell's experiment to demonstrate, 163–164, 203–204. *See also* vector potential

*Encyclopedia Britannica*: Faraday's reading in, 11; Maxwell's scientific editorship of, 53, 206

energy, 28–29; conservation of, 30, 169, 295–297, 310, 362–363; dissipation as heat, 171, 208–209, 258, 286; distributed in field, 103, 143, 318, 327–328, 333; of flowing current, 329, 388; in fluid flow model, 104, 106, 130, 175; kinetic, 29, 260, 245–246, 290–291, 312; in Lagrangian formulation, 303, 310–313, 369–372, 388–390; of Leyden jar, 108; in Maxwell's vortex theory, 160, 167, 199, 203, 210, 214–215, 242–246; mechanical, in interchange with field, 47, 113, 206, 267, 314; of photon, 206, 289; potential, 28–29, 102, 108, 131, 328, 363; of propagating electromagnetic field, 256, 259, 261, 286, 290, 344, 363, 400; relation to force, 107, 114, 173, 313, 332, 392–393; as scalar quantity, 131, 173; stored in capacitor, 136; of sunlight, 286–287, 299, 344–345, 400–403; of system of charges, 103–104, 215, 275, 281; and term "dynamical," 253, 298; total, in electromagnetic field, 271–278, 298, 325–326, 388–390; of unit cell, 93, 112, 131–133; as univocal term, 276, 306, 328

energy, distribution in fields: of circular current, 113*fig*2.16; of current in uniform magnetic field, 114*fig*2.17; of parallel currents, 115*fig*3.10; of parallel plate capacitor, 95*fig*2.8

equation: mixed, expressing an analogy, 99–100; as vehicle for ideas, 124

equilibrium condition: and disequilibrium, 184; in Maxwell's vortex

theory, 176–177, 183–184, 240–242; of stress, 148–149; thermal, 120–122

equipotential lines, 94, 122, 131, 320; of circular current, 113*fig*2.16; of parallel plate capacitor, 94*fig*2.7, 95*fig*2.8; of two like charges, 93*fig*2.6

equipotential surface, 135, 260, 271–272, 320–321; Maxwell's "soap bubble," of circular current, 321*fig*4.10

equivalence: of current and dielectric displacement, 215, 294; of current and magnetic shell, 111*fig*2.14; of cutting surface distribution and sources, 84, 90*fig*2.5, 98*fig*2.11, 104; of magnet and dipole, 110, 110*fig*2.14

error: in science, 33, 109; of sign, in Maxwell's argument, 215, 384

ether, 125, 166, 206, 231, 340, 256–257, 268, 289–291, 362; Maxwell's attempted measurement of motion through, 290

Euclid, 88*fig*2.4

existence, of action-at-a-distance, 255; of Ampère's current elements, 110; of curved lines of force, 47; denied, for model fluid, 86, 141–142; of displacement current, 292; of electric charge, 25, 34, 99, 169, 254, 293, 386; of the electromagnetic field, 276; of electromagnetic waves, 281, 300, 361–362; of the electrotonic state, 158; of the ether, 172, 206, 256–257, 259, 268, 298–290, 311–312, 363; Faraday on "real," 15, 139, 300, 361; and Lagrangian theory, 303; of the lines of force, 12; of lines within a bar magnet; of magnetic poles, 250, 391; of normal vibrations, 285, 343; of physical lines of force, 41, 139–140; of the vortices of "Physical Lines," 140, 142–144, 156–157, 165–167, 174; problem of, 180–181

experiment: of Coulomb, 26,
28*fig*1.23; of Hertz, on displacement
current, 295; on "isolated" poles, 36;
of Kohlrausch and Weber, on ratio
of units, 216, 261–262, 280, 284,
341; of Maxwell, on mass of electric
current, 107–109; of Maxwell, on
motion through the ether, 329, 334;
of Maxwell, on ratio of units, to determine *v*, 353–356, 356*fig*4.23; of
Maxwell, with "electromagnetic
top," 207–208, 207*fig*3.20; of Maxwell, with "spinning coil," 350–353,
352*fig*4.20; in Maxwell's scientific
method, 79, 119, 125, 143–144, 164,
170, 172, 269, 276, 280, 284–285,
296, 314; of Oersted, 50; on velocity
of light, 284, 341
*Experimental Researches in Electricity*
(Faraday), 1, 6, 9, 11–12, 15, 32,
70–77, 269, 316, 407n1.17
explanation, mechanical, 140, 153,
157, 256, 281; not equivalent to illustration, 276, 328; partial or premature, 55, 252; in science, 15, 28,
51, 56, 118, 143, 168–169, 343. *See
also entries under* mechanical
exploring wire, 106, 326–327, 391; circuits, 315*fig*4.7, 316*fig*4.8
extension: as Cartesian matter, 125

Faraday, Michael, 9–16, 10*fig*1.8,
11*fig*1.9, 17*fig*1.12, 361; on absolute
charge, 25, 34–35, 95, 105; and Ampère, 27; charged cage experiment, 35,
95, 135; coil used by, 49*fig*1.38; concept of science, 12–16; and continuous
rotations, 19, 46–47; on Coulomb's
law, 32–33; on diamagnetism and
paramagnetism, 38–41, 71, 106–107,
190; on electromagnetic induction, 47,
76; on electrostatic induction, 32–34,
36, 70, 104; on the electrotonic state,
86, 175, 203, 265, 273; and formal
mathematics, 7, 12–13, 27–28, 46, 59,
77, 83, 140, 407n2.1; on the identity

of electricities, 213; on the lines of
force, 10–12, 36–37, 41, 70–71, 106,
115, 131, 135, 173–174; and Maxwell,
1, 6, 10, 16, 36, 41, 47, 51, 57, 70–71,
76–78, 83–85, 99, 104, 139, 144, 173,
262, 269, 292, 299–301, 307,
316–318, 365, 407n2.1; on the moving wire, 135, 316*fig*4.8, 408n4.1; on
the propagation of magnetism, 252,
300, 360–361; religious faith, 13; and
speculation, 15; on the sphondyloid of
power, 40, 132; and Thomson, 36
fertility of method, 56, 119
field: of circular current, 113*fig*2.16;
concept of, 81, 89, 94, 103, 107,
139–140, 143; connectedness of, in
Maxwell's sketches, 296*fig*4.3; of
current in uniform field, 114*fig*2.17;
electromagnetic, 11, 28, 46–47, 106,
112–114, 139, 142, 175, 190, 206,
255–258, 260–268, 363–367,
388–393; electrostatic, 35, 100,
103–105, 131, 135; energy in,
388–390, 400–403; forces exerted
by, 189*fig*3.10, 194*fig*3.13; magnetic, 21, 37–40, 45, 71, 77, 106,
115, 143, 206; of parallel circular
currents, 115*fig*2.18; of parallel plate
capacitor, 94*fig*2.7, 95*fig*2.8,
130*fig*2.26; of two like electric
charges, 93*fig*2.6, 96*fig*2.9; velocity
of propagation, 350–368; of vortices, 146, 151–156, 163–169, 180,
182, 188–194, 199, 202–204, 207,
214–215, 248–251
fluid, imaginary: arbitrariness of unit,
99; as caloric, 118; definition of, 60;
divergence as measure of source,
188*fig*3.9; and electromotive force,
74; energy dissipation in, 95*fig*2.8,
104; flow from parallel plate
sources, 94*fig*2.7, 95*fig*2.8,
130*fig*2.25, flow from two sources,
93*fig*2.6; flow from a point source,
97*fig*2.10, 133–134, 133*fig*2.26;
flow model that does not work,

fluid, imaginary (*continued*)
123*fig*2.22; law of resistance to
flow, 63, 119; lines and tubes of motion, 60–61, 84*fig*2.3; as a mathematical idea, 58–60, 71, 77, 81–82,
124, 126; as massless, 63–64, 86,
91, 107, 116, 141; pressure in fluid
flow, 63–69, 73, 92–104, 96*fig*2.9,
109, 112, 127–134, 133*fig*2.26,
136–137; sources and sinks, 57,
62–71, 85, 96–116, 129–130,
133–134; theory of motion, 59–68,
116, 127–129. *See also* analogy; caloric; cells; tubes of flow; unit cells
flux: of energy, 132, 401; magnetic,
112, 315, 380
force: in Newton's laws of motion,
228–231. *See also* centrifugal force;
electrodynamic force, between parallel currents; electromagnetic force;
electrostatic force; gravitational force
Fourier, Jean-Baptiste: *Analytical Theory of Heat,* 120; solution of equilibrium temperature problem, 82*fig*2.2,
121*fig*2.20, 122*fig*2.21; theory of
heat flow, 82, 91–92, 120–123
Fourier series, 82, 120–121, 126; in solution for equilibrium temperature
over heated bar, 82*fig*2.2,
122*fig*2.21; superposition of terms
in, 121*fig*2.20
Franklin, Benjamin, 18, 342; *Experiments and Observations in Electricity,* 406n1.8
French tradition in analytic mathematics, 2, 7, 36, 408n4.5. *See also* Coulomb; Fourier; Lagrange; Laplace

galvanometer, 45, 47, 106, 109, 284, 315,
359–360; ballistic, 315; Thomson's,
9*fig*1.7; Weber's tangent, 45, 45*fig*1.3
Gauss, Karl Friedrich, 21, 348; magnetic map of the earth, 24*fig*1.20;
magnetic observations, 23*fig*1.19
generalized quantities, in mechanics. *See*
Lagrangian theory; reduced momentum

geometrical ideas, 2, 58–59, 61, 83–84,
125, 144, 174–175, 265, 289, 312,
314, 320–321. *See also* cylinder;
fluid, imaginary; line of sight
glass: apparent surface charge of, 104,
386; comparison of optical and dielectric properties of, 344*fig*4.16; as
dielectric, 108, 117, 136, 299; electric displacement in, 259; insulating
power of, 73, 108, 170–171; magnetic rotation of plane of polarized
light in, 291; specific inductive capacity, 104, 343. *See also* refraction
gradient, in vector calculus: defined,
381; of potential, 278, 329–330,
375, 381, 389; of pressure, 97–98,
112, 129–131, 133, 195
gravitational force, 26, 30, 34, 41, 51,
56, 81, 104, 143, 146, 173, 186, 206,
231, 300; as model, 129–131, 319;
problem of lines of force for, 173,
179, 211; unified theory of, 361

harmonics. *See* overtones
heat: radiant, 245, 255–262, 289; transformation of energy into, 159, 167,
171, 209, 255, 268, 308, 310, 328
heat flow: and analogy to laws of attraction, 7, 56–57, 81, 83–85,
131–132, 136, 211–212; theory of,
57, 81, 92, 97, 120–123, 347. *See
also* Fourier; Fourier series
Helmholtz, Hermann: on conservation
of energy, 295; on electromagnetic
theory, 260, 295; on fluid motion,
169, 211–212; proposed test of Maxwell's theory, 285
Henry, Joseph: possible discovery of
electromagnetic induction, 407n1.21
Hertz, Heinrich: experiment to demonstrate the displacement current,
294*fig*4.2; on the electric charge,
292–293, 293*fig*4.1; judgment of
Maxwell's theory, 384, 387; production of electromagnetic radiation,
295, 362

Hooke's law, 123, 212, 215. *See also* elastic medium

Hopkins, William, 4

Huygens, Christiaan: wave theory of light, 80*fig*2.1, 118–119, 118*fig*2.19

hypothesis: absence in dynamical theory, 176, 312, 329, 363; in science, 35, 55, 79, 281, 363; in vortex theory, 151–152, 157–160, 163, 167–178, 180, 183, 191, 196–197, 208, 314

idler wheels: in Maxwell's vortex theory, 156–158, 166–170, 194, 198, 198*fig*3.14, 205, 208, 210, 212. *See also* electric current; vortices

illustration, 58, 79, 143–144, 154, 194, 263, 354, 367–369, 377. *See also* rhetoric

imaginary fluid. *See* fluid, imaginary

impulse, in mechanics: as measure of momentum, 204–205, 263

induction. *See* coefficient of mutual induction; electromagnetic induction; electrostatic induction; electrotonic state; magnetic induction; self-induction

inertia, 63–64, 91, 116, 180, 208; of coupled circuits, 329, 386, 388; of electric current, Maxwell's experiment on, 286, 329, 388, 398*fig*4.26; in Lagrangian theory, 306, 309, 373; Maxwell's search for inertial coupling, 207*fig*3.20. *See also* Lagrangian theory; matter; moment of inertia

intelligibility, as goal in science, 3, 5, 51, 60, 87, 139–140, 174, 250, 252, 347, 366; of physics of Descartes, 406n1.10

intensity: of current, 41–42, 74, 109; of electric field, 214, 278, 297, 375; of electromotive force, 74, 109, 112; of magnetic force, 75–77, 151–153, 186, 188, 190, 192, 195–196, 207, 250, 277–279, 327–327, 329, 332, 340, 354, 374–375, 389, 391; of magnetic pole, 249, 251; of stress, 145; of sunlight, 344–346, 402–403

interference: wave theory of, 118*fig*2.19(b), 119. *See also* optics; Young

interpretation, 16, 33, 49, 59, 99, 104, 106, 136, 167, 193–195, 308; of Faraday's lines, 10, 57; of Lagrangian terms, 324; of models, 129–130, 134–136, 151, 155, 163, 173, 185–186, 203, 247–251, 314, 329

interstices, 59, 85

intuition, 7, 85, 87, 130, 160, 204, 217–218, 252, 300, 395. *See also* mind's eye

inverse square law, 32, 57, 59, 85, 124, 403; of electrostatic attraction, 26, 32; for fluid flow, 85, 97*fig*2.26, 124, 134; for gravity, 143; of magnetic attraction, 35; radial variation of pressure or potential, under, 133*fig*2.26. *See also* Coulomb's law

iron filings, 12, 36, 40, 143, 145, 174, 292, 315, 317; patterns of, 13*fig*1.10, 14*fig*1.11. *See also* magnetic lines of force

isobars, 93, 129–132; between parallel plane sources, 94*fig*2.7; between two sources, 93*fig*2.6

isothermal lines, 122, 130

isotropic media, 271, 275, 375, 378

kinetic energy, 29, 160, 245–246, 303, 312, 328–330, 372; as "actual" energy, 290–291

Kirchoff, Gustav, 107

Kohlrausch, Friedrich, 172, 216, 261, 280, 284, 341, 347–348, 353, 360

Lagrange, Joseph-Louis, 7, 405n1.2; *Traité Analytique,* 302

Lagrangian theory, 302–306, 364; Maxwell and, 346–367; in Maxwell's illustration of the bell ringers, 372; as rhetorical instrument, 306; and Thomson and Tait, *Treatise on Natural Philosophy,* 408n4.5. *See also* generalized quantities; reduced momentum

Laplace, Pierre-Simon, 7, 405n1.2

law, in science, 25–27, 32–36, 45, 49–50, 55, 60, 68, 71, 85–87, 91, 106, 143–144. *See also* laws of force; laws of motion

law of sines. *See* refraction

laws of force. *See* Ampère's law; Coulomb's law; electrodynamic force; electromagnetic force; electrostatic force; inverse square law; Ohm's law

laws of motion: Aristotelian, 91, 116; of fluids, 92, 97, 123, 128, 131; of heat (caloric), 92. *See also* Newton's laws of motion

Leyden jar, 27*fig*1.22, 108, 108*fig*2.13, 110, 170. *See also* capacitor

light, 15, 310; analogy to particle motion, 56, 80*fig*2.1(a), 117, 119; as vibrations of elastic fluid, 58, 81, 118–119, 170; wave theory, Huygens, 80*fig*2.1(b), 118*fig*2.19(a); wave theory, Young, 118*fig*2.19(b). *See also* arc lamp; electromagnetic theory of light; refraction; sunlight

limit, in mathematics, 61, 97, 100–101, 110, 121, 219–226, 229, 233, 235, 240–241, 248, 322. *See also* calculus; differential

line of sight, 113–114. *See also* geometrical ideas

lines of fluid motion, 58, 60–61, 64, 84, 88, 125–126, 130, 132, 169, 206; physical, 15–16, 174

lines of force, 68, 174, 217; curved, 33, 34*fig*1.27, 109; Faraday on "vibration" of, 300; for gravity, 173; iron filing patterns of, 13*fig*1.10, 14*fig*1.11, 98*fig*2.11; and lines of fluid motion, 84*fig*2.3, 70–71, 78, 83–84, 134; and lines of heat flow, 131, 156–157, 159, 166; and mechanical force, 101, 316; physical, 15–16, 73–74, 139–251, 252–253, 292–294, 312, 314, 328, 347, 366–367; polarity of, 177–178, 180; in vortex theory, 167, 169, 177, 179,

196. *See also* electric lines of force; field; magnetic lines of force

magnetic compass, 20–21, 36, 45, 215, 315–316; in tracing lines of force, 21, 35, 315–317

magnetic dipole, 110

magnetic induction, 36–38, 37*fig*1.28; 71, 76, 143, 150–156, 161, 174, 187, 190, 247–250, 271, 275, 326, 338, 389–392. *See also* coefficient of magnetic induction; electromagnetic induction; magnetic inductive capacity; self-induction

magnetic inductive capacity, 153–154, 157, 166, 169, 187, 190, 262. *See also* coefficient of magnetic induction; magnetic induction

magnetic intensity. *See* intensity

magnetic lines of force, 8, 10–11, 15, 36, 39–40, 53, 58, 106, 187, 191, 316–317; conduction of, 71–72, 190; as diagrams, 40, 173; and the electrotonic state, 164; exploring with coil or wire, 269–271, 292, 315–317, 315*fig*4.7, 316*fig*4.8; and lines of fluid motion, 78; solenoidal, 71, 105–106, 112; tracing with compass, 21, 36, 315–317; tracing with iron filings, 12, 36, 40, 143–145, 174, 292, 315, 317. *See also* electromagnetic induction; Faraday; field; iron filings; lines of force; Maxwell

magnetic pole, 35–37, 58, 70–71, 99, 105, 113–116, 143–146, 150–151, 154, 156, 190, 196, 277, 321–327; in closed ring, 163–164; directionality of stress, 177–178; in dynamical theory, 318–320, 331–332, 390–391; force on, 249–251, 261, 271–273, 278, 389–390; like dielectric induction, 171, 258; and magnetic induction, 275, 326–327, 390–391; and magnetic intensity, 391, 393; and magnetic potential, 271–273, 278, 389–390; never isolated, 250, 278, 318, 331–332, 365, 383, 390–391; strength of, 35, 187–188, 277, 331–332, 390–391; unit, 191–192, 194, 271, 331, 333, 391. *See also* magnetic induction; magnetic lines of force; potential; solenoidal

magnetic potential. *See* potential

magnetic shell: equivalence to current, 75, 97, 111–112, 111*fig*2.15

Magnetic Union (Magnetische Verein), 21; magnetic map of the earth, 24*fig*1.10; observation station, 23*fig*1.19. *See also* Gauss; Weber

magnetism. *See* bar magnet; diamagnetism; magnetic induction; magnetic pole; magnetic shell; magnets, permanent; paramagnetism

magnets: Ampère's theory of, 50–51, 319–320; composed of shells, 105–106; in dynamical theory, 313–314, 318; in fluid flow analogy, 70–71; permanent, 68, 89–99. *See also* bar magnet; magnetic shell

mass: absence of, in imaginary fluid of "Faraday's Lines," 64, 139, 179, 228, 230; and absolute measurement, 351; of electric current, 388; of ether, 206–207, 231, 289, 312; of electron, 210, 387; gravitational, 26, 206, 289; inertial, in Newtonian theory, 87, 125, 141, 160, 228, 235, 304, 351, 368; negligible, in particles of "Physical Lines," 158, 199, 209–210, 229, 328; proportionality of inertial and gravitational, 206, 289; rest, of photon, 207; virtual, in dynamical theory, 253, 303–306, 312, 369; of vortices of "Physical Lines," 141–142, 147, 160, 171, 200, 228, 230. *See also* gravitational force; inertia

*Mathematical Principles of Natural Philosophy* ("*Principia*") (Newton), 2, 33, 79, 81, 87, 91, 124–125, 181, 230–231, 233, 364, 405n1.1. *See also* Newton

mathematical theories, 31–32, 45, 52, 55–56, 124, 190, 300–301; Ampère on, 27–28, 74, 78–79, 81, 109–110, 119; Faraday and, 13, 15–16, 18, 28, 32–33, 41, 46–47, 57–58; Maxwell and, 15–16, 51, 55–60, 71–79, 82–83, 107–110, 119, 124, 140, 144–147, 162, 168–169, 196, 211, 242–243, 252–254, 300–301, 325–327, 331, 365, 388; Newton and, 79, 124, 181; Thomson and, 6, 8, 57, 349. *See also* mathematics

mathematics: in Cambridge curriculum, 1–4, 126, 367; Faraday and, 4, 6–7, 12–13, 16, 32, 46, 57–58, 77, 83; mathematical mind, 57, 82–83, 119, 124–125, 139–140, 181, 196, 324, 365; mathematical rhetoric, 242, 253, 327, 331, 387; Maxwell and, 4, 8, 181; survey of analytic mathematics, 217–228; *ta mathemata* as the knowable, 2, 83. *See also* mathematical theories

matter: elastic, in "Physical Lines," 215, 262; and electricity, 34–35, 68, 77, 107, 157, 170, 214; of electromagnetic medium, 33, 165, 255–256; and ether, 125, 166, 206, 255–256, 262, 289–291, 300; as extension, in Descartes, 125, 365, 406n1.10; gravitational, 206; and imaginary fluid, 86–87, 116, 125; and magnetism, 38–40; and Newton's laws of motion, 228, 364; and vortices, 169–170, 175; *See also* inertia; mass; momentum; physical concepts

Maxwell, James Clerk: on action-at-a-distance (attraction), 143, 156, 165–166, 173; on Ampère, 57, 409n4.6; birthplace, 5*fig*1.3; and the concept of the field, 146–147, 163, 165–166, 199, 206; concept and goal of science, 139–140; on contiguity,

Maxwell, James Clerk (*continued*)
158; correspondence with Faraday,
19–20, 407n2.1; and Descartes,
159–160; on the displacement current, 171, 215; "A Dynamical Theory of the Electromagnetic Field,"
252–403; early biography, 1–7; on
the electromagnetic theory of light,
172, 216; and the experiment to
measure the inertia of electric current, 389*fig*4.26; and Faraday, 139,
407n2.1; on Faraday's electronic
state, 161–166, 201–203; as fellow
of Cambridge, 1, 3–7, 1*fig*1.1; and
formal mathematics, 141, 160; and
geometry, 144; and Helmholtz, 169,
211; on hypothesis in science, 156,
163, 167, 175, 208, 217; on imaginary systems, 168, 211–212; on interpretation of analytic results, 150–156,
186–196, 247, 249–251; at King's College, London, 142; and Lagrangian
theory, 165, 205, 253; on lines of
force, 143–144, 155–159, 191; and
mathematical rhetoric, 242–247; and
*Matter and Motion*, 7, 160, 206; on
mechanisms and mechanical explanation, 144–146, 153, 156–157, 159,
165–168; on method in science,
143–144, 156–160, 167–169,
180–182, 200, 206, 211–212,
252–253; on the nature of electric
charge, 157, 167, 170, 195, 199, 210,
213–214; "On Faraday's Lines of
Force," 63, 187–188; "On Physical
Lines of Force," 139–251; on physical
analogy, 169; on physical interpretation, 151–155, 163; and the question
of production of electromagnetic radiation, 215, 285, 362, 410n4.16; and the
ratio-of-units experiment to determine
*v*, 353–356, 356*fig*4.23, 409n4.9,
409n4.10, 410n4.13; on real existence
in nature, 167, 174; scientific editorship of *Encyclopedia Britannica*, 61,
206; and the spinning coil determination of the ohm, 346*fig*4.18,
350–353, 352*fig*4.20, 409n4.9,
409n4.10, 410n4.14; on the size of
the vortices, 166, 174; style, 114,
140, 181; on the term "physical,"
174; on the theory of Weber, 57,
252; *Treatise on Electricity and
Magnetism*, 61; and the *Treatise on
Natural Philosophy* of Thomson and
Tait, 253, 408n4.5; and vector
mathematics, 141, 2201
measurement: absolute, 46*fig*1.36, 334,
337, 350–353; Ampère on, 27; Ampère's, of force between conductors,
394; Maxwell's, of momentum of
electric current, 329; Maxwell's, of
possible motion relative to ether,
290; Maxwell's, of size and momentum of vortices, 207–208, 266; as
universal analogy, 103–104. *See
also* balance; ratio of units; standards; systems of units
mechanical analogy, 269, 276, 301,
305–306
mechanical conception, 78, 167, 288, 350
mechanical and dynamical, compared,
217, 266, 332–335
mechanical explanation, 17, 144, 146,
153, 156–159, 255, 328
mechanical illustrations, 144, 276, 328,
367
mechanical interaction with the field, 45,
47, 112–114, 144–146, 154, 206–210,
260–261, 268–270, 277–281,
289–290, 306–307, 312–315, 337
mechanical interpretation, 163
mechanical model, 231, 367–369
mechanical reasoning, 196, 269
mechanical system: complete, 266,
305–306, 314, 367–368; of
Descartes, 406n1.10
mechanical theory, 143–145, 260
medium, 26, 46, 51, 70, 144, 157, 172,
175–177, 189–190, 210; conducting
power of, 38–41, 71–73; as explaining attraction, 39; as filled with vor-

tices, 156–158; and force in curved lines, 33–35; of light, as elastic vibrations in, 56, 117, 170; and plenum of Descartes, 406n1.10; stress in, 188

metaphor, 78–79, 81, 100–105, 124, 131, 138, 253, 304–307, 310, 328–239, 247, 365–367. *See also* analogy; rhetoric

method: of Ampère, 119; of analogy, 55–56, 79, 87, 124, 139; Faraday's, Maxwell's use of, 15–16, 57–58, 139, 316; of Fourier, 120; Lagrangian, 302, 372; mathematical, 30, 32, 58, 83, 119–120, 124, 160, 252, 302, 314; Maxwell's, 37, 55–59, 67–69, 77–79, 83, 124, 139, 252–253, 297, 363–364; of Newton, 102; speculative versus analytic, 112, 252; and truth, 59, 124; vividness and fertility as goals of, 56, 124. *See also* analogy; physical analogy; rhetoric; speculation; theory

Millikan, Robert A., 107

mind's eye, 33, 53, 83–84, 89, 132, 321

model: calibration of, 103, 127, 131, 136–138, 314; fluid, 104, 196–197, 216, 231, 351; geometrical, 58, 84–86, 103–104, 116; gravitational, 31, 129, 319, 329; Maxwell's, of connected mechanical system, 368*fig*4.24; mechanical, 231, 314, 367; thermal, 85, 349; vortex, 181, 196–197, 216, 231, 251

moment of inertia, 200, 207

momentum, 116, 161, 175, 199–208, 228–230, 301, 304, 307, 310, 328, 338, 368, 392; defined, 228; generalized, 332; of electric current, 276, 307, 329, 388; and impulse, 204–205; of vortices, 163, 199–200, 204, 207–208, 210, 216. *See also* electromagnetic momentum; reduced momentum

motion: analytic description of, 218–222, 219*fig*3.24; 221*fig*3.25; 222*fig*3.26. *See also* calculus

motor, electric: 19, 22*fig*1.17, 34, 47, 113, 311

moving wire: Faraday on, 292, 316–317. *See also* magnetic lines of force

music: as a liberal art, 126; tone, 120–121. *See also* Fourier series

Newton, Isaac: and action-at-a-distance, 25, 36, 45, 81; Ampère on, 27, 49–51; Faraday and, 32, 36, 46; geometrical style, 2, 181, 405n1.1; and hypothesis, 79; law of universal gravitation, 26, 45, 81, 206, 304; method, 124; Newtonian tradition in England, 2, 25, 32, 253, 364, 408n4.5

Newton's laws of motion, 45, 48, 91, 139–142, 228–231; and concept of force, 48–49, 51; and concept of mass, 87, 125, 142, 200; and centrifugal force, 231–233; first law, 91–92, 228; second law, 91, 199–200, 204, 228–230, 301, 304, 322, 369; third law, 49, 178, 210, 230

Oersted, Hans Christian, 44, 47

Oersted effect, 44–47, 45*fig*1.35, 379; discovery of, 407n1.20; law relating force and distance, 355*fig*4.22. *See also* electromagnetic force

ohm, 43; standard, of British Association for the Advancement of Science, 345*fig*4.17, 350–353, 356–357, 359–360, 410n4.13; as a velocity, 357–358. *See also* resistance; standards; unit

Ohm, George Simon, 42, 72, 407n1.19

Ohm's law, 42–43, 107, 109, 297, 308–309, 351, 375, 381; analogy to fluid flow, 109; in dimensional form, 357. *See also* dimensional analysis; resistance

"On Faraday's Lines of Force" (Maxwell), 53–138; sent by Maxwell to Faraday, 16

open circuits, 79, 110, 386. *See also* capacitor

operators, in mathematics, 101, 127,
226–227, 377–379, 382. *See also*
calculus
optics. *See* electromagnetic theory of
light; interference; light; refraction;
sunlight
orthogonal sets of lines, 130, 135
overtones: in Fourier analysis, 120,
121*fig*2.20. *See also* Fourier series;
music

parallel connection, of circuits, 44,
44*fig*1.34
parallels in geometry, 88. *See also* cylinder
paramagnetism, 37–40, 39*fig*1.30, 154,
257, 361; in Ampère's theory, 51;
Faraday's interpretation of, 38–40,
40*fig*1.31; in fluid flow analogy, 40,
71–72, 106–107; and magnetic inductive capacity, 190; in Weber's
theory, 51
physical analogy, 100, 119–121, 127,
169, 196, 216, 367; defined, as a
method in science, 56, 79. *See also*
physical concepts
physical concepts: concerning actions,
16; and existence, 12, 56–57, 125,
142–143, 289–292, 298, 334, 347,
362–365; ideas of, 55–56, 140; interpretation in terms of, 127, 150–151,
155, 161, 193–194, 247, 250–251;
nature of, 41, 49; and physical explanation, 124, 211; signification of,
15, 59; and truth, 56–57, 298. *See
also* matter; physical hypothesis;
physical lines of force
physical hypothesis, 79, 175, 197, 217;
and blindness to facts, 55; contrasted
with mathematical formulas, 55–56,
58, 79
physical lines of force, 15, 41, 47,
139–251. *See also* lines of force;
magnetic lines of force
polarization: electric, 33–35, 71, 108,
214, 258, 261, 276, 280, 292–293,

375, 386; of light, 119, 145,
256–257, 284, 291, 300, 340–341;
magnetic, 105–107, 171, 276. *See
also* electromagnetic theory of light;
wave equation; wave motion
polarization, magnetic, 36, 71–72, 105,
107, 145–146, 177, 180, 257, 276
potential: analogy to fluid pressure, 43,
73, 94, 101–104, 130–131, 135–138;
analogy to temperature in heat flow,
57, 82–83, 82*fig*2.2, 121–122; electric, 28–31, 29*fig*1.24, 41–43,
69–72, 100–104, 108, 135–138, 358,
378; magnetic, 129, 260, 271–272,
277, 319–321, 321*fig*4.10; 329–332,
389–390; measurement of, 102, 104;
of a point source, 133*fig*2.26; of a
system on itself, 102–104; superposition of, 96*fig*2.9, 100–102, 121*fig*2.20,
131–133; in topographic model,
30*fig*1.25, 121; in vortex theory,
170, 208. *See also* equipotential
lines; equipotential surface
potential energy, 29–31, 108, 256, 259,
276, 291, 328, 372; total, of a system
of charges, 102–104. *See also* potential
power: definition, 132, 310; extraction
from field, 45; in fluid motion, 109,
132; in mutual induction, 310, 312;
sphondyloid of, 40–41, 132; transformation to heat, 208
pressure: additive, as scalar, 94, 98,
133; in fluid flow, 43–44, 63–69, 73,
92, 95–96, 105, 109, 112, 127–137;
hydrostatic, 178–179, 185*fig*3.6,
237, 239–240; in medium, 188–191;
principal, 177; in vortex theory,
143–149, 153, 156–157, 165–166,
169–171, 175–180, 185, 195–197,
203, 206, 214, 231–234, 237–241
Priestly, Joseph, 18; *History and Present State of Electricity,* 406n1.8
primary circuit, 47–48, 49*fig*1.38
*Principia Mathematica: See Mathe-
matical Principles of Natural
Philosophy*

Ptolemy, 127, 227, 391; and Fourier series, 126

quadrivium, 126
quantity of electricity: relation of current and static quantities, 74, 93, 267, 395, 408n3.2
quantum theory, 119, 228, 230, 289, 346, 364–366

radian measure of angles, 233
radio. *See* electromagnetic waves
ratio of units, in electrostatic and electromagnetic systems: force between conductors, in two systems, 394*fig*4.28; Maxwell's experiment to measure, 353–356, 356*fig*4.23, 359–360, 409n4.9; measurement by Kohlrausch and Weber, 172, 216, 261–262, 280, 284, 341, 347, 353, 360, 363–364; and velocity of light, 279–281, 334–337, 357–359, 393–396
ratios, in mathematics: style of Euclid and Newton, used by Maxwell, 181
reduced momentum, 165, 205, 253, 263–266, 301–307, 311, 372. *See also* Lagrangian theory
refraction: as dielectric effect, 344*fig*4.16; law of sines, 117–118, 148; particle theory of (Descartes), 56, 80*fig*2.1, 119; wave theory of (Huygens), 80*fig*2.1
religious faith, and Faraday's scientific work, 12
repulsion, magnetic, 190
resistance: dimensional equivalence to velocity, 357; in dynamical theory, 257–258, 261–264, 274, 308–310, 372–375; electrical, 42, 44, 109; in fluid medium, 43, 63–74, 92, 104, 128, 131–132; in mechanical model, 368–369; in vortex theory, 158, 161–163, 209
rhetoric, 124, 297, 306; in mathematics, 242–243, 297, 306, 311, 325, 331,

377. *See also* analogy; physical analogy; illustration; intuition; metaphor; mind's eye; speculation; vividness
rotation, mechanical. *See* continuous rotation
Royal Institution (London): coil used by Faraday, 49*fig*1.38; Faraday's charged cage experiment in, 11, 95; Faraday's time of propagation experiment at, 360–361; great battery of, 43*fig*1.38; great electromagnet of, 35*fig*1.29, 45. *See also* Faraday

saddle point, geometric, 94, 136
Sandeman, Robert, 12
scalar quantities, 93–94, 98–100, 102, 131–133, 137, 173, 176, 237, 246, 378
secondary circuit, 47–48, 49*fig*1.38
self-induction, 262, 373. *See also* electromagnetic induction
self-consistency, as criterion for science, 62, 91
series circuit, 43*fig*1.33, 44, 44*fig*1.34, 109
shear, 177*fig*3.1, 186*fig*3.7, 187*fig*3.8
shell. *See* magnetic shell
Skilling, Hugh: notion of the "curl meter," 201
solenoidal, 134–135, 143
source: charge as, 136–137, 298; in fluid theory, 62–71, 85, 96–106, 129, 223; as inferred from cutting surfaces, 62, 90*fig*2.5, 98*fig*2.11, 132–133; magnetic, 187, 195, 247–250, 331–332, 360, 390; measured by divergence of flow, 188*fig*3.9
space, 289–294; displacement current in, 215, 312–314; in dynamical theory, 255, 268, 276, 312–314, 319–321, 364–366; energy in, 327–330; and the ether, 125, 289–294; Faraday on, 15, 33–35, 39–40; as filled with imaginary fluid, 58–62, 83–85, 88–89, 95, 103,

space (*continued*)
116; as filled with physical lines, 139–140, 143, 173; as filled with vortices, 179–182, 196–198, 202, 210, 215
specific inductive capacity, 33, 70, 104, 261–262, 286, 343. *See also* electrostatic induction
specific magnetic capacity. *See* coefficient of magnetic induction
speculation, 16, 51, 55, 59, 78–79, 86–87, 124, 143, 174, 250–252, 255, 312
sphondyloid of power, 40, 132
standards, 20–21, 41, 44, 255, 298, 348–351; and cable industry, 20, 348–349; of electrical resistance, 263, 350–353, 356, 359; magnetic, 21; physical, 126–127, 350–353; role in dynamical theory, 298; standard ohm, 345*fig*4.17; 350–351. *See also* British Association for the Advancement of Science; calibration; Magnetic Union; measurement; unit
stellar aberration, 341, 342*fig*4.12
Stokes, George Gabriel: and method of vector potential, 160–161
stress, 145, 156, 175–179, 184–185, 197, 203, 350; concept, 237–240; elastic, 212–213; equilibrium of in vortex theory, 240–242; isotropic, 145, 176–180, 179*fig*3.2, 271; tension and shear, 177*fig*3.1, 186*fig*3.7, 187*fig*3.8; in a vortical medium, 139–251
style: Faraday's, 9–10, 12–16, geometric, 2; Maxwell's, 88, 353
summation operation ($\Sigma$), 101
sunlight: mechanical value of, 286–287, 299, 344–346, 400–402. *See also* light
superconductivity, 209
superposition: of potentials, 96*fig*2.9, 100, 102, 131–133, 136; of pressures in fluid flow, 94, 98, 132–133, 137; of sources in fluid flow, 63–65,

96*fig*2.9; of terms in Fourier series, 121*fig*2.20
surface: cutting, 60–65, 89–90, 94, 260, 270; equipotential, 94, 135, 260, 271–272, 320–321; equivalence of distribution on, to sources, 62, 90*fig*2.5, 98*fig*2.11, 132–133; isobaric, 65, 129–133
systems of units, 298; arbitrariness, for imaginary fluid, 126, 161–163; electromagnetic, 261, 279–280, 332–335, 393–396; electrostatic, 255, 261, 280, 333, 393–396; foundations, 394*fig*4.28; practical, 43, 48, 255. *See also* measurement; ratio of units; standards

Tait, Peter Guthrie, 302*fig*4.4; *Treatise on Natural Philosophy,* with Thomson, 253, 372, 408n4.5. *See also* Lagrangian theory
tangent galvanometer. *See* galvanometer
telegraph, 8, 19–20, 21*fig*1.15, 41, 45; operating room, 20*fig*1.15; Wheatstone's, 19*fig*1.14
temperature: as analogy for potential, 81–82, 120–121, 131, 224; distribution over heated bar, 82*fig*2.2, 121*fig*2.20, 122*fig*2.21; in electrical calibration, 209; in thermal flow, 212
tension: and electrotonic state, 86, 202–203; electrical "tension," 73–76, 170, 213; in "Physical Lines," 143–149, 152–153, 166, 177–180, 187, 196–197, 213, 237–328, 250
theory: Ampère's account of, 27, 50–51, 110–111; Faraday's skepticism of, 12–15, 32–33, 36, 83; Lagrangian, 306–307, 371; mature, 71; Maxwell on, 55–59, 72, 78, 107, 124, 139, 144, 156, 166–169, 172, 175, 196–197, 211–214, 250–261, 276–277, 290–296, 300–301, 324–325, 347, 363–367; Newtonian, 51; and speculation, 101; Thomson's

role in, 8, 36; Weber's role in,
21–22, 51, 254, 287–288, 363
Thomson, J. J.: editor of Maxwell's
*Treatise on Electricity and Magnet-
ism,* 345, 367, 402; experiments on
cathode rays, 107
Thomson, William (Lord Kelvin), 7–9,
8*fig*1.6, 180; analogy of heat and
electrostatics, 7, 57, 81, 122; and the
Atlantic cable, 20, 349–350; on the
ether, 256–257, 286–287; instruments, 9*fig*1.7, 45; interpretation of
Faraday, 57; and the magnetic compass, 21; on magnetism, 40; on mechanical analogies, 100; on the
standard of resistance, 349–351,
353; theory of electromagnetic induction, 260, 295; theory of electrostatic induction, 32, 36, 51, 139;
*Treatise on Natural Philosophy,*
with Tait, 253, 372
time: in absolute measurement, 351,
357, 364, 366, 395; in connected systems, 116, 126; delay term in Weber's theory, 362–363; of
electromagnetic propagation, 256,
282, 300, 360–362; of electromagnetic vibrations, Faraday on, 287,
299, 345–346, 403; in fluid motion,
107, 127; and geometry, 125, 127;
as stopped, in "Faraday's Lines,"
103, 116, 138
trigonometry: functions, illustrated,
227*fig*3.27; introduction to,
227–228; series, 120–121. *See also*
Fourier series
truth: and criterion of self-consistency,
91; as criterion in science, 33, 56, 75,
91, 119, 252–253, 300; empirical, 350,
365–366; and Lagrangian theory, 303,
366; Maxwell on, 55–56, 75, 77; and
Newton, 228; relation to "physical"
concepts, 12, 77; and role of imaginary fluid, 91. *See also* method
tubes of fluid flow, 34*fig*1.27, 58–70,
85, 88–90, 126–132, 128*fig*2.24;

unit tube, 60–68, 79, 89, 93*fig*2.6,
112, 126–128, 127*fig*2.23. *See also*
unit cells
Tyndall, John, 107

uniqueness: in definition of magnetic
potential, 320; of pressure or potential distribution, 104
unit: of charge, 100, 102, 107–108,
333–335; of current, 73–75, 125,
155, 192, 335; of electrical resistance,
350–351, 356; of electromotive force,
214, 280; of line of magnetic force,
75, 144; of magnetic induction
(flux), 154, 271, 273, 327, 390–391;
of magnetic intensity, 151, 186, 192,
273, 318; of "magnetic matter," 151,
248, 390; pole, 154, 272, 277, 319,
331–332, 390–391; of potential,
100, 102; of power, 132, 310; of
quantity of imaginary fluid, 60, 87;
of work and energy, 131–132,
289–290. *See also* measurement; ratio of units; standards
unit cells, 65, 71, 93, 131; and energy
distribution in the field, 103,
131–132; in relation to force exerted
by field, 189*fig*3.10. *See also* field
unit tube. *See* tubes of fluid flow

vector: calculus, 184; and induction in
curved lines, 34*fig*1.27; Maxwell
and, 141; quantities, 29, 93–94, 98,
173, 324–325; and tensor, 237–239;
unit, 183
vector potential, 384. *See also* electromagnetic momentum; electrotonic
state
velocity of electromagnetic propagation, 118, 172, 216, 262, 284–286,
289, 300, 341–343, 347, 350; and dimensions of electrical resistance,
357–360; Faraday on, 361. *See also*
electromagnetic theory of light; ratio
of units
viscous fluid, 78, 86

vitreous electricity, 180. *See also* electric charge

vividness of conception, in the sciences, 56, 119. *See also* method; rhetoric

volt, 43, 299, 343–345, 358, 400–403

voltaic battery, 43–45, 73, 109, 199, 309; of Daniell cells, 42*fig*1.32; Gassiot's, 359–360, 409n4.11; great battery of the Royal Institution, 43*fig*1.33

voltmeter, 43–44, 44*fig*1.34, 109, 208–209, 305–306

vortices, 139–251, 182*fig*3.3, 295, 328, 332 ; centrifugal forces in, 146–150, 233–237, 234*fig*3.9; electromagnetic induction in, 203*fig*3.17, 205*fig*3.19; electromagnetic propagation in, 204*fig*3.18; elasticity of figure and electric induction in, 214*fig*3.18; field of, 146; hexagonal, 183*fig*3.4; idler wheels between, in Maxwell's theory, 156–158, 166–170, 194, 198, 198*fig*3.14, 205, 208, 210, 212; pressure and tension in, 185*fig*3.6, 186*fig*3.7; size of, Maxwell on, 166, 231. *See also* experiment; measurement

Watts, Isaac, 12

wave equation, 339, 396–400; advancing wave, illustrated, 399*fig*4.29

wave motion: absent in fluid of "Faraday's Lines," 116; question of normal vibrations in, 340; sound, 300. *See also* electromagnetic theory of light; polarization; wave equation

wave theory of light, 117–119; interference in, 116*fig*2.19(b); refraction in, 80*fig*2.1; wave front in, 116*fig*2.19(a). *See also* Huygens; Young

Weber, Wilhelm, 21; electromagnetic theory, 51, 139, 287–288, 362; establishment of standards, 348, 406n1.9; and Gauss, 406n1.9; magnetic observations, with Gauss, 23*fig*1.19; Maxwell on his theory, 51, 252, 254–255; ratio of units experiment, with Kohlrausch, 172, 216, 261–262, 280, 284, 341, 347, 353, 360, 363–364; tangent galvanometer, 45, 46*fig*1.36

Whewell, William, 407n1.17

work: around closed loop, and curl operator, 191*fig*3.12; definition, 100, 131; and electric potential, 28–29, 71–72, 100–102, 131; in fluid model, 65, 70, 76, 103, 112–113, 131–132; and potential energy, 28–30. *See also* energy

Young, Thomas, 119; wave theory of interference, 118*fig*2.19

# About the Author

Thomas K. Simpson retired recently from teaching the "Great Books" program at St. John's College. With a background in both engineering and the classics, he holds a doctorate in the history of science from the Johns Hopkins University. He has written numerous articles in the sciences and the humanities and is currently an enthusiastic consultant to schools and science museums.